CHAPMAN & HALL/CRC
Monographs and Surveys in
Pure and Applied Mathematics   142

QUASILINEAR

HYPERBOLIC SYSTEMS,

COMPRESSIBLE FLOWS,

AND WAVES

CHAPMAN & HALL/CRC
Monographs and Surveys in
Pure and Applied Mathematics   **142**

QUASILINEAR

HYPERBOLIC SYSTEMS,

COMPRESSIBLE FLOWS,

AND WAVES

Vishnu D. Sharma

CRC Press
Taylor & Francis Group
Boca Raton London New York

CRC Press is an imprint of the
Taylor & Francis Group, an **informa** business

A CHAPMAN & HALL BOOK

CRC Press
Taylor & Francis Group
6000 Broken Sound Parkway NW, Suite 300
Boca Raton, FL 33487-2742

First issued in paperback 2019

© 2010 by Taylor and Francis Group, LLC
CRC Press is an imprint of Taylor & Francis Group, an Informa business

No claim to original U.S. Government works

ISBN-13: 978-1-4398-3690-3 (hbk)
ISBN-13: 978-0-367-38415-9 (pbk)

Library of Congress Cataloging-in-Publication Data

Sharma, Vishnu D.
　　Quasilinear hyperbolic systems, compressible flows, and waves / Vishnu D. Sharma.
　　　　p. cm. -- (Monographs and surveys in pure and applied mathematics ; 142)
　　Includes bibliographical references and index.
　　ISBN 978-1-4398-3690-3 (hardcover : alk. paper)
　　　　1. Wave equation--Numerical solutions. 2. Differential equations,
Hyperbolic--Numerical solutions. 3. Quasilinearization. I. Title.

QC174.26.W28S395 2010
515'.3535--dc22
　　　　　　　　　　　　　　　　　　　　　　　　　　　　　　　　　　2010008125

Visit the Taylor & Francis Web site at
http://www.taylorandfrancis.com

and the CRC Press Web site at
http://www.crcpress.com

Contents

Preface

The material in this book has evolved partly from a set of lecture notes used for topic-courses and seminars at IIT Bombay and elsewhere during the last few years; much of the material is an outgrowth of the author's collaborative research with many individuals, whom the author would like to thank for their scientific contributions.

The book provides a reasonably self-contained discussion of the quasi-linear hyperbolic equations and systems with applications. Entries from the bibliography are referenced in the body of the text, but these do not seriously affect the continuity of the text, and can be omitted at first reading. The aim of this book is to cover several important ideas and results on the subject to emphasize nonlinear theory and to introduce some of the most active research areas in this field following a natural mathematical development that is stimulated and illustrated by several examples. As the book has been written with physical applications in mind, the author believes that the analytical approach followed in this book is quite appropriate, and is capable of providing a better starting point for a graduate student in this fascinating field of applied mathematics. In fact, the book should be particularly suitable for physicists, applied mathematicians and engineers, and can be used as a text in either an advanced undergraduate course or a graduate level course on the subject for one semester. Care has been taken to explain the material in a systematic manner starting from elementary applications, in order to help the reader's understanding, progressing gradually to areas of current research. All necessary mathematical concepts are introduced in the first three chapters, which are intended to be an introduction both to wave propagation problems in general, and to issues to be developed throughout the rest of the book in particular. The remaining chapters of the book are devoted to some of the recent research work highlighting the applications of the characteristic approach, singular surface theory, asymptotic methods, self-similarity and group theoretic methods, and the theory of generalized functions to several concrete physical examples from gasdynamics, radiation gasdynamics, magnetogasdynamics, nonequilibrium flows, and shallow water theory. A few general remarks have been included at the end of sections or chapters with the hope that they will provide useful source material for ideas beyond the scope of the text.

Chapter 1 provides a link between continuum mechanics and the quasi-linear partial differential equations (PDEs); it begins with a discussion of the conditions necessary for such systems to be hyperbolic. Several examples are considered, that illustrate the ideas and show some of the peculiarities that may arise in the classification of systems.

Chapter 2 introduces the scalar conservation laws. The main nonlinear feature is the breakdown of smooth solutions, leading to the notion of weak solutions, the loss of uniqueness, entropy conditions, and shocks. These are all presented in detail with the aid of examples.

Chapter 3 is devoted to hyperbolic systems in two independent variables; notions of genuine nonlinearity, k-shock, contact discontinuity, simple waves and Riemann invariants are introduced. Special weak solutions, namely, the rarefaction waves and shocks are discussed; these special solutions are particularly useful in solving Riemann problems, which occur frequently in physical applications. All these ideas are applied to solve the Riemann problem for shallow water equations with arbitrary data.

Chapter 4 presents the evolutionary behavior of weak and mild discontinuities in a quasilinear hyperbolic system, using the method of characteristics and the singular surface theory. Local and global behavior of the solution of transport equation, describing the evolution of weak discontinuities, is studied in detail, modifying several known results in the literature. Weak nonlinear waves, namely, the rarefaction waves, compression waves, and shocks are studied in the one-dimensional motion of an ideal plasma permeated by a transverse magnetic field. The method of relatively undistorted waves is introduced to study high frequency waves in a relaxing gas. The problem of interaction between a weak discontinuity wave and a shock, which gives rise to reflected and transmitted waves, is studied in detail.

Chapter 5 deals with weakly nonlinear geometrical optics (WNGO). It is an asymptotic method whose objective is to understand the laws governing the propagation and interaction of high frequency small amplitude waves in hyperbolic systems. The procedure is illustrated for nonequilibrium and stratified gas flows. Expressions for the energy dissipated across shocks, and the evolution equations describing mixed nonlinearity are derived. Singular ray expansions and resonantly interacting waves in hyperbolic systems are briefly discussed.

Chapter 6 demonstrates the power, generality, and elegance of the self-similar and basic symmetry (group theoretic) methods for solving Euler equations of gasdynamics involving shocks. An exact self-similar solution and the results of interaction theory are used to study the interaction between a weak discontinuity wave and a blast wave in plane and radially symmetric flows. The method of Lie group invariance is used to determine the class of self-similar solutions to a relaxing gas flow problem involving shocks of arbitrary strength. The method yields a general form of the relaxation rate for which the self-similar solutions are admitted. A particular case of collapse of an imploding shock is worked out in detail for radially symmetric flows. Numerical calculations have been performed to describe the effects of relaxation and the ambient density on the self-similar exponent and the flow pattern. With the help of canonical variables, associated with the group generators that leave the first order system of PDEs invariant, the system of PDEs is reduced to an autonomous system whose simple solutions provide nontrivial solutions of the original system. We remark that one of the special solutions of the Euler system, discussed here, is precisely the blast wave solution known in the literature.

Chapter 7 contains a discussion of the kinematics of a shock of arbitrary strength in three dimensions. The dynamic coupling between the shock front and the rearward flow is investigated by considering an infinite system of transport equations that hold on the shock front. At the limit of vanishing shock strength, the first order truncation approximation leads to an exact description of acceleration waves. Asymptotic decay law for weak shocks and rearward precursor disturbances are obtained as special cases. At the strong shock limit, the first order approximation leads to a propagation law that has a structural resemblance with Guderley's exact similarity solution. Attention is drawn to the connection between the transport equations along the rays obtained here and the corresponding results obtained by an alternative method using the theory of distributions. Finally, the procedure is used to describe the behavior of a bore of arbitrary strength as it approaches the shoreline on a sloping beach. The evolutionary behavior of weak and strong bores described by the lowest order truncation approximation is in excellent agreement with that predicted by the characteristic rule. Strangely enough, in the characteristic rule (CCW approximation), the critical values of the bore strength for which the bore height and the bore speed attain extreme values, remain uninfluenced by the initial bore strength and the undisturbed water depth. The present method takes care of this situation in a natural manner, and the first few approximations yield results with reasonably good accuracy. It may be remarked that even the lowest order approximation describes the evolutionary behavior of a bore with good accuracy.

Acknowledgments

I have been helped in the present venture by many colleagues and former students. I am indebted to all of them; in particular, I mention Dr. R. Radha, Dr. G. K. Srinivasan, and Dr. T. Rajasekhar for their cooperation and help. I am thankful to C. L. Antony, Pradeep Kumar and Ashish Mishra for their persistent support, which was essential for the completion of this book. During the work on the manuscript, I have benefited from contact with Prof. J. B. Keller, Prof. T. P. Liu, Prof. T. Ruggeri, Prof. D. Serre, and Prof. C. Dafermos, to whom I express my most sincere appreciation. I am greatly indebted to Prof. A. Jeffrey for being a constant source of inspiration and encouragement over the years.

I acknowledge the continuous support from IIT Bombay and the Department of Mathematics, in particular, for giving me time and freedom to work on this book. I also acknowledge the support of the Curriculum Development Programme at IIT Bombay in writing the book. I express my sincere thanks to the personnel of the CRC Press, Dr. Sunil Nair and his editorial staff, for their kind cooperation, help, and encouragement in bringing out this book. In particular, I thank Karen Simon who painstakingly went through the draft of the entire manuscript.

I express my deep gratitude to my family members for providing me much needed moral support without which this work would have remained only in my mind. Finally, I dedicate this book to my teacher Prof. Rishi Ram who not only introduced me to this difficult area of mathematics but also guided me through the tortuous paths of life; to him I owe more than words can express.

V. D. Sharma

About the Author

Dr. V.D. Sharma has been at the Indian Institute of Technology, Bombay (IITB) as a professor since 1988. Presently, he holds the position of Institute Chair Professor in the Department of Mathematics at IIT Bombay, and is President of the Indian Society of Theoretical and Applied Mechanics. He is a Fellow of the National Academy of Sciences India, and of the Indian National Science Academy. Professor Sharma has published numerous research articles in the area of hyperbolic systems of quasilinear partial differential equations and the associated nonlinear wave phenomena. He is also a member of the editorial board of the *Indian Journal of Pure and Applied Mathematics*. He has visited several universities and research institutions such as the University of Maryland at College Park, the Mathematics Research Center at the University of Wisconsin, and Stanford University. He was the head of the Department of Mathematics at IITB from 1996 to 2000 and from 2003 to 2006. Professor Sharma received IITB's awards for excellence in research in basic sciences in 2005, and for excellence in teaching in 1998 and in 2003. He received the C.L. Chandna Mathematics Award from the Canadian World Education Foundation for distinguished and outstanding contributions to mathematics research and teaching in 1999. He also received the M.N. Saha Award for Research in Theoretical Sciences from the University Grants Commission, Government of India in 2001.

Chapter 1

Hyperbolic Systems of Conservation Laws

Any mathematical model of a continuum is given by a system of partial differential equations (PDEs). In continuum mechanics, the conservation laws of mass, momentum and energy form a common starting point, and each medium is then characterized by its constitutive laws. The conservation laws and constitutive equations for the field variables, under quite natural assumptions, reduce to field equations, i.e., partial differential equations, which, in general, are nonlinear and nonhomogeneous. For nonlinear problems, neither the methods of their solutions nor the main characteristics of the motion are as well understood as in the linear theory. Before we proceed to discuss mathematical concepts and techniques to understand the phenomena from a theoretical standpoint and to solve the problems that arise, we introduce here the hyperbolic systems of conservation laws in general (see Benzoni-Gavage and Serre [13]), and then present some specific examples, for application or motivation, which are of universal interest.

1.1 Preliminaries

A large number of physical phenomena are modeled by systems of quasilinear first order partial differential equations that result from the balance laws of continuum physics. These equations, expressed in terms of divergence, are commonly called conservation laws. The general form of the system of conservation laws, in its differentiated form, is given by

$$\frac{\partial \mathbf{u}}{\partial t} + \frac{\partial}{\partial x_i} \mathbf{f}_i(\mathbf{u}) = \mathbf{g}(\mathbf{u}, \mathbf{x}, t), \quad 1 \le i \le m, \tag{1.1.1}$$

where $\mathbf{u}(\mathbf{x}, t) = (u_1, u_2, \ldots, u_n)^{t_r}$ is a vector of conserved quantities, dependent on $\mathbf{x} = (x_1, x_2, \ldots, x_m)^{t_r} \in \mathbb{R}^m$ and $t \in [0, \infty)$, and the superscript t_r denotes transposition. The vector fields $\mathbf{f}_i(\mathbf{u}) = (f_{1i}, f_{2i}, \ldots, f_{ni})^{t_r}$ and $\mathbf{g} = (g_1, g_2, \ldots, g_n)^{t_r}$ represent, respectively, the flux and production densities, which are assumed to be smooth functions of their arguments. Here, and throughout, summation convention on repeated indices is automatic unless otherwise stated.

System (1.1.1) arises in the study of nonlinear wave phenomena, when weak dissipative effects such as chemical reaction, damping, stratification, relaxation etc., are taken into account. It expresses that the time variation

1

of the total amount of the substance $\int_D \mathbf{u} dx$, contained in any domain D of $\mathbb{R}^m$, is equal to the flux of the vector fields $\mathbf{f}_i$ across the boundary of D, plus contribution due to the sources or sinks distributed in the interior of D; system (1.1.1) with $\mathbf{g} = \mathbf{0}$ is said to be in strict conservation from. When the differentiation in (1.1.1) is carried out, the following quasilinear system of first order results

$$\mathbf{u}_{,t} + \mathbf{A}_i(\mathbf{u})\mathbf{u}_{,i} = \mathbf{g}, \qquad (1.1.2)$$

where $\mathbf{A}_i = (\partial f_{ji}/\partial u_k)_{1 \leq j,k \leq n} \equiv \nabla \mathbf{f}_i$ is the $n \times n$ Jacobian matrix with $\nabla = (\partial/\partial u_1, \partial/\partial u_2, \ldots, \partial/\partial u_n)$ as the gradient operator with respect to the elements of $\mathbf{u}$, and a comma followed by t (respectively, an index i), denotes partial differentiation with respect to time t (respectively, the space variable x_i).

The system (1.1.1) is called hyperbolic if for each $\mathbf{x}, t$ and $\mathbf{u}$, and the unit vector $\boldsymbol{\xi} = (\xi_1, \xi_2, \ldots, \xi_m)^{tr} \in \mathbb{R}^m$, the $n \times n$ matrix $\xi_j A_j$ has n real eigenvalues $\lambda_1, \lambda_2, \ldots, \lambda_n$ with linearly independent eigenvectors $\mathbf{r}_1, \mathbf{r}_2, \ldots, \mathbf{r}_n$; if, in addition, the eigenvalues are all distinct, the system (1.1.2), and hence (1.1.1), is called strictly hyperbolic. However, if the algebraic multiplicity of an eigenvalue is greater than its geometrical multiplicity, the system is referred to as nonstrictly hyperbolic and cannot be diagonalized (see Li et al. [107] and Zheng [215]).

It may be remarked that the eigenvalues of the system (1.1.2) are the same as those of (1.1.1), and for a smooth solution the two forms (1.1.1) and (1.1.2) are equivalent. Working with (1.1.1) allows us to consider discontinuous solutions as well, so that the equation is interpreted in some generalized sense. It is our goal here to study certain problems involving these equations. Before proceeding with the discussion of some of the consequences of such equations, it will be useful to have some specific physical examples which illustrate their occurrence.

1.2 Examples

1.2.1 Traffic flow

The simplest example of a nonlinear conservation law in one space dimension is a first order partial differential equation

$$u_{,t} + (f(u))_{,x} = 0, \qquad (1.2.1)$$

where u is the density function and $f(u)$, a given smooth nonlinear function, is the flux function. Such an equation appears in the formulation of traffic flow where $u(x, t)$ denotes the density, the number of cars passing through the position x at time t on a highway, and the function $f(u) = uv$ denotes the flux of cars with v being the average (local) velocity of the cars, which is assumed to be a given function of u. This assumption seems to be reasonable since the drivers are supposed to increase or decrease their speed as the density decreases or increases, respectively. The simplest model is the linear relation described by the equation $v = v_{\max}(1 - (u/u_{\max}))$, which shows that the maximum value of v occurs when $u = 0$, and when u is maximum, $v = 0$. On setting $\tilde{u} = u/u_{\max}$ and $\tilde{x} = x/v_{\max}$, the resulting normalized conservation law, after suppressing the overhead tilde sign, reduces to the form

$$u_{,t} + (u(1-u))_{,x} = 0. \tag{1.2.2}$$

Further details on traffic modeling and analyses may be found in Whitham [210], Goldstein [64], Sharma et al. [181], and Haberman [67].

1.2.2 River flow and shallow water equations

For river flows, a rectangular channel of constant breadth and inclination is considered, and it is assumed that the disturbance is roughly the same across the breadth. If $h(x,t)$ be the depth and $u(x,t)$ the mean velocity of the fluid in the channel, then the governing equations for the river flow can be written in the conservation form (see Whitham [210] and Ockendon et al. [132])

$$\mathbf{u}_{,t} + \mathbf{f}_{,x} = \mathbf{b}, \tag{1.2.3}$$

where
$\mathbf{u} = (h, hu)^{tr}, \mathbf{f} = (hu, hu^2 + (gh^2 \cos\alpha)/2)^{tr}$, and $\mathbf{b} = (0, gh\sin\alpha - C_f u^2)^{tr}$; α is the angle of inclination of the surface of the river, g the acceleration due to gravity, and C_f the friction coefficient that appears in the expression for the friction force of the river bed. As the Jacobian matrix $\mathbf{A} = \nabla \mathbf{f}$ has distinct real eigenvalues $u \pm \sqrt{gh \cos\alpha}$, the system (1.2.3) is strictly hyperbolic. Equations (1.2.3) may be simplified to become the equivalent pair in nonconservative form.

$$h_{,t} + uh_{,x} + hu_{,x} = 0,$$
$$u_{,t} + uu_{,x} + g\cos\alpha\, h_{,x} = g\sin\alpha - C_f^2(u^2/h).$$

In the shallow water theory, where the height of the water surface above the bottom is small relative to the typical wave lengths, usually the slope and friction terms are absent in (1.2.3), and so the governing system of equations is in strict conservative form, and assumes on simplification the following form

$$c_{,t} + uc_{,x} + (cu_{,x}/2) = 0,$$
$$u_{,t} + uu_{,x} + 2cc_{,x} = 0, \tag{1.2.4}$$

where $c = \sqrt{gh}$. It may be noticed that the above system (1.2.4) describes the flow over a horizontal level surface; for a nonuniform bottom, there is an additional term in the horizontal momentum equation due to the force acting on the bottom surface, and so the corresponding shallow water equations can be written in the following nonconservative form

$$c_{,t} + uc_{,x} + (cu_{,x}/2) = 0,$$
$$u_{,t} + uu_{,x} + 2cc_{,x} = g(dh_0/dx), \qquad (1.2.5)$$

where the function $-h_0(x)$ describes the bottom surface relative to an origin located on the equilibrium surface of the water along which lies the x-axis.

1.2.3   Gasdynamic equations

One of the simplest examples that appears to be fundamental in the study of gasdynamics is provided by unsteady compressible inviscid gas flow; the Euler equations in the Cartesian coordinate system $x_i, 1 \leq i \leq 3$, can be written in the following conservative form.

$$\rho_{,t} + (\rho u_i)_{,i} = 0,$$
$$(\rho u_i)_{,t} + (\rho u_i u_j + p\delta_{ij})_{,j} = 0, \quad i,j = 1,2,3 \qquad (1.2.6)$$
$$(\rho E)_{,t} + ((\rho E + p)u_j)_{,j} = 0,$$

where ρ is the gas velocity, u_i the ith component of gas velocity vector $\mathbf{u}$, p the gas pressure, $E = e + (|\mathbf{u}|^2/2)$ the total energy per unit mass with e as the internal energy, and δ_{ij} the Kronecker δ's. The equation of state can be taken in the form $p = p(\rho, e)$. Identifying (1.2.6) with (1.1.1), we see that these equations are in conservative form with $\mathbf{u} = (\rho, \rho u_1, \rho u_2, \rho u_3, \rho E)^{tr}$, $\mathbf{f}_1 = (\rho u_1, p + \rho u_1^2, \rho u_1 u_2, \rho u_1 u_3, (\rho E + p)u_1)^{tr}$, $\mathbf{f}_2 = (\rho u_2, \rho u_1 u_2, p + \rho u_2^2, \rho u_2 u_3, (\rho E + p)u_2)^{tr}$, $\mathbf{f}_3 = (\rho u_3, \rho u_1 u_3, \rho u_2 u_3, p + \rho u_3^2, (\rho E + p)u_3)^{tr}$, and $\mathbf{b} = \mathbf{0}$.

The eigenvalues of the matrix $\xi_j \mathbf{A}_j$, where $\mathbf{A}_j$ is the Jacobian matrix of $\mathbf{f}_j$ and $|\boldsymbol{\xi}| = 1$, are easily found to be $u_i\xi_i - a, u_i\xi_i + a$ and $u_i\xi_i$; the first two eigenvalues are simple and the remaining one, i.e., $u_i\xi_i$, is of multiplicity three having three linearly independent eigenvectors associated with it. Thus, all the five eigenvalues of $\xi_j \mathbf{A}_j$ are real, but not distinct, and the eigenvectors span the space $\mathbb{R}^5$, the system (1.2.6) is hyperbolic. Conservative forms (1.2.6) are obtained from the corresponding integral representation forms, and are needed for the treatment of shocks. But for other purposes, the equations may be simplified. The energy equation (1.2.6)$_3$ can be written in various forms; using the other two equations (1.2.6)$_1$ and (1.2.6)$_2$, it takes an alternative form

$$de/dt - (p/\rho^2)(d\rho/dt) = 0, \qquad (1.2.7)$$

where $d/dt = \partial/\partial t + u_i\partial/\partial x_i$ denotes the time derivative following an individual particle. Further, using the thermodynamical relations $TdS = de + pd(1/\rho) = dh - (1/\rho)dp$, where $T(p,\rho)$, $S(p,\rho)$ and $h(p,\rho)$ denote, respectively, the absolute temperature, the specific entropy and the specific enthalpy, (1.2.7) reduces to

$$T(dS/dt) = 0 \quad \text{or} \quad dh/dt - (1/\rho)dp/dt = 0. \qquad (1.2.8)$$

Since the expression for S in terms of p and ρ may be solved in principle as $p = p(\rho, S)$, an equivalent form of $(1.2.8)_1$ is $dp/dt - a^2 d\rho/dt = 0$, where $a = \sqrt{(\partial p/\partial \rho)_S}$ is the local speed of sound.

Thus, the following alternative formulation of the Euler equations (1.2.6), though not in conservative form, will be convenient for future reference.

$$d\rho/dt + \rho u_{i,i} = 0, \quad du_i/dt + (1/\rho)p_{,i} = 0, \quad dS/dt = 0. \tag{1.2.9}$$

For a polytropic gas, we have

$$e = p/(\rho(\gamma - 1)), S = C_v \ln(p/\rho^\gamma) + \text{ constant}, \quad a^2 = \gamma p/\rho, \tag{1.2.10}$$

where $\gamma > 1$ is the specific heat ratio for the gas.

If we assume that the flow has some symmetry, one can reduce the number of space variables. For instance for a one-dimensional unsteady compressible inviscid gas flow with plane, cylindrical or spherical symmetry, system (1.2.6) can be written [96]

$$
\begin{aligned}
&(x^m \rho)_{,t} + (x^m \rho u)_{,x} = 0, \\
&(x^m \rho u)_{,t} + (x^m (p + \rho u^2))_{,x} = mpx^{m-1}, \\
&(x^m \rho(e + \frac{u^2}{2}))_{,t} + (x^m \rho u(e + \frac{p}{\rho} + \frac{u^2}{2}))_{,x} = 0,
\end{aligned}
\tag{1.2.11}
$$

where $u = u(x,t), \rho = \rho(x,t), p = p(x,t)$ and $e = e(x,t)$ are, respectively, the gas velocity, density, pressure and internal energy with x as the spatial coordinate being either axial in flows with planar ($m = 0$) geometry, or radial in cylindrically symmetric ($m = 1$) and spherically symmetric ($m = 2$) configurations. System (1.2.11) can be written as

$$\mathbf{u}_{,t} + \mathbf{f}_{,x} = \mathbf{b}, \tag{1.2.12}$$

where $\mathbf{u}, \mathbf{f}$ and $\mathbf{b}$ are column vectors, which can be read off by inspection of (1.2.11); with the usual equation of state, the system (1.2.12) is strictly hyperbolic as the Jacobian matrix of $\mathbf{f}$ has real and distinct eigenvalues $u \pm a$ and u.

1.2.4 Relaxing gas flow

As a result of high temperatures attained by gases in motion, the effects of non equilibrium thermodynamics on the dynamics of gas motion can be important. Assuming that the departure from equilibrium is due to vibrational relaxation, and the rotational and translational modes are in local thermodynamical equilibrium throughout, the governing system of equations for an unsteady flow in the absence of viscosity, heat conduction and body forces, is obtained by adjoining to the gasdynamic equations (1.2.6) the rate equation (see Vincenti and Kruger [207])

$$(\rho\sigma)_{,t} + (\rho\sigma u_j)_{,j} = \rho\Phi, \tag{1.2.13}$$

where σ is the vibrational energy per unit mass, and Φ is the rate of change of vibrational energy, which is assumed to be a known function of p, ρ and σ, given by $\Phi = (\sigma^* - \sigma)/\tau$; here, σ^* is the equilibrium value of σ given by $\sigma^* = R\Theta_v/(\exp(\Theta_v/T) - 1)$ and τ^{-1} is the relaxation frequency given by $\tau^{-1} = k_1 p \exp(-k_2/T^{1/3})$, where R is the gas constant, Θ_v the characteristic temperature of the molecular vibration, $T = p/(\rho R)$ the gas temperature, and k_1, k_2 the positive constants depending on the physical properties of the gas. Using the continuity equation $(1.2.6)_1$, equation $(1.2.13)$ can be put in a simpler form $d\sigma/dt = \Phi$, where $d/dt = \partial/\partial t + u_i \partial/\partial x_i$ is the material derivative.

To close the system of equations, we need to add an equation of state which can be written in the form

$$e = \sigma + (\gamma - 1)^{-1}(p/\rho), \tag{1.2.14}$$

where γ is the frozen specific heat ratio.

Thus, in case of smooth flows, we may write the governing system of equations in nonconservative form

$$\begin{aligned} d\rho/dt + \rho u_{i,i} = 0, \quad du_i/dt + (1/\rho)p_{,i} = 0, \\ dp/dt + \rho a_f^2 u_{i,i} + (\gamma - 1)\rho\Phi = 0, \quad d\sigma/dt = \Phi, \end{aligned} \tag{1.2.15}$$

where $a_f^2 = \gamma p/\rho$ is the frozen speed of sound.

System $(1.2.15)$ may be written in the form $\mathbf{u}_{,t} + \mathbf{A}_i \mathbf{u}_{,i} = \mathbf{b}$, where $\mathbf{u}$ is the six-dimensional column vector having components $\rho, \mathbf{u}, p$ and σ; the column vector $\mathbf{b}$ and 6×6 matrices $\mathbf{A}_i$ can be read-off by inspection of $(1.2.15)$. Indeed the matrix $A_i \xi_i = (\alpha_{IJ})_{1 \leq I, J \leq 6}$ with $\alpha_{11} = \alpha_{22} = \alpha_{33} = \alpha_{44} = \alpha_{55} = \alpha_{66} = u_i \xi_i, \alpha_{12} = \rho \xi_1, \alpha_{13} = \rho \xi_2, \alpha_{14} = \rho \xi_3, \alpha_{25} = \xi_1/\rho, \alpha_{35} = \xi_2/\rho, \alpha_{45} = \xi_3/\rho, \alpha_{52} = \rho a_f^2 \xi_1, \alpha_{53} = \rho a_f^2 \xi_2, \alpha_{54} = \rho a_f^2 \xi_3$, and the remaining entries being zeros, has eigenvalues $u_i \xi_i + a_f$, $u_i \xi_i - a_f$ and $u_i \xi_i$. The first two eigenvalues are simple, whilst the third one, $u_i \xi_i$, is of multiplicity four, having four linearly independent eigenvectors associated with it. Thus, the system $(1.2.15)$ is hyperbolic.

For a one-dimensional motion with plane, $(m = 0)$, cylindrical $(m = 1)$ or spherical $(m = 2)$ symmetry, the rate equation $(1.2.13)$ can be written as

$$(x^m \rho\sigma)_{,t} + (x^m \rho\sigma u)_{,x} = x^m \rho\Phi. \tag{1.2.16}$$

Equations $(1.2.11)$ together with $(1.2.14)$ and the rate equation $(1.2.16)$ describe the one-dimensional motion of a vibrationally relaxing gas with plane, cylindrical or spherical symmetry; these equations may be simplified to assume the following nonconservative form

$$\begin{aligned} \rho_{,t} + u\rho_{,x} + \rho u_{,x} + m\rho u/x = 0, \quad u_{,t} + uu_{,x} + (1/\rho)p_{,x} = 0, \\ p_{,t} + up_{,x} + \rho a_f^2(u_{,x} + mux^{-1}) + (\gamma - 1)\rho\Phi = 0, \quad \sigma_{,t} + u\sigma_{,x} = \Phi. \end{aligned} \tag{1.2.17}$$

System $(1.2.17)$ is hyperbolic as the Jacobian matrix has four real eigenvalues, but not all distinct, and the eigenvectors span the space $\mathbb{R}^4$.

It is sometimes of interest to examine the entropy production in nonequilibrium flows, where there is a transfer of energy from one internal mode of gas molecules to another, and one needs to supplement the gasdynamic equations with the rate equation (see Chu [36] and Clarke and McChesney [37])

$$(\rho q)_{,t} + (\rho q u_i)_{,i} = \rho w(p, S, q), \quad \text{or} \quad dq/dt = w, \qquad (1.2.18)$$

where q is the progress variable characterizing the extent of internal transformation in the fluid, and w is the rate of internal transformation which is assumed to be a known function of p, S and q. Then by specifying the specific enthalpy $h = h(p, S, q)$, through the equation of state and by invoking the Gibbs relation, $T dS = dh - (1/\rho)dp + \alpha dq$, where T denotes the temperature and α the affinity of internal transformation characterized by the variable q; all other variables such as ρ, T and α are known functions of p, S and q, given by

$$\rho^{-1} = \partial h/\partial p, \quad T = \partial h/\partial S, \quad \alpha = -\partial h/\partial q. \qquad (1.2.19)$$

When the internal transformation attains a state of equilibrium

$$w(p, S, q) = 0 = \alpha(p, S, q) \Rightarrow q = q^*(p, S), \qquad (1.2.20)$$

where q^* is the equilibrium value of q evaluated at local p and S. When the Gibbs relation is applied to the changes following a fluid element, and use is made of the energy equation $dh/dt = (1/\rho)dp/dt$ along with the rate equation $(1.2.18)_2$, we obtain $dS/dt = w\alpha/T$. This gives the rate of entropy production following a fluid element which, in general, is not zero in a nonequilibrium flow. When $\rho = \rho(p, S, q)$ is substituted into the equation of continuity and use is made of $(1.2.18)_2$ along with the entropy equation $dS/dt = w\alpha/T$, we obtain an equivalent form of the continuity equation as

$$dp/dt + \rho a_f^2 u_{i,i} = -w a_f^2 (\partial \rho/\partial q + (\alpha/T)\partial \rho/\partial S),$$

where a_f is the frozen speed of sound given by $a_f^2 = (\partial p/\partial \rho)_{S,q}$.

Thus, in the absence of external body forces, the equations governing an unsteady nonequilibrium flow of an inviscid and non-heat conducting gas, which has only one lagging internal mode, can be written in the following nonconservative form

$$\begin{aligned} &dp/dt + \rho a_f^2 u_{i,i} = -w a_f^2 (\partial \rho/\partial q + (\alpha/T)\partial \rho/\partial S), \\ &du_i/dt + (1/\rho)p_{,i} = 0, \quad dS/dt = \alpha w/T, \quad dq/dt = w. \end{aligned} \qquad (1.2.21)$$

One can easily check that the system (1.2.21) is hyperbolic; regarding some of the results concerning (1.2.21), the reader is referred to Singh and Sharma [186].

1.2.5 Magnetogasdynamic equations

Magnetogasdynamics is concerned with the study of interaction between magnetic field and the gas flow; the governing equations, thus, consist of

Maxwell's equations and the gasdynamic equations of motion. The effect of magnetic field on the conventional gasdynamic momentum equation is to add a term corresponding to the Lorentz force which, with the aid of Maxwell's equations, may be written as the divergence of a tensor. For a perfectly conducting inviscid gas motion in a magnetic field $\mathbf{H} = (H_1, H_2, H_3)$, the equations of magnetogasdynamics can be written in the conservative form (see Jeffrey and Tanuiti [79] and Cabannes [27])

$$\rho_{,t} + (\rho u_i)_{,i} = 0;$$
$$(\rho u_i)_{,t} + (\rho u_i u_j + p\delta_{ij} - \mu(H_i H_j - \tfrac{1}{2}|\mathbf{H}|^2 \delta_{ij}))_{,j} = 0;$$
$$(\rho e + \tfrac{1}{2}\rho|\mathbf{u}|^2 + \tfrac{1}{2}\mu|\mathbf{H}|^2)_{,t} + (\rho u_i(e + \tfrac{p}{\rho} + \tfrac{1}{2}|\mathbf{u}|^2) + \mu(|\mathbf{H}|^2 u_i - u_j H_j H_i))_{,i} = 0;$$
$$(H_i)_{,t} + u_j H_{i,j} - H_j u_{i,j} + H_i u_{j,j} = 0; \qquad H_{i,i} = 0,$$

$$(1.2.22)$$

where μ is the magnetic permeability and other symbols have their usual meaning; the suffixes i, j run over 1,2,3. The thermal pressure p is related to the internal energy e through the gamma-law equation of state $p = \rho e(\gamma - 1)$. Equation $(1.2.22)_4$ which is a consequence of Maxwell's equations and Ohm's law, is called the induction equation or the field equation. The magnetic field vector $\mathbf{H}$, by virtue of the induction equation and initial conditions, satisfies the equation $(1.2.22)_5$, which is a statement of the experimental fact that the magnetic charges do not exist in nature; in fact, it may be regarded as a restriction on $\mathbf{H}$, and may be conveniently used in place of one of the projections of the induction equation. Although the electric field $\mathbf{E}$ and the current density $\mathbf{J}$ do not explicitly appear in equations (1.2.22), they may be obtained through the magnetic field and flow velocity: $\mathbf{E} = \mu \mathbf{H} \times \mathbf{u}$ and $\mathbf{J} = \mu \ \mathbf{curl} \ \mathbf{H}$ under the approximation of ideal magnetogasdynamics. The theory of non-linear system (1.2.22) is basically similar to that of ordinary compressible gasdynamics; nevertheless, in detail, the magnetogasdynamic phenomena are decidedly more complex than the gasdynamic ones. The momentum and energy equations in (1.2.22) can be simplified with the help of the remaining equations, and can be written in the following nonconservative form

$$u_{i,t} + u_j u_{i,j} + (1/\rho)p_{,i} + (\mu/\rho)H_j(H_{j,i} - H_{i,j}) = 0,$$
$$p_{,t} + u_i p_{,i} + \rho a^2 u_{i,i} = 0,$$

$$(1.2.23)$$

where $a = \sqrt{(\gamma p/\rho)}$ is the sound speed in the medium. The system of equations described by $(1.2.22)_1$, $(1.2.22)_4$ and (1.2.23) is hyperbolic (see Singh and Sharma [185] and Sharma ([178], [176])).

When the flow and field variables depend only on the Cartesian coordinate x and on time t, H_1 turns out to be a constant, and the governing system of equations, represented by $(1.2.22)_1$, $(1.2.22)_4$ and (1.2.23), for the one-dimensional motion becomes $\mathbf{u}_{,t} + \mathbf{A}\mathbf{u}_{,x} = \mathbf{0}$, where $\mathbf{u} = (\rho, u_1, u_2, , u_3, H_2, H_3, p)^{tr}$ and $\mathbf{A} = (A_{IJ})_{1 \leq I, J \leq 7}$ is a 7×7 matrix with $A_{11} = A_{22} = A_{33} = A_{44} = A_{55} = A_{66} = A_{77} = u_1, A_{12} = \rho, A_{25} = \mu H_2/\rho, A_{26} = \mu H_3/\rho, A_{27} = 1/\rho, A_{35} = -\mu H_1/\rho = A_{46}, A_{52} = H_2, A_{53} = A_{64} = -H_1, A_{62} = H_3, A_{72} = \rho a^2$ and the remaining entries being zeros. When $H_1 \neq 0$, the eigenvalues of A are real and distinct, namely, $u_1 \pm c_f, u_1 \pm c_s, u_1 \pm c_a$ and u, where c_f, c_s and c_a are respectively the fast magnetoacoustic speed, slow magnetoacoustic speed, and Alfven speed given by $c_f^2 = \frac{1}{2} \left\{ a^2 + b^2 + \sqrt{(a^2 + b^2)^2 - 4a^2 c_a^2} \right\}^{1/2}$, $c_s^2 = \frac{1}{2} \left\{ a^2 + b^2 - \sqrt{(a^2 + b^2)^2 - 4a^2 c_a^2} \right\}^{1/2}$, and $c_a^2 = \mu H_1^2/\rho$ with $b^2 = \mu |\mathbf{H}|^2/\rho$. The corresponding eigenvectors span $\mathbb{R}^7$ and so the system is strictly hyperbolic.

For a one-dimensional planar and cylindrically symmetric motion of a plasma, which is assumed to be an ideal gas with infinite electrical conductivity and to be permeated by a magnetic field orthogonal to the trajectories of gas particles, the governing equations can be easily derived from the equations $(1.2.22)_1$, $(1.2.22)_4$ and $(1.2.23)$. We assume that the vector $\mathbf{H}$ is directed perpendicular to the trajectories of gas particles. In the cylindrical case, the field vector $\mathbf{H}$ may, in particular, be directed along the axis of symmetry, along tangents to concentric circles with center on the symmetry axis, or in the general case, it may create a screw field with components H_ϕ and H_z. Thus, the basic equations can be written down in the following nonconservative form (see Korobeinikov [96])

$$\rho_{,t} + (\rho u)_{,r} + m\rho u/r = 0,$$
$$\rho(u_{,t} + uu_{,r}) + (p + (\mu/2)H^2)_{,r} + (n\mu/r)H^2 = 0,$$
$$H_{,t} + (uH)_{,r} + m(1 - n)uH/r = 0, \qquad (1.2.24)$$
$$p_{,t} + up_{,r} + \rho a^2(u_{,r} + mu/r) = 0,$$

where H is the transverse magnetic field, which in a cylindrically symmetric $(m = 1)$ motion is either axial $(n = 0)$ with $H = H_z$ or annular $(n = 1)$ with $H = H_\phi$; for a planer $(m = 0)$ motion, $H = H_z$ and the constant n assumes the value zero.

For a screw field, the vectors $\mathbf{H}$ and $\mathbf{u}$ have components $(0, H_\phi, H_z)$ and $(u, 0, 0)$ along the axes associated with the cylindrical coordinates (r, ϕ, z), and the basic equations become

$$\rho_{,t} + (\rho u)_{,r} + \rho u/r = 0,$$
$$\rho(u_{,t} + uu_{,r}) + (p + (\mu/2)(H_\phi^2 + H_z^2))_{,r} + (\mu/r)H_\phi^2 = 0,$$
$$H_{z,t} + (uH_z)_{,r} + uH_z/r = 0, \qquad (1.2.25)$$
$$H_{\phi,t} + (uH_\phi)_{,r} = 0,$$
$$p_{,t} + up_{,r} + \rho a^2(u_{,r} + u/r) = 0.$$

Both the systems described above by $(1.2.24)$ and $(1.2.25)$ are hyperbolic, but not strictly hyperbolic; for further details regarding some of the results concerning $(1.2.24)$, the reader is referred to Menon and Sharma [122].

1.2.6   Hot electron plasma model

We consider a homogeneous unbounded plasma consisting of mobile electrons in a neutralizing background of stationary ions of number density n_o. Assuming that the processes taking place in a plasma are such that the heat conduction, viscosity and energy interchange due to collisional interactions are all negligible, the evolution of the electron density n and mean electron velocity u for a one component warm electron fluid, in a one-dimensional configuration, are then given by

$$n_{,t} + un_{,x} + nu_{,x} = 0,$$
$$u_{,t} + uu_{,x} + (nm)^{-1}p_{,x} = -eE/m, \qquad (1.2.26)$$

where m is the electron mass, $-e$ is the electron charge, E is the electric field, and p is the electron pressure, the evolution of which is determined from the approximate moment of the Vlasov equation; in the one-dimensional case, the evolution of p is governed by the following equation (see Davidson [46])

$$p_{,t} + up_{,x} + 3pu_{,x} = 0, \qquad (1.2.27)$$

which, in view of the continuity equation $(1.2.26)_1$ yields the familiar adiabatic gamma-law (for $\gamma = 3$)

$$p = (\text{const})n^3. \qquad (1.2.28)$$

The electric field E in $(1.2.26)_2$, in view of Poisson's equation $E_{,x} = -4\pi e(n - n_o)$, and the Maxwell equation in the electrostatic approximation, $E_{,t} = 4\pi enu$, is determined from the equation

$$E_{,t} + uE_{,x} = 4\pi eun_o. \qquad (1.2.29)$$

It may be recalled that it is possible to heat a plasma due to collisional effects by periodically compressing and decompressing it. This compression can be one, two, or three-dimensional depending on the geometrical arrangement. The thermodynamic state of charged particles subject to adiabatic relation, $pn^\gamma = \text{const}$, where $\gamma = 1 + (2/\delta)$ with δ being the number of degrees of freedom affected by the compression. For a one-dimensional compression, $\delta = 1$, we have $\gamma = 3$, and hence the adiabatic relation (1.2.28) (see Sharma et al. [180]). Equations (1.2.26), (1.2.27) and (1.2.29), thus constitute a closed one-dimensional description of a warm electron fluid in the electrostatic approximation; it may be noticed that the description of this model is equivalent to the single-water bag model of a Vlasov plasma (see Bertrand and Feix [14]). This system can be expressed in the form $\mathbf{u}_{,t} + \mathbf{A}\mathbf{u}_{,x} = \mathbf{b}$, where the matrix $\mathbf{A}$ has eigenvalues $u \pm (3p/n)^{1/2}$ and u (multiplicity two) with four linearly independent associated eigenvectors; hence it is hyperbolic. For further details regarding some of the results concerning this system, the reader is referred to [180].

1.2.7 Radiative gasdynamic equations

In radiative gasdynamics, the basic equations governing the flow form a system of coupled integro-differential equations of considerable complexity. The consequence of these complexities has been to stimulate a search for approximate formulation of the equation of radiative transfer which leads merely to a system of nonlinear differential equations. Thus, using a general differential approximation for the equation of radiative transfer, which is applicable over the entire range from the transparent limit to the optically thick limit, the set of fundamental equations governing an unsteady flow of a thermally radiating inviscid gas can be written in the form (see Vincenti and Kruger [207])

$$
\begin{aligned}
&\rho_{,t} + u_i\rho_{,i} + \rho u_{i,i} = 0; \quad \rho u_{i,t} + \rho u_j u_{i,j} + p_{,i} + (1/3)E_{R,i} = 0; \\
&\rho T(S_{,t} + u_i S_{,i}) + (4/3)E_R u_{i,i} + u_i E_{R,i} - k_p(cE_R - 4\sigma T^4) = 0; \quad (1.2.30) \\
&E_{R,t} + q_{i,i} = -k_p(cE_R - 4\sigma T^4); \quad (1/c)q_{i,t} + (c/3)E_{R,i} = -k_p q_i,
\end{aligned}
$$

where $1 \leq i,j \leq 3$; u_i are the gas velocity components, ρ the density, $p = p(\rho, S)$ the gas pressure with S being the entropy, E_R is the radiative energy density, $T = T(\rho, S)$ the temperature, c the velocity of the light, k_p the Plank's absorption coefficient, σ the Stefan's constant, and q_i the components of the radiative flux vector. The system of equations can be written in the form $\mathbf{u}_{,t} + \mathbf{A}_i \mathbf{u}_{,i} = \mathbf{b}$, where $\mathbf{u}$ is the column vector with nine components $\rho, u_1, u_2, u_3, S, E_R$, and $\mathbf{q}$; the column vector $\mathbf{b}$ can be read off by inspection, and $\xi_i \mathbf{A}_i = (\alpha_{IJ})_{1 \leq I, J \leq 9}$ is a 9×9 matrix with $\alpha_{11} = \alpha_{22} = \alpha_{33} = \alpha_{44} = \alpha_{55} = u_i\xi_i$, $\alpha_{12} = \rho\xi_1$, $\alpha_{13} = \rho\xi_2$, $\alpha_{21} = a^2\xi_1/\rho$, $\alpha_{25} = p_{,S}\xi_1/\rho$, $\alpha_{26} = \xi_1/(3\rho)$, $\alpha_{31} = a^2\xi_2/\rho$, $\alpha_{35} = p_{,S}\xi_2/\rho$, $\alpha_{36} = \xi_2/(3\rho)$, $\alpha_{41} = a^2\xi_3/\rho$, $\alpha_{45} = p_{,S}\xi_3/\rho$, $\alpha_{46} = \xi_3/(3\rho)$, $\alpha_{52} = 4E_R\xi_1/(3\rho T)$, $\alpha_{53} = 4E_R\xi_2/(3\rho T)$, $\alpha_{54} = 4E_R\xi_3/(3\rho T)$, $\alpha_{56} = u_i\xi_i/(\rho T)$, $\alpha_{67} = \xi_1$, $\alpha_{68} = \xi_2$, $\alpha_{69} = \xi_3$, $\alpha_{76} = c^2\xi_1/3$, $\alpha_{86} = c^2\xi_2/3$ and $\alpha_{96} = c^2\xi_3/3$ and the remaining entries being all zeros. The eigenvalues of $\xi_i \mathbf{A}_i$ are real: $\pm c/\sqrt{3}$, $u_i\xi_i$ (multiplicity three), $u_i\xi_i \pm (\partial p/\partial \rho + (4E_R/3\rho^2 T)\partial p/\partial S)^{1/2}$ and zero (multiplicity two), and the corresponding eigenvectors span $\mathbb{R}^9$; hence the system is hyperbolic. For further details regarding some of the consequences of (1.2.30), we refer to Helliwell [70], Sharma and Shyam [167], and Sharma [182].

1.2.8 Relativistic gas model

A model for one-dimensional, isentropic, barotropic fluid in special relativity is given by the following conservation laws of mass and momentum (see Smoller and Temple [188])

$$
\begin{aligned}
&\frac{\partial}{\partial t}\left(\rho + \frac{(p + \rho c^2)v^2}{c^2(c^2 - v^2)}\right) + \frac{\partial}{\partial x}\left(\frac{(p + \rho c^2)v^2}{c^2 - v^2}\right) = 0, \\
&\frac{\partial}{\partial t}\left(\frac{(p + \rho c^2)v}{c^2 - v^2}\right) + \frac{\partial}{\partial x}\left(p + \frac{(p + \rho c^2)v^2}{c^2(c^2 - v^2)}\right) = 0,
\end{aligned}
\qquad (1.2.31)
$$

where v is the particle speed, c the velocity of the light, ρ the density and $p(\rho)$ the pressure which is a smooth function of ρ. For more general models, the reader is referred to Taub [195]. The above system is strictly hyperbolic as the eigenvalues of the above system are real and distinct, i.e.,

$$\lambda_1 = c^2(v-a)/(c^2 - av) \text{ and } \lambda_2 = c^2(v+a)/(c^2 + av),$$

where $a = \sqrt{dp/d\rho}$ is the sound speed.

1.2.9 Viscoelasticity

Within the framework of the extended thermodynamics, a set of balance laws describing viscoelastic materials has been proposed in Lebon et al. [104] by introducing the stress tensor and the heat flux vector among the state variables; this model includes, as particular cases, a number of classical models such as the ones of Maxwell, Kelvin-Voigt and Poynting-Thomson. Assuming slab symmetry, the balance equations for a one-dimensional case become

$$\begin{aligned}
&\epsilon_{,t} + v\epsilon_{,x} - \epsilon v_{,x} = 0, \\
&v_{,t} + vv_{,x} + (\epsilon/\rho_o)\sigma_{,x} = 0, \\
&e_{,t} + ve_{,x} + (\sigma\epsilon/\rho_o)v_{,x} = 0, \\
&\sigma_{,t} + v\sigma_{,x} = (\phi_1/\tau)(\epsilon_{,t} + v\epsilon_{,x}) + (\phi_2/\tau)(T_{,t} + vT_{,x}) + \phi_3,
\end{aligned} \qquad (1.2.32)$$

where ϵ is the strain, v is the velocity, ρ_o is the reference density, T denotes the absolute temperature, σ is the stress, $e = e(\epsilon, T, \sigma)$ represents the internal energy, τ is the relaxation time and ϕ_1, ϕ_2, ϕ_3 are the material response functions dependent upon ϵ, T and σ.

The group invariance of the above system gives rise to a complete characterization of the material response functions (see Fusco and Palumbo [57] and Jena and Sharma [86]). Using an invariance analysis for $\tau = \text{const}$, the Maxwell model of viscoelastic media is characterized by the response functions $\phi_1 = \psi T^{-2\alpha/\gamma}, \phi_2 = 0$ and $\phi_3 = -\Lambda\sigma T^{-1/\gamma}$, where ψ, Λ, α and γ are constants with $\Lambda/\psi \leq 0$; the invariant system of basic equations can be written in the form $\mathbf{u}_{,t} + \mathbf{A}\mathbf{u}_{,x} = \mathbf{b}$ where $\mathbf{u} = (\epsilon, v, T, \sigma)^{tr}, \mathbf{b} = (0, 0, \Lambda\sigma e_{,\sigma} T^{-1/\gamma}/(\tau e_{,T}), -\Lambda\sigma T^{-1/\gamma}/\tau)^{tr}$, and $\mathbf{A} = (A_{IJ})_{1 \leq I, J \leq 4}$ is a 4×4 matrix with $A_{11} = A_{22} = A_{33} = A_{44} = v, A_{12} = -\epsilon, A_{24} = \epsilon/\rho_o, A_{42} = -\psi T^{-2\alpha/\gamma}\epsilon/\tau, A_{32} = (\epsilon/e_{,T})(e_{,\epsilon} + \rho_o^{-1}\sigma + \psi\tau^{-1}T^{-2\alpha/\gamma}e_{,\sigma})$ and the remaining entries being all zeros. Matrix $\mathbf{A}$ admits eigenvalues $v \pm \epsilon\Gamma T^{-\alpha/\gamma}$ and v (multiplicity two) with $\Gamma = (-\psi/(\rho_o\tau))^{1/2}$; hyperbolicity of the system is assured by the condition $\psi < 0$.

1.2.10 Dusty gases

The study of a two-phase flow of gas and dust particles has been of great interest because of many applications to different engineering problems. Gas flows, which carry an appreciable amount of solid particles, may exhibit significant relaxation effects as a result of particles being unable to follow rapid changes of the velocity and temperature of the gas. When the mass concentration of the particles is comparable with that of the gas, the flow properties become significantly different from that of a pure gas. Here, we consider a mixture of a perfect gas and a large number of small dust particles of uniform spherical shape. The viscosity and heat conductivity of the gas are neglected except for the interaction with the dust particles, which do not interact among themselves. The thermal motion of dust particles is also assumed negligible. Let p, ρ, T and u be respectively the pressure, density, temperature and velocity of the gas, and k, θ, v the mass concentration, temperature and velocity of dust particles. Then, assuming slab symmetry, the equations of mass, momentum and energy, for a two-phase flow can be written in the following form (see Marble [118], Rudinger [156], Pai et al. [139], and Sharma et al. [177])

$$\rho_{,t} + (\rho u)_{,x} = 0; \quad \rho(u_{,t} + uu_{,x}) + k(v_{,t} + vv_{,x}) + p_{,x} = 0,$$

$$(\rho(\frac{u^2}{2} + c_v T) + k(\frac{v^2}{2} + c_m \theta))_{,t} + (\rho u(\frac{u^2}{2} + c_p T) + kv(\frac{v^2}{2} + c_m \theta))_{,x} = 0,$$

$$k_{,t} + (kv)_{,x} = 0; \quad v_{,t} + vv_{,x} = (u - v)/\tau_v, \tag{1.2.33}$$

$$\theta_{,t} + v\theta_{,x} = (T - \theta)/\tau_T,$$

where c_p, c_v and c_m are specific heats of the gas and dust particles; τ_v and τ_T are the velocity and temperature equilibrium times, which are assumed to be constant. The equation of state of a perfect gas is taken to be $p = \rho RT$ with R as the gas constant. The above system, after simplification, can be written as $\mathbf{u}_{,t} + \mathbf{A}\mathbf{u}_{,x} = \mathbf{b}$, where $\mathbf{u} = (\rho, u, T, k, v, \theta)^{tr}$,

$$\mathbf{b} = \left(0, -\frac{k}{\rho}\frac{u - v}{\tau_v}, \frac{k}{\rho c_v}(\frac{(u - v)^2}{\tau_v} - \frac{c_m(T - \theta)}{\tau_v}), 0, \frac{(u - v)}{\tau_v}, \frac{(T - \theta)}{\tau_T}\right)^{tr},$$

and $\mathbf{A} = (A_{IJ})_{1 \leq I, J \leq 6}$ is a 6×6 matrix with $A_{11} = A_{22} = A_{33} = u, A_{44} = A_{55} = A_{66} = v, A_{12} = \rho, A_{21} = RT/\rho, A_{23} = R, A_{32} = (\gamma - 1)T, A_{45} = k$ and the remaining entries being all zeros. The eigenvalues of A are real : $u \pm a, u, v$ (multiplicity three) with $a = (\gamma p/\rho)^{1/2}$, but the corresponding eigenvectors span $\mathbb{R}^5$; thus the system is nonstrictly hyperbolic.

1.2.11 Zero-pressure gasdynamic system

The following two-dimensional convective system, known as the pressureless gasdynamic system, arises in the study of adhesion particle dynamics to describe the motion of free particles, which stick under collision, and explains the formation of the large scale structures in the universe. (see Weinan et al. [209])

$$\rho_{,t} + (\rho u_i)_{,i} = 0, \quad (\rho u_j)_{,t} + (\rho u_i u_j)_{,i} = 0,$$

where i, j take values 1 and 2. This system has a repeated eigenvalue $\lambda = \xi_1 u_1 + \xi_2 u_2$, in the direction of unit vector $\boldsymbol{\xi} = (\xi_1, \xi_2)$ with an incomplete set of right eigenvectors; its one-dimensional version takes the form

$$\rho_{,t} + (\rho u)_{,x} = 0, \quad (\rho u)_{,t} + (\rho u^2)_{,x} = 0,$$

which again has a single eigenvalue whose geometrical multiplicity is less than its algebraic multiplicity. Again, each of these systems is not hyperbolic, and is referred to as nonstrictly hyperbolic.

Remarks 1.2.1:

(i) There is no general theory to solve globally in time the initial value problem for the system of PDEs (1.1.1); indeed, it is usual to consider discontinuous weak solutions satisfying admissibility criteria, but not much is known about their existence (see [45] and [164]).

(ii) The study of nonstrictly hyberbolic systems is far from complete with respect to the existence and uniqueness of their solutions that may comprise waves such as undercompressive shocks, overcompressive shocks, delta shocks, and oscillations (see Li et al. [107], Zheng [215], and Dafermos [45]).

Chapter 2

Scalar Hyperbolic Equations in One Dimension

It is well known that the Cauchy problem for the system (1.1.1), satisfying the initial condition $\mathbf{u}(\mathbf{x}, 0) = \mathbf{u}_o(\mathbf{x}), \mathbf{x} \in \mathbb{R}^m$ does not have, in general, a smooth solution beyond some finite time interval, even when $\mathbf{u}_0$ is sufficiently smooth. For this reason, we study the weak solutions containing the discontinuities; these weak solutions are, in general, not unique, and one needs an admissibility criterion to select physically relevant solutions. Here, we study the one-dimensional scalar case to highlight some of these fundamental issues in the theory of hyperbolic equations of conservation laws.

2.1 Breakdown of Smooth Solutions

Let us set $m = n = 1, \mathbf{g} = \mathbf{0}$, and write (1.1.1) as

$$u_{,t} + (f(u))_{,x} = 0, x \in \mathbb{R}, t > 0 \tag{2.1.1}$$

satisfying the initial condition

$$u(x, 0) = u_0(x). \tag{2.1.2}$$

We write (2.1.1) as $u_{,t} + a(u)u_{,x} = 0$, where $a(u) = f'(u)$; the basic idea underlying the hyperbolicity of an equation like (2.1.1) is that the Cauchy problem, in which appropriate data are prescribed at some initial time, should be well posed for it. Since (2.1.1) may be interpreted as an ordinary differential equation (ODE) along any member of the characteristic family, determined by the solution of $dx/dt = a(u)$, the notion of hyperbolicity for (2.1.1) is synonymous with the existence of real characteristic curves.

Let us assume that $a'(u) \neq 0$ for all u; this means that $f''(u) \neq 0$, i.e., the flux function $f(u)$ is either convex ($f''(u) > 0$) or concave ($f''(u) < 0$). The convexity condition $a(u) \neq 0$ corresponds to the notion of genuine non-linearity introduced by Lax [102]. For convenience, we may take $a'(u) > 0$; it then immediately follows that along the characteristic curves $dx/dt = a(u), u$ remains constant, and so the characteristics are straight lines. The solution

$u = u(x,t)$ of (2.1.1) is then implicitly given by

$$u = u_0(x - ta(u)). \tag{2.1.3}$$

The above solution (2.1.3) represents a wave, with shape determined by the function $u_0(x)$, moving to the right with speed $a(u)$; the dependence of a on u produces a typical nonlinear distortion of the wave as it propagates. If u_0 is differentiable and $u_0'(x) \geq 0$, it follows from (2.1.1) on using the implicit function theorem that $u_{,t}$ and $u_{,x}$ stay bounded for all $t > 0$, and the solution u exists for all time. On the other hand, if $u_0' < 0$ at some point, both $u_{,t}$ and $u_{,x}$ become unbounded as $1 + u_0' a_0'(u_0)t$ approaches zero; this corresponds to the situation when the characteristics, emanating from $(x_1, 0)$ and $(x_2, 0)$ with $x_1 < x_2$ and $f'(u_0(x_1)) > f'(u_0(x_2))$, intersect at a point (x_c, t_c) in the domain $t > 0$, where a continuous solution is overdetermined. Thus, the initial value problem (2.1.1), (2.1.2) has a continuous single valued solution, called a simple wave if $u_0(x)$ is nondecreasing (see Courant and Friedrichs [41]). A simple wave of the kind in which the characteristics emanating from the initial line fan out in the positive t-direction is called an expansion (or rarefaction wave); see Figure 2.1.1(a). However, if the characteristics of different slopes fan out from a single point, the wave is called a centered simple wave; see Figure 2.1.1(b). But, if $u_0(x)$ is monotonically decreasing so that the characteristics emanating from the initial line converge in the positive t-direction, and the simple wave region tends to be narrowed (see the region: $t \leq x \leq \alpha, 0 \leq t < \alpha$ in Figure 2.1.2(a)), then the simple wave is called a compression wave (or, a condensation wave). In fact, any compressive part of the wave, where the propagation velocity is a decreasing function of x, ultimately breaks to give a triple-valued solution for $u(x,t)$; the time $t = t_c > 0$ of first breaking is given by

$$t_c = -(\min_{\xi \in \mathbb{R}} \phi'(\xi))^{-1}; \quad \phi(\xi) = a(u_0(\xi)). \tag{2.1.4}$$

Example 2.1.1: Consider the following initial value problem (IVP) for the inviscid Burgers equation

$$\begin{aligned} u_{,t} + u u_{,x} &= 0, \quad x \in \mathbb{R}, t > 0 \\ u(x,0) &= 1/(1+x^2). \end{aligned} \tag{2.1.5}$$

Here, the characteristic speed $a(u) = u$ and $u_0'(x) = -2x/(1+x^2)^2$, we expect that the wave will break for $x > 0$, and the solution $u(x,t)$ will become multivalued. The breaking time t_c in (2.1.4) requires finding the most negative value of $\phi'(\xi) = u_0'(\xi) = -2\xi/(1+\xi^2)^2$, which is attained at $\xi = 1/\sqrt{3}$. The breaking time t_c is, then, given by $t_c = 8/3^{3/2}$.

The differential equation (2.1.5)$_1$ implies that along the characteristics, $x = \xi + tu_0(\xi)$, which meet the line $t = 0$ at $x = \xi$, the function $u(x,t)$ has the value $u_0(\xi)$. For $t < t_c$, these characteristics do not intersect, and the solution of the Cauchy problem is given implicitly by

$$u(x,t) = u_0(x - tu). \tag{2.1.6}$$

On the other hand, when $t \geq t_c$ the characteristic lines begin to intersect; indeed, in this case the Jacobian of the transformation $(\xi, t) \rightarrow (x, t)$ is zero, i.e., $(\partial x / \partial \xi)|_t = 0$, and (2.1.6) no longer defines a single valued solution to the Cauchy problem.

The foregoing solution can be saved for all times $t \geq 0$ by allowing discontinuities into the solution; this we shall discuss in the following subsection.

2.1.1 Weak solutions and jump condition

Let us consider the conservation law (2.1.1). The function $u(x, t)$ is called a generalized or a weak solution of (2.1.1), if it satisfies the integral form of the conservation law, i.e.,

$$\frac{d}{dt} \int_{x_1}^{x_2} u(x, t) dx = f(u(x_1, t)) - f(u(x_2, t)), \qquad (2.1.7)$$

holds for any fixed interval $x_1 \leq x \leq x_2$ and time t.

It may be noticed that the integral form (2.1.7) is more fundamental than the differential form (2.1.1) in the sense that unlike (2.1.1), it makes sense even for piecewise continuous functions; under the additional assumption of smoothness, these two forms are equivalent.

An alternative integral formulation of the weak solution by integrating (2.1.1) against smooth test functions with compact support, namely

$$\int_R \int (u\phi_{,t} + f(u)\phi_{,x}) dx dt = 0, \qquad (2.1.8)$$

where R is an arbitrary rectangle in (x, t) plane, ϕ is an arbitrary test function with continuous first derivatives in R and $\phi = 0$ on the boundary of R, has a similar feature; indeed, there is a direct connection between (2.1.7) and (2.1.8), and it can be shown that the two integral forms are mathematically equivalent (see LeVeque [106]).

We shall now consider solution of (2.1.1) in the sense of distributions that are only piecewise smooth and, thus, admit discontinuities. Let $C : x = X(t)$ be a smooth curve representing the line of discontinuity in the (x, t) plane such that $x_1 < X(t) < x_2$. Suppose u and f, and their first derivatives are continuous in $x_1 \leq x < X(t)$ and in $X(t) < x \leq x_2$, and have finite limits as $x \rightarrow X(t)^\pm$. Then (2.1.7) may be written

$$
\begin{aligned}
f(u(x_1, t)) - f(u(x_2, t)) &= \frac{d}{dt} \left(\int_{x_1}^{X(t)} + \int_{X(t)}^{x_2} \right) u(x, t) dx \\
&= \int_{x_1}^{X(t)} u_{,t} dx + \int_{X(t)}^{x_2} u_{,t} dx + (u_l - u_r) \frac{dX}{dt},
\end{aligned}
$$

where $u_l(t) = \lim\limits_{x \rightarrow X(t)^-} u(x, t)$ and $u_r(t) = \lim\limits_{x \rightarrow X(t)^+} u(x, t)$; now, if we let

$x_2 \to X(t)$ and $x_1 \to X(t)$, the integrals tend to zero in the limit as $u_{,t}$ is bounded in each of intervals separately, and we are left with the following Rankine-Hugoniot (R-H) jump condition, expressing the velocity of propagation of the discontinuity in terms of the solution on either side

$$s = dX/dt = [f(u)]/[u], \qquad (2.1.9)$$

where $[u] = u_l - u_r$ and $[f(u)] = f(u_l) - f(u_r)$ denote the jumps in u and $f(u)$ across $x = X(t)$.

If we write the Burgers equation $(2.1.5)_1$, as a conservation law with flux function $f(u) = u^2/2$, i.e.,

$$u_{,t} + (u^2/2)_{,x} = 0, x \in \mathbb{R}, t > 0 \qquad (2.1.10)$$

then the R-H condition (2.1.9) yields

$$s = \frac{u_l + u_r}{2}. \qquad (2.1.11)$$

Example 2.1.2 (Rarefaction Wave): Consider the Burgers equation (2.1.10) with the initial data

$$u_o(x) = \begin{cases} 0, & x \le 0 \\ x/\alpha, & 0 < x \le \alpha \\ 1, & x > \alpha. \end{cases} \qquad (2.1.12)$$

As the vertical characteristics issuing from $x \le 0$ carry the value $u = 0$ from the initial data, we obtain the solution $u(x, t) = 0$ for $x \le 0$. Similarly, the solution for $x > \alpha$ is $u(x, t) = 1$. It may be noticed that the characteristics issuing from the interval $0 < x \le \alpha$ fan out; indeed, the solution is smooth everywhere and has the parametric form

$$x = \xi + tu_0(\xi), \quad u(x, t) = u_0(\xi), \quad 0 \le \xi \le \alpha. \qquad (2.1.13)$$

Equation $(2.1.13)_1$ implies that $\xi = \alpha x/(\alpha + t)$, which together with $(2.1.13)_2$, shows that $u(x, t) = x/(\alpha + t)$ in the region $R : 0 < x \le \alpha, t > 0$ (see Figure 2.1.1(a)). Thus, we have obtained a weak solution of the initial value problem (2.1.10) and (2.1.12) in the form

$$u(x, t) = \begin{cases} 0, & x \le 0 \\ x/(\alpha + t), & 0 < x \le \alpha + t \\ 1, & x > \alpha + t. \end{cases} \qquad (2.1.14)$$

It may be noticed that u is continuous along the rays $t > 0, x = 0$ and $x = t + \alpha$, but $u_{,x}$ and $u_{,t}$ are not. In this solution, known as the expansion or rarefaction wave, the wave amplitude u increases from 0 to 1 on an x−interval whose length increases with an increase in t; this wave does not break and the solution is valid for all time.

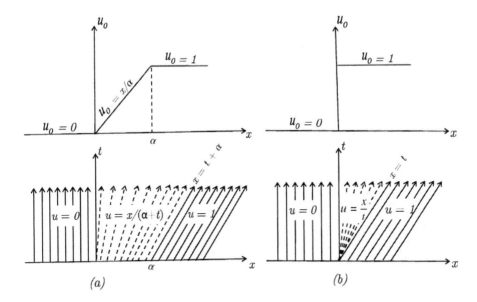

Figure 2.1.1: (a) Solution of (2.1.10) with initial data (2.1.12); (b) Solution of (2.1.10) with initial data (2.1.15).

It may be remarked that in the limit $\alpha \to 0$, the initial function in (2.1.12) has a jump discontinuity at $x = 0$, i.e.,

$$u_o(x) = \begin{cases} 0, & x \leq 0 \\ 1, & x > 0, \end{cases} \tag{2.1.15}$$

and the corresponding solution is obtained from (2.1.14) by letting α tend to zero (see Figure 2.1.1(b)), i.e.,

$$u(x,t) = \begin{cases} 0, & x \leq 0 \\ x/t, & 0 < x \leq t \\ 1, & x > t. \end{cases} \tag{2.1.16}$$

Example 2.1.3 (Compression Wave): Consider the Burgers equation (2.1.10) with the initial data

$$u_0(x) = \begin{cases} 1, & x \leq 0 \\ 1 - (x/\alpha), & 0 < x \leq \alpha \\ 0, & x > \alpha. \end{cases} \tag{2.1.17}$$

Again, in view of (2.1.13), the initial value problem has the solution (see Figure 2.1.2(a))

$$u(x,t) = \begin{cases} 1, & x \leq t \\ (x - \alpha)/(t - \alpha), & t < x \leq \alpha \quad \text{for } 0 < t < \alpha. \\ 0, & x > \alpha. \end{cases} \tag{2.1.18}$$

In this solution, known as compression wave, the wave amplitude decreases from 1 to 0 on an x−interval whose length tends to zero as t increases to α. At the point (α, α) a continuous solution is overdetermined, because the characteristic lines issuing from the interval $0 \le x \le \alpha$ meet at (α, α), and each carries different values of u. Hence, no solution exists in the classical sense for $t \ge \alpha$. Indeed, the solution can be extended beyond $t = \alpha$ by introducing a discontinuity at (α, α), and for this, we use (2.1.11); since $u_l = 1$ and $u_r = 0$, we get $s = 1/2$. So, for $t \ge \alpha$, there is a discontinuous solution, given by

$$u(x,t) = \begin{cases} 1, & x \le (t+\alpha)/2 \\ 0, & x > (t+\alpha)/2. \end{cases} \tag{2.1.19}$$

Thus, the solution of the initial value problem has two distinct time intervals $0 < t < \alpha$ and $t \ge \alpha$, describing two different states, given by (2.1.18) and (2.1.19), respectively.

In the limit as $\alpha \to 0$, the initial profile (2.1.17) has a jump discontinuity at $x = 0$, i.e.,

$$u_0(x) = \begin{cases} 1, & x \le 0 \\ 0, & x > 0, \end{cases} \tag{2.1.20}$$

and the corresponding solution is obtained from (2.1.19) by letting α tend to zero (see Figure 2.1.2(b)).

$$u(x,t) = \begin{cases} 1, & x \le t/2 \\ 0, & x > t/2. \end{cases} \tag{2.1.21}$$

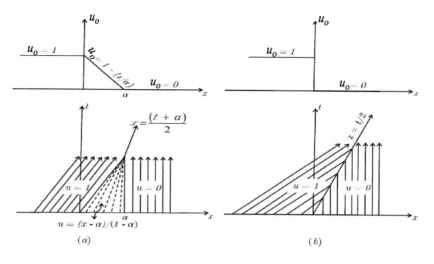

Figure 2.1.2: (a) Solution of (2.1.10) with initial data (2.1.17); (b) Solution of (2.1.10) with initial data (2.1.20).

2.1.2   Entropy condition and shocks

Introducing generalized or weak solutions makes it possible for an IVP to possess many different solutions; this can be illustrated with the help of scalar convex equation (2.1.10) and the initial data (2.1.15). For this IVP, we have obtained a rarefaction wave solution (2.1.16). Now we check that there are many other weak solutions of this IVP. In fact, the condition (2.1.11) suggests that we obtain a weak solution to this problem by introducing a discontinuity, propagating with $s = (u_l + u_r)/2 = 1/2$, which gives (see Figure 2.1.3)

$$u(x,t) = \begin{cases} 0, & x < t/2 \\ 1, & x > t/2. \end{cases} \tag{2.1.22}$$

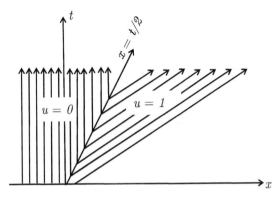

Figure 2.1.3: A discontinuous solution of (2.1.10) with initial data (2.1.15).

It is easy to check that (2.1.22) satisfies (2.1.10) and (2.1.15) along with (2.1.9), and therefore, it is also a weak solution of (2.1.10), (2.1.15). In fact, it can be shown that there are infinite number of weak solutions to this IVP. For instance, if β is any number, satisfying $0 < \beta < 1$, then the function (see Figure 2.1.4)

$$u(x,t) = \begin{cases} 0, & x \leq \beta t/2 \\ \beta, & \beta t/2 < x \leq (\beta+1)t/2 \\ 1, & x > (\beta+1)t/2, \end{cases} \tag{2.1.23}$$

is also a weak solution with two discontinuities propagating with speeds $\beta/2$ and $(\beta+1)/2$. Thus, there is a need to have some mechanism to decide about the admissible weak solution. A criterion that selects the right solution was proposed by Lax [101], according to which a discontinuity in the solution is admissible only if it prevents the intersection of characteristics coming from the points of the initial line on the two sides of it, i.e.,

$$u_l(t) > s(t) > u_r(t). \tag{2.1.24}$$

It may be noticed that the condition (2.1.24) is violated in the solutions given by (2.1.22) and (2.1.23), which show that the characteristics originating from the discontinuity line diverge into the domain on the two sides of it as t increases. This is clearly a nonphysical situation, since the solution should depend upon the initial conditions, not on conditions at the discontinuity; and, therefore, the solutions (2.1.22) and (2.1.23) have to be discarded. The only admissible solution is the continuous one, namely (2.1.16).

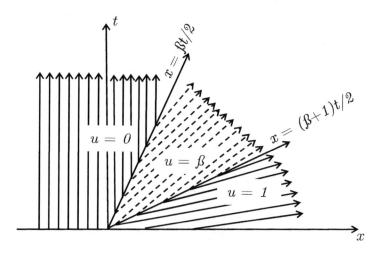

Figure 2.1.4: Weak solutions of (2.1.10) with initial data (2.1.15).

For the general scalar conservation law (2.1.1) with convex or concave f, a generalized or weak solution is called admissible if for each discontinuity, propagating with speed s defined by (2.1.9), the inequality

$$f'(u_l) > s > f'(u_r), \qquad (2.1.25)$$

is satisfied. This is known as the Lax's entropy condition for the flux equation $f''(u) \neq 0$; for convex f, the condition (2.1.25) requires that $u_\ell > u_r$, whereas for concave f, the requirement is $u_\ell < u_r$. In the context of gasdynamics, the corresponding condition excludes discontinuities across which the entropy does not increase.

Definition 2.1.1 A discontinuity satisfying the jump condition (2.1.9) and the entropy condition (2.1.25) is called a shock wave.

It may be remarked that shocks are not associated with only discontinuous initial data; indeed, Example 2.1.1 illustrates that even with smooth initial data, the continuity of the solution breaks down after a finite time $t_c = 8/3^{3/2}$.

2.1.3   Riemann problem

The conservation law together with piecewise constant data having a single discontinuity is known as the Riemann problem; as an example, consider the conservation law (2.1.1) with piecewise constant initial data

$$u(x,0) = \begin{cases} u_l, & x \leq 0 \\ u_r, & x > 0, \end{cases} \tag{2.1.26}$$

where u_l and u_r are constants.

It may be noticed that the stretching transformation $x \to ax, t \to at$ leaves both the equation (2.1.1) and the initial data in (2.1.2) invariant. Thus, if $u(x,t)$ is an entropy solution of (2.1.1), (2.1.2), then for every constant $a > 0$, the function $u_a(x,t) = u(ax, at)$ is also a solution. Since the entropy solution for a given initial condition is unique, we choose $a = 1/t$, and obtain $u(x,t) = u(x/t, 1)$. We, therefore, look for self-similar solutions of the form $u(x,t) = \phi(x/t)$, and consider the Riemann problem (2.1.1), (2.1.26) with the uniformly convex flux function, and $u_l < u_r$; then the convexity implies $f(u_l) < f(u_r)$ and the rarefaction wave is given by

$$u(x,t) = \begin{cases} u_l, & x \leq tf'(u_l) \\ \phi(x/t), & tf'(u_l) < x \leq tf'(u_r) \\ u_r, & x > tf'(u_r), \end{cases} \tag{2.1.27}$$

where $\phi(x/t)$ solves $f'(\phi(x/t)) = x/t$. However, if $u_l > u_r$, then the weak solution, satisfying the entropy condition, is a shock wave, and is given by

$$u(x,t) = \begin{cases} u_l, & x < st \\ u_r, & x > st, \end{cases} \tag{2.1.28}$$

where $s = (f(u_l) - f(u_r))/(u_l - u_r)$.

Thus, the Riemann problem (2.1.1), (2.1.26) with $f'' > 0$ (respectively, $f'' < 0$) is solved uniquely by a rarefaction wave if $u_l < u_r$ (respectively, $u_l > u_r$), and a shock wave if $u_l > u_r$ (respectively, $u_l < u_r$). The solutions of the Riemann problem for a scalar equation, namely shocks and rarefaction waves, are called elementary waves.

Example 2.1.4 (Traffic flow): Let us consider the traffic flow problem described by (1.2.2), in which the flux function $f(u) = u(1 - u)$ is concave, i.e., $f''(u) < 0$ for all u, together with the initial data (2.1.26).

Case 1: Let $u_l = 1/2$ and $u_r = 1$, so that $u_l < u_r$. The characteristics issuing from the negative half-line, entering into the region $t > 0$, are vertical lines along which $u = 1/2$, whereas the characteristics starting from $x > 0$ are straight lines with negative slope, -1, along which $u = 1$. Consequently, the characteristics intersect right at the origin, and there is a need to introduce a shock through the point (0,0) satisfying the R-H jump condition (2.1.9), namely $s = 1 - (u_l + u_r) = -1/2$, and the stability criterion (2.1.24). Thus, the resulting weak solution to the initial value problem is $u(x,t) = 1/2$ for $x \leq -t/2$ and $u(x,t) = 1$ for $x > -t/2$ (see Figure 2.1.5). This solution, in which the density of cars jumps from $u_l = 1/2$ to $u_r = 1$ at the shock, corresponds to the situation where the cars moving at speed $v_l = (1 - u_l) = 1/2$

encounter a traffic jam, and suddenly have to reduce their velocity to $v_r = (1 - u_r) = 0$.

Case 2: Let $u_l = 1$ and $u_r = 1/2$, so that $u_l > u_r$. This initial data may model the situation where a traffic light turns green at $t = 0$. Cars to the left, leaving the traffic light, are initially stationery, $v_l = (1 - u_l) = 0$, with density high ($u_l = 1$), while on the other side of the light, there is a small constant density. Indeed, the weak solution to this initial value problem is $u(x, t) = 1$ for $x < -t$, $u(x, t) = \frac{1}{2}(1 - \frac{x}{t})$ for $-t < x \le 0$, and $u(x, t) = 1/2$ for $x > 0$; see Figure 2.1.6. The solution shows that the cars which were stationary initially, begin to accelerate when the cars in front of them start moving; and, thus, the spreading out of characteristics leads to a spreading out of cars, accelerating out of the high density region into the low density region.

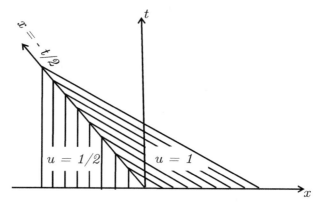

Figure 2.1.5: Solution of (1.2.2) with initial data (2.1.26); $u_l = 1/2$, $u_r = 1$.

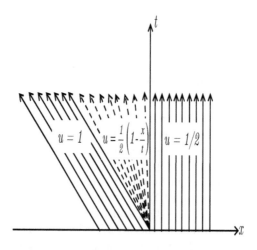

Figure 2.1.6: Solution of (1.2.2) with initial data (2.1.26); $u_l = 1$, $u_r = 1/2$.

2.2 Entropy Conditions Revisited

We have already seen in the preceding subsections that the weak solutions of (2.1.1), in general, are not unique, and there is a need to appeal to some extra admissibility conditions in order to decide which of the many possible weak solutions is relevant to the physical situation; we have already noticed that for a convex flux function $f(u)$, a weak solution of (2.1.1) is admissible only if the Lax's entropy condition (2.1.24) is satisfied across it. For nonconvex flux functions, which appear in several applications, we need more general admissibility conditions (see Holden and Risebro [73] and LeFloch [105]).

2.2.1 Admissibility criterion I (Oleinik)

It was Oleinik [133], who first formulated the generalized entropy condition, which applies to nonconvex scalar functions as well. According to this, the function $u(x,t)$ is an entropy solution of (2.1.1), if a discontinuity propagating with speed s, defined by (2.1.9), satisfies the condition

$$(f(u) - f(u_l))/(u - u_l) \geq s \geq (f(u) - f(u_r))/(u - u_r), \qquad (2.2.1)$$

for all u between u_l and u_r. This condition reduces to the compressibility condition (2.1.25) when the flux function is strictly convex. For a general flux, by taking the limits $u \to u_l$ and $u \to u_r$ in (2.2.1), it implies the weak compressibility condition

$$f'(u_l) \geq s \geq f'(u_r). \qquad (2.2.2)$$

The inequality (2.2.2) has a simple geometric interpretation by noticing that $f'(u_l)$ and $f'(u_r)$ are slopes of the graph of $f(u)$ at the points $(u_l, f(u_l))$ and $(u_r, f(u_r))$, while s is the slope of the chord that connects these two points. In the nonconvex case, the characteristic may be tangent to a discontinuity; in fact, the discontinuity propagating with speed s is called a semi-characteristic shock to the left or to the right according as $f'(u_l) = s \geq f'(u_r)$ or $f'(u_l) \geq s = f'(u_r)$, respectively. However, if $f'(u_l) = s = f'(u_r)$, i.e., the states on both sides propagate with the same speed as the discontinuity, the discontinuity is called a contact discontinuity or a characteristic shock according as the flux function, between u_l and u_r, is affine or not affine (see Serre [164] and Boillat [20]).

2.2.2 Admissibility criterion II (Vanishing viscosity)

In the vanishing viscosity approach, we replace the conservation law (2.1.1) by the higher order equation

$$u_{,t} + (f(u))_{,x} = \epsilon u_{,xx}, \qquad (2.2.3)$$

in which the various dissipative effects make their appearance by the presence of the viscous term $\epsilon u_{,xx}$ with $\epsilon > 0$. Assuming that the viscous equation (2.2.3), satisfying the initial data (2.1.2), has smooth solution $u_\epsilon(x,t)$ and $u = \lim_{\epsilon \to 0} u_\epsilon(x,t)$, then u is called the physically relevant weak solution of (2.1.1). Unfortunately, it is often not so easy to work with it as the solutions of (2.2.3) are not often known for a general $f(u)$; we, therefore, look for simpler conditions, which can be more easily verified in practice. But, before we proceed to the next selection principle, we would like to remark that (2.2.3) with initial data (2.1.2) can be solved explicitly when $f(u) = u^2/2$, using the Cole-Hopf transformation (see Whitham [210]). It can be shown that the exact solution does indeed tend to the weak solution of (2.1.1), which we have discussed in preceding subsections, using characteristics and shock fitting, obeying (2.1.11); indeed, with initial data (2.1.26), the vanishing viscosity solution in the case $u_l < u_r$ is the rarefaction wave, and in the case $u_l > u_r$, it is the shock wave solution, in agreement with the Lax's shock condition (2.1.24).

2.2.3 Admissibility criterion III (Viscous profile)

Another method, based on specific physical arguments, is the viscous profile approach, in which the scalar conservation law (2.1.1) is again replaced by (2.2.3), so that the physical problem has some diffusion. For (2.2.3), a viscous profile is a traveling wave solution $u = u_\epsilon(x - st) = u(\frac{x-st}{\epsilon})$ for some constant $s \in \mathbb{R}$, which is to be determined, and the solution u is required to approach (i) the constant states u_l and u_r as $x - st \to -\infty$ and $+\infty$, respectively, and (ii) the shock wave as $\epsilon \to 0^+$. If a discontinuity has no such profile, it is not admissible. It is easy to see that no smooth traveling wave can exist when $\epsilon = 0$, but for $\epsilon > 0$, we obtain

$$u' = f(u) - su + \alpha, \qquad (2.2.4)$$

where the prime denotes differentiation with respect to the argument function $x - st, \alpha$ is a constant of integration, and the flux function f is assumed to be convex. Thus, if the requirement (i) is imposed, then $\lim_{x-st \to \pm\infty} u' = 0$, and (2.2.4) implies that $\alpha = su_l - f(u_l) = su_r - f(u_r)$, which gives us the R-H condition (2.1.9). In particular, we say that a discontinuity has a viscous profile, or it satisfies the traveling wave entropy condition if the following boundary value problem holds across the discontinuity

$$u' = g(u); \ u(+\infty) = u_r, \ u(-\infty) = u_l,$$

where $g(u) = -s(u - u_r) + f(u) - f(u_r)$, and $s = (f(u_l) - f(u_r))/(u_l - u_r)$. It may be noticed that u_l and u_r are zeros of $g(u)$, and therefore the above ODE has a solution if the function $g(u)$ is negative (respectively, positive) for $u_l > u > u_r$ (respectively, $u_l < u < u_r$). Thus, if $g(u) < 0$ for $u_r < u < u_l$, then we have $u' < 0$, and consequently the graph of $f(u)$ must be below the straight line joining the points $(u_l, f(u_l))$ and $(u_r, f(u_r))$, i.e.,

$$f(u) < f(u_r) + s(u - u_r), \qquad (2.2.5)$$

for all $u \in (u_l, u_r)$. The solution is as shown in Figure 2.2.1 with u increasing monotonically from u_r at $+\infty$ to u_l at $-\infty$. In fact, as $\epsilon \to 0$, the viscous profile in Figure 2.2.1 tends to the discontinuous hyperbolic shock of (2.1.1) – a step function increasing from u_r to u_l, and traveling with speed s; this is because the viscous shock solution is a function of $(x - st)/\epsilon$, which implies that the shock layer is of width ϵ. This discontinuous shock solution is in agreement with Lax's shock inequality (2.1.25). Similarly, if $u_l < u < u_r$, then u' must be positive, and the solution increases from u_l at $-\infty$ to u_r at $+\infty$; indeed, the graph of $f(u)$ must be above the straight line joining the points $(u_l, f(u_l))$ and $(u_r, f(u_r))$, i.e.,

$$f(u) > f(u_l) + s(u - u_l), \qquad (2.2.6)$$

for all $u \in (u_l, u_r)$. As $\epsilon \to 0$, the solution tends to the weak solution of (2.1.1) and (2.1.26), satisfying (2.1.25). When the above two cases are combined, it can be concluded, that the viscous profile or traveling wave entropy condition is equivalent to

$$(f(u) - f(u_l))/(u - u_l) > s > (f(u_r) - f(u))/(u_r - u), \qquad (2.2.7)$$

for all u between u_l and u_r. It may be noticed that the condition (2.2.7) is the strict version of Oleinik's condition. By taking the limits $u \to u_l$ and $u \to u_r$ in (2.2.7), one obtains the Lax's condition (2.1.25), i.e.,

$$f'(u_l) > (f(u_l) - f(u_r))/(u_l - u_r) > f'(u_r).$$

It may be remarked that the converse implication does not hold in the case of a general (nonconvex) flux function f.

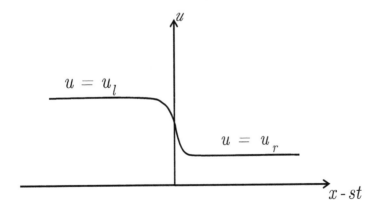

Figure 2.2.1: Viscous profile with shock structure.

It may be noticed that the condition (2.2.1) implies that

$$s|k - u_l| \leq (k - u_l)(f(k) - f(u_l)),$$

for all k between u_l and u_r; in fact, an identical inequality holds with u_l replaced by u_r. These inequalities motivate another entropy condition due to Kruzkov [97], which is often more convenient to work with.

2.2.4 Admissibility criterion IV (Kruzkov)

In this approach to the entropy condition, a continuously differentiable function $\eta(u)$ is called an entropy function for the conservation law (2.1.1) with entropy flux $q(u)$, provided $\eta(u)$ is convex (i.e., $\eta''(u) > 0$), as is the case with the physical entropy in gasdynamics, and

$$\eta'(u)f'(u) = q'(u). \tag{2.2.8}$$

The pair (η, q) is called an entropy/entropy-flux pair for the conservation law (2.1.1). It may be observed that if u is a smooth solution of (2.1.1), then (2.2.8) implies that

$$(\eta(u))_{,t} + (q(u))_{,x} = 0. \tag{2.2.9}$$

In other words, for a smooth solution to (2.1.1), not only the quantity u is conserved, but the additional conservation law (2.2.9) also holds. However, when u is discontinuous, the operations performed above in arriving at (2.2.9) are not valid, i.e., a weak solution of (2.1.1), in general, does not provide a weak solution to (2.2.9). In order to study the behavior of η for the vanishing viscosity weak solution, we deal with the corresponding dissipative equation (2.2.3), and then let the viscosity tend to zero. In fact, it turns out that given a convex entropy η for (2.1.1), with entropy flux q, the weak solution u of (2.1.1) is an entropy solution if it satisfies the following entropy inequality in the sense of distribution (see Evans [54])

$$(\eta(u))_{,t} + (q(u))_{,x} \leq 0. \tag{2.2.10}$$

Following the arguments used in deriving the R-H condition for the conservation law (2.1.1), we have an equivalent form of the entropy condition (2.2.10), namely

$$s(\eta(u_l) - \eta(u_r)) \leq q(u_l) - q(u_r), \tag{2.2.11}$$

where u_l and u_r are respectively the states on the left and right sides of the discontinuity, propagating with speed s. Let us now consider the entropy/entropy-flux pair of Kruzkov

$$\eta(u) = |u - k|, \ q(u) = \text{sgn}(u - k)(f(u) - f(k)),$$

where k is a parameter taking values in $\mathbb{R}$. One can easily check that η, q satisfy (2.2.8) at every $u \neq k$, and with the pair (η, q), the inequality (2.2.11) yields

$$s(|u_l - k| - |u_r - k|) \leq \text{sgn}(u_l - k)(f(u_l) - f(k)) - \text{sgn}(u_r - k)(f(u_r) - f(k)),$$
$$(2.2.12)$$

for all $k \in \mathbb{R}$, which is known as the Kruzkov entropy condition. It can be easily shown that for $u_l < u_r$ (respectively, $u_l > u_r$) and k lying between u_l and u_r, the Kruzkov entropy condition (2.2.12) implies the Oleinik entropy condition (2.2.1).

Here, we consider the case $u_l < u_r$. Let us choose $k \geq u_r$ and $k \leq u_l$, successively; then the entropy condition (2.2.12), with these choices, yields two inequalities, which together express the R-H condition (2.1.9). Finally, if $u_l < k < u_r$, then from (2.2.12), we have $s(u_r + u_l - 2k) \geq f(u_r) + f(u_l) - 2f(k)$. But $(u_r - u_l)s = f(u_r) - f(u_l)$, with the result that $s(u_r - k) \geq f(u_r) - f(k)$ and $s(u_l - k) \geq f(u_l) - f(k)$, which are equivalent to the Oleinik condition (2.2.1).

2.2.5   Admissibility criterion V (Oleinik)

If the flux function in (2.1.1) is uniformly convex, i.e., there exists a number $\theta > 0$ such that $f''(u) \geq \theta > 0$ for all u, $f(0) = 0$ and $u_0(x)$ is bounded and integrable, then we have an explicit form of the solution $u(x,t)$ of (2.1.1), (2.1.2) known as Lax-Oleinik formula, which satisfies another form of entropy condition given below.

According to this criterion, the function $u(x,t)$ is the entropy solution of (2.1.1) and (2.1.2), if there exists a constant $c > 0$ such that for all $h > 0, t > 0$ and $x \in \mathbb{R}$,

$$\frac{u(x+h,t) - u(x,t)}{h} \leq \frac{c}{t}. \tag{2.2.13}$$

The above form (2.2.13) turns out to be useful in studying numerical methods; for details, see Smoller [189], LeVeque [106], and Evans [54].

Remarks 2.2.1:

(i) In order to realize (2.2.13), we notice that for smooth initial data $u_0(x)$, with $u_0' > 0$, the solution (2.1.3) of the initial value problem (2.1.1), yields $u_{,x} = u_0'/(1 + tu_0' f''(u_0))$. This shows that if f is uniformly convex, i.e., $f''(u) \geq \theta > 0$ for all u, then $u_{,x} < c/t$, where $c = 1/\theta$; this is indeed, (2.2.13). Furthermore, if we fix $t > 0$, and choose x and $x + h$ on the left and right sides of the discontinuity, where h is a measure of the discontinuity width, much smaller than the jump in u, then simple reasoning suggests that $u_r - u_l \leq ch/t \approx 0$; this implies the entropy condition $u_l \geq u_r$, i.e., if we let x go from $-\infty$ to $+\infty$, then the quantity u can only jump down. For details, we refer to Smoller [189] and Evans [54].

(ii) We may recall (2.1.27), as a solution of the Riemann problem (2.1.1), (2.1.26) with a uniformly convex flux function f and $u_l < u_r$. In the fan-like region $f'(u_l) < \frac{x}{t} < f'(u_r)$, the solution $u = \phi(x/t)$ is given by solving the equation $f'(\phi(\frac{x}{t})) = \frac{x}{t}$ for ϕ. The solution is smooth everywhere in $t > 0$, but is only Lipschitz continuous on the two lines $x = tf'(u_l)$ and $x = tf'(u_r)$. It may be noticed that in this region, the solution satisfies $u(x + h, t) - u(x, t) = \phi(\frac{x+h}{t}) - \phi(\frac{x}{t}) \leq \text{Lip} (\phi)\frac{h}{t}$, if $tf'(u_l) < x < x + h < tf'(u_r)$; here Lip (ϕ) is a Lipschitz constant of ϕ, and is finite. This inequality implies that u satisfies the entropy condition (2.2.13).

(iii) For global existence of weak solutions to the initial value problem, (2.1.1), (2.1.2) with a smooth convex flux, and bounded initial data with bounded total variation, the reader is referred to Chang and Hsiao [31].

2.3   Riemann Problem for Nonconvex Flux Function

We have already seen in subsection 2.1.3 that when the flux function f in (2.1.1) is convex, the solution to the Riemann problem (2.1.1), (2.1.26) is always either a shock or a rarefaction wave. For nonconvex f, the Oleinik condition (2.2.1) suggests that the entropy solution to the Riemann problem might involve both a shock and a rarefaction wave. In fact, the entropy solution to the Riemann problem (2.1.1), (2.1.26) can be determined from the graph of f in a simple manner. Assume that $f \in C^2$ has only one inflection point between u_l and u_r, and consider the case

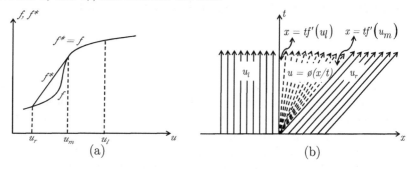

Figure 2.3.1: (a) Concave hull of f; (b) Solution of (2.1.1) and (2.1.26) with nonconvex f.

(i) $u_l > u_r$: In order to solve this problem, we consider the smallest concave function f^* such that $f^*(u) \geq f(u)$ for all $u \in [u_r, u_l]$; indeed, an elastic string stretched above the graph from $(u_r, f(u_r))$ to $(u_l, f(u_l))$ will follow the graph of f^*. We, thus, observe that f^*, called concave hull of f, consists of the straight line segment, connecting the two points $(u_r, f(u_r))$ and $(u_m, f(u_m))$, which is tangent to the graph of f at the point $(u_m, f(u_m))$, together with the portion of f over $[u_m, u_l]$; see Figure 2.3.1(a). It turns out that the entropy solution consists of a semi-characteristic shock joining u_r to u_m, where u_m can be calculated from $f'(u_m) = \{f(u_m) - f(u_r)\}/(u_m - u_r)$, which is precisely the shock speed, and a centered simple wave joining u_m and u_l, i.e.,

$$
u(x, t) = \begin{cases} u_l, & x < f'(u_l) \\ \phi(x/t), & tf'(u_l) \le x \le tf'(u_m) \\ u_r, & x > tf'(u_m), \end{cases} \tag{2.3.1}
$$

where ϕ solves $f'(\phi(x/t)) = x/t$; it may be noticed that the solution is continuous across the line $x = tf'(u_l)$, whereas it is discontinuous across the line $x = tf'(u_m)$ with left and right states u_m and u_r, respectively (see Figure 2.3.1(b)).

(ii) $u_l < u_r$: In this case, we consider the largest convex function f_* such that $f_*(u) \le f(u)$ for all $u \in [u_l, u_r]$; observe that f_*, called convex hull of f, consists of the portion of f over $[u_l, u_m]$ together with the straight line segment, connecting the points $(u_m, f(u_m))$ and $(u_r, f(u_r))$, which is tangent to the graph of f at $(u_m, f(u_m))$. The entropy solution consists of a centered simple wave joining u_l and u_m, and a shock that connects u_m and u_r, where u_m is given by $f'(u_m) = \{f(u_m) - f(u_r)\}/(u_m - u_r)$; see Figure 2.3.2.

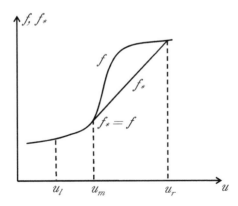

Figure 2.3.2: Convex hull of f.

2.4   Irreversibility

We have already noticed some interesting results about the solution of non-linear equations that have no counterpart for linear equations. For instance, unlike linear equations, not only that the breakdown in the solution of nonlinear equations can happen with smooth initial data (see Example 2.1.1), but also a continuous solution can have discontinuous initial data (see Example 2.1.2). Furthermore, the physical processes described by smooth solutions of hyperbolic equations are reversible in time, in the sense that if we know the solution at $t = t_0$, then the solution at any $t > t_0$ or $t < t_0$ can be determined uniquely. But, if the process is described by a discontinuous solution, then there is a high degree of irreversibility in the sense that the solution may not be traced back uniquely in time. We shall illustrate this by means of an example.

Example 2.4.1: Consider the conservation law (2.1.10) with the initial data

$$u(x,0) = \begin{cases} 1, & -\infty < x \le -1 \\ 0, & -1 < x \le 0 \\ 2, & 0 < x \le 1 \\ 0, & x > 1. \end{cases} \qquad (2.4.1)$$

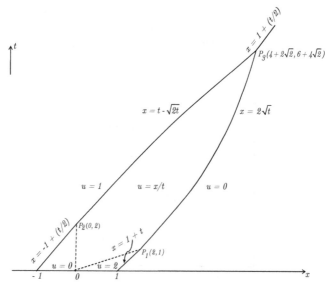

Figure 2.4.1: Solution of (2.1.10) with initial data (2.4.1).

The compressive parts of the wave at $x = -1$ and $x = 1$ break immediately and require the introduction of shocks, originating from $(-1, 0)$ and $(1, 0)$, with speeds $s = 1/2$ and 1, respectively. The equations of shock fronts originating from $(-1, 0)$ and $(1, 0)$ are $x = -1 + (t/2)$ and $x = t + 1$, respectively; the shock paths are straight as long as they continue to have $u = 1$ (respectively, $u = 2$) on the left and $u = 0$ (respectively, $u = 0$) to the right; see Figure 2.4.1. Since the rarefaction wave originating from $(0, 0)$, overtakes the shock $x = t + 1$ at point $P_1(2, 1)$, and it is overtaken by the left shock $x = -1 + (t/2)$ at $P_2(0, 2)$, the shapes and speeds of the two shocks get altered beyond P_1 and P_2, which we obtain by invoking the R-H condition (2.1.11). Indeed, the shock paths beyond P_1 and P_2 turn out to be parabolic, namely, $x = t - \sqrt{2t}$ and $x = \sqrt{2t}$, respectively.

At $P_3(4 + 2\sqrt{2}, 6 + 4\sqrt{2})$, the two shocks eventually collide and merge into a single shock $x = (t/2) + 1$, leading to an entirely different state. The stability criterion (2.1.24) ensures that the solution has four distinct time intervals, describing four different states:

$$u(x, t) \;=\; \begin{cases} 1, & x \leq -1 + (t/2) \\ 0, & -1 + (t/2) < x \leq 0 \\ x/t, & 0 < x \leq 2t \\ 2, & 2t < x \leq t + 1 \\ 0, & x > t + 1, \end{cases} \qquad \text{for } 0 < t < 1;$$

$$u(x, t) \;=\; \begin{cases} 1, & x \leq -1 + (t/2) \\ 0, & -1 + (t/2) < x \leq 0 \\ x/t. & 0 < x \leq 2\sqrt{t} \\ 0, & x > 2\sqrt{t} \end{cases} \qquad \text{for } 1 \leq t < 2; \tag{2.4.2}$$

$$u(x, t) \;=\; \begin{cases} 1, & x \leq t - 2\sqrt{t} \\ x/t, & t - 2\sqrt{t} < x \leq 2\sqrt{t} \\ 0 & x > 2\sqrt{t} \end{cases} \qquad \text{for } 2 \leq t < 6 + 4\sqrt{2};$$

$$u(x, t) \;=\; \begin{cases} 1, & x \leq 1 + (t/2), \\ 0, & x > 1 + (t/2) \end{cases} \qquad \text{for } t \geq 6 + 4\sqrt{2}.$$

Example 2.4.2: The unique admissible weak solution of (2.1.10), satisfying the initial data

$$u(x, 0) = \begin{cases} 1, & x \leq 1 \\ 0, & x > 1, \end{cases} \tag{2.4.3}$$

is

$$u(x, t) = \begin{cases} 1, & x \leq 1 + (t/2) \\ 0, & x > 1 + (t/2) \end{cases} \quad \text{for } t \geq 0. \tag{2.4.4}$$

It may be noticed that the two different initial value problems, in the above two examples, represent the same state, namely

$$u(x, t) = \begin{cases} 1, & x \leq 1 + (t/2) \\ 0, & x > 1 + (t/2) \end{cases} \quad \text{for } t \geq 6 + 4\sqrt{2}.$$

In fact, the solution at time $t \geq 6 + 4\sqrt{2}$ forgets the initial details, and it is not possible to trace it back uniquely in time in the sense that it corresponds to two different initial states (2.4.1) and (2.4.3). It is in this sense that solutions of nonlinear conservation laws are irreversible.

2.5   Asymptotic Behavior

Let us consider the conservation law (2.1.10) with the initial data $u(x, 0) = u_0(x), x \in \mathbb{R}, t > 0$ where $u_0(x)$ is bounded, integrable, and represents a single hump disturbance with $u_0(x) \equiv 0$ outside the range $x_1 < x < x_2$, and $u_0(x) > 0$ in the range. Then the integral form of the conservation laws implies that

$$\int_{-\infty}^{\infty} u(x,t)dx = \int_{-\infty}^{\infty} u_0(x)dx, \quad \text{for all } t > 0. \qquad (2.5.1)$$

Since $\int_{x_1}^{x_2} u(x,t)dx = \int_{x_1}^{x_2} u_0(x)dx \equiv \Delta$, where Δ is the area under the initial hump (see Figure 2.5.1(a)), holds for both the discontinuous solution in Figure 2.5.1(b) and the multivalued solution in Figure 2.5.1(c), we must insert the shock into the profile at a position so as to give equal areas $\Delta_1 = \Delta_2$ for the two lobes, as shown in Figure 2.5.1(d). This is known as the equal area rule, and the procedure, known as shock fitting, is applicable when there is only one shock; for its implementation and justification, see Whitham [210].

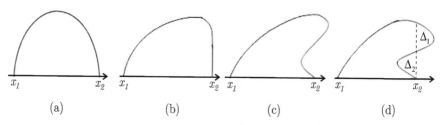

| (a) | (b) | (c) | (d) |

Figure 2.5.1: Distortion of the initial profile and shock fitting.

The behavior of the solution as $t \to \infty$ can be studied without going through the above construction in detail. Let us consider the following examples.

Example 2.5.1: Consider the solution of the conservation law (2.1.10) with the initial data

$$u_o(x) = \begin{cases} 0, & x < -1 \\ -A & -1 < x < 0 \\ A, & 0 < x < 1 \\ 0, & x > 1, \end{cases} \qquad (2.5.2)$$

where A is a positive constant.

We notice that the compressive parts of the wave at $x = -1$ and $x = 1$ break immediately and require the introduction of shocks; see Figure 2.5.2(b). The shock path originating at $x = -1$ (respectively, $x = 1$) is given by $x = -(At/2) - 1$ (respectively, $x = (At/2) + 1$) with the state $u = 0$ (respectively, $u = A$) to the left of it and $u = -A$ (respectively, $u = 0$) to the right. The centered rarefaction wave $u = x/t$ originates from $(0,0)$ with $u = -A$ to the left of it and $u = A$ to the right; the trailing (respectively, leading) front of the rarefaction wave overtakes the left (respectively, right) running shock at the point P_1: $x = -2$, $t = 2/A$ (respectively, P_2: $x = 2$, $t = 2/A$). The shock path beyond P_1 (respectively, P_2) turns out to be parabolic, namely $x = -\sqrt{2At}$ (respectively, $x = \sqrt{2At}$) which separates a state $u_l = 0$ (respectively, $u_l = \sqrt{2A/t}$) from the state $u_r = -\sqrt{2A/t}$ (respectively, $u_r = 0$). Thus, the solution has two distinct time intervals describing two different states, namely,

$$
u(x,t) \;=\; \begin{cases}
0, & x < -1 - (At/2) \\
-A, & -1 - (At/2) < x < -At \\
x/t, & -At < x < At \\
A, & At < x < 1 + (At/2) \\
0, & x > 1 + (At/2),
\end{cases}
\qquad \text{for } 0 < t < 2/A
$$

$$
u(x,t) \;=\; \begin{cases}
0, & x < -\sqrt{2At} \\
x/t, & -\sqrt{2At} < x < \sqrt{2At} \\
0, & x > \sqrt{2At},
\end{cases}
\qquad \text{for } t \geq 2/A.
$$

It may be noticed that the area of the pulse on the right, as well as the left, of $x = 0$ in Figures 2.5.2(a) – (d) remains equal to the constant A, i.e., the area under the initial curve is conserved; this is in agreement with the property (2.5.1). Moreover, the initial profile with positive and negative parts tends to a wave pattern, which has the shape of inverted N, called an N wave, and the shock amplitude $u_l - u_r$ decays to zero like $t^{-1/2}$ as $t \to \infty$. Indeed, it is only the area under the initial pulse that appears in the asymptotic solution. In fact, unlike the linear case, all other initial details are completely lost.

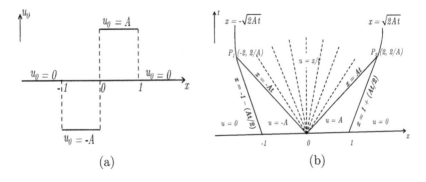

(a)   (b)

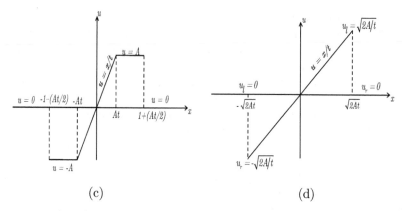

(c)   (d)

Figures 2.5.2(a–d): Solution of (2.1.10) with initial data (2.5.2), showing that the area under the initial curve is conserved.

Example 2.5.2: Consider the following initial value problem in $\mathbb{R} \times (0, \infty)$

$$u_{,t} + (u^2/2)_{,x} = 0, \quad u(x,0) = \begin{cases} 2, & |x| \geq 2 \\ -x, & -2 < x < -1 \\ x + 2, & |x| \leq 1 \\ 4 - x, & 1 < x < 2. \end{cases} \quad (2.5.3)$$

We examine the pattern of characteristics and associated values of u that emerge from different portions of the x-axis.

Starting with $x \leq -2$, we have $u = 2$ on the family of straight lines $x = 2t + \xi, -\infty < \xi \leq -2$, and thus $u = 2$ for $x \leq 2t - 2$.

The next segment of x-axis has a compression wave $u = x/(t - 1)$ for $2(t-1) < x \leq t-1$; in this triangle shaped region u decreases from 2 to 1 on the x-interval, whose length tends to zero as t increases to 1; see Figure 2.5.3(a). At the point $(0, 1)$, a continuous solution is overdetermined; in fact, the graph of u steepens as $t \uparrow 1$. Thus, there is a need to introduce a shock, originating from $(0, 1)$ into the region $t > 1$, by invoking the R-H condition (2.1.11). As the next segment of x-axis has an expansion wave $u = (x + 2)/(t + 1)$ for $t - 1 \leq x \leq 3t + 1$, the shock originating from $(0, 1)$ has $u = 2$ to the left and $u = (x + 2)/(t + 1)$ to the right; its trajectory is given by $x = 2t - \sqrt{2(t + 1)}$, which separates a state $u_l = 2$ from a state $u_r = (x + 2)/(t + 1) = 2 - \sqrt{2/(t + 1)}$; see Figure 2.5.3(b).

The next segment of x-axis has another compression wave $u = (x-4)/(t-1)$ for $3t+1 < x < 2t+2$, across which u decreases from 3 to 2 on the x-interval, whose length tends to zero as $t \uparrow 1$. Again, at the point $(4, 1)$, a continuous solution is overdetermined, and there is a need to introduce a shock by invoking the R-H condition. As the segment of x-axis past $x = 2$ has $u = 2$ for $x \geq 2t+2$, the shock emanating from $(4, 1)$ has $u = (x + 2)/(t + 1)$ to the left and $u = 2$ to the right; its trajectory in the region $t > 1$ is given by $x = 2t + \sqrt{2(t + 1)}$, which separates a state $u_l = (x + 2)/(t + 1) = 2 + \sqrt{2/(t + 1)}$ from the state

$u_r = 2$. Thus, the solution has two distinct time intervals describing two different states, namely,

$$u(x,t) = \begin{cases} 2, & x \le 2t - 2, \\ x/(t-1), & 2(t-1) < x < t-1, \\ (x+2)/(t+1), & t-1 \le x \le 3t+1, \\ (x-4)/(t-1), & 3t+1 < x < 2(t+1), \\ 2, & x \ge 2(t+1), \end{cases} \quad \text{for } 0 < t < 1$$

$$u(x,t) = \begin{cases} 2, & x \le 2t - \sqrt{2(t+1)}, \\ (x+2)/(t+1), & 2t - \sqrt{2(t+1)} < x < 2t + \sqrt{2(t+1)}, \\ 2, & x \ge 2t + \sqrt{2(t+1)}. \end{cases} \quad \text{for } t \ge 1$$

The wave form for $t \ge 1$ is an N wave; see Figure 2.5.3(b); the shock amplitude, $u_l - u_r$, decays to zero like $t^{-1/2}$, as $t \to \infty$.

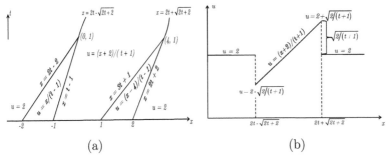

(a)   (b)

Figure 2.5.3: (a) Characteristics curves in the $x-t$ plane and the development of shocks; (b) Graph of the solution of (2.5.3) for $t \ge 1$

Remarks 2.5.1: For the general scalar conservation law (2.1.1) with initial condition (2.1.2), we have the following decay estimate:

If $u_0(x)$ is a bounded, integrable function having compact support, and the flux function f is smooth and uniformly convex with $f(0) = 0$, then there exists a constant k such that $|u(x,t)| \le kt^{-1/2}$ for all $x \in \mathbb{R}$ and $t > 0$.

However, it is interesting to note that the asymptotic behavior of shocks that originated from periodic data, the shock strength, $u_l - u_r$, decays like t^{-1} for large t, rather that $t^{-1/2}$, i.e., the solution of a periodic wave would decay faster than the solution in a single hump; see Lax [100], Whitham [210], Smoller [189], and Evans [54].

Chapter 3

Hyperbolic Systems in One Space Dimension

In this chapter, we study certain properties which ensure that the Cauchy problem for a first order system of conservation laws is well posed. We begin with the study of weak solutions, satisfying the system of conservation laws in the distributional sense, and then outline the admissibility conditions which ensure uniqueness of solutions. Notions of simple waves and Riemann invariants are introduced, and some general properties of shocks and rarefaction waves are noted. The Riemann problem for shallow water equations is presented with some general remarks given at the end.

3.1   Genuine Nonlinearity

We begin by considering the one-dimensional system of conservation laws

$$\mathbf{u}_{,t} + (\mathbf{f}(\mathbf{u}))_{,x} = 0, \; x \in \mathbb{R}, \; t \geq 0 \tag{3.1.1}$$

where $\mathbf{u} = \mathbf{u}(x,t)$ and $\mathbf{f}(\mathbf{u})$ are n vectors, representing the density and the flux functions. Assume that $\mathbf{f}(\mathbf{u})$ is a C^2 smooth function, and let $\mathbf{A}(\mathbf{u}) = (\partial f_i / \partial u_j)_{1 \leq i,j \leq n}$ be the $n \times n$ Jacobian matrix. The system (3.1.1) is called hyperbolic, if for each $\mathbf{u}$ the matrix $\mathbf{A}(\mathbf{u})$ is diagonalizable with real eigenvalues $\lambda_1 \leq \lambda_2 \leq \ldots \leq \lambda_n$; if all the eigenvalues are distinct, the system is called strictly hyperbolic. Let $\mathbf{r}_i(\mathbf{u})$ be a right eigenvector corresponding to the eigenvalue $\lambda_i(\mathbf{u})$, referred to as the ith characteristic field. The system is said to be genuinely nonlinear (convex) in the ith characteristic field if, and only if

$$\nabla \lambda_i(\mathbf{u}) \cdot \mathbf{r}_i \neq 0 \; \text{ for all } \mathbf{u}, \tag{3.1.2}$$

where ∇ is the gradient operator with respect to the elements of $\mathbf{u}$; the system is called genuinely nonlinear, if it is so in all the characteristic fields. This implies that by virtue of the continuity of first derivatives, $\lambda_i(\mathbf{u})$ is monotonically increasing or decreasing as $\mathbf{u}$ varies along the integral curve of the vector field $\mathbf{r}_i(\mathbf{u})$, which is everywhere tangent to $\mathbf{r}_i(\mathbf{u})$; for a scaled $\mathbf{r}_i$, the nonvanishing condition (3.1.2) takes the form

$$\nabla \lambda_i \cdot \mathbf{r}_i = 1 \; \text{ for all } \mathbf{u}. \tag{3.1.3}$$

If instead

$$\nabla \lambda_i \cdot \mathbf{r}_i \equiv 0 \ \text{ for all } \mathbf{u}, \tag{3.1.4}$$

then the ith characteristic field is said to be linearly degenerate or equivalently, the system is called linearly degenerate in the ith characteristic field; indeed, $\lambda_i = \lambda_i(\mathbf{u})$ is constant along every integral curve of $\mathbf{r}_i$.

Example 3.1.1: For $n = 1$, (3.1.1) represents a single conservation law and $\lambda_1(u) = f'(u)$, while $r_1(u) = 1$ for all u. The condition (3.1.2) reduces to the convexity requirement $f''(u) \neq 0$, which corresponds to the notion of genuine nonlinearity. On the other hand if $\lambda_1(u) = $ constant, this corresponds to the linearly degenerate case of a linear equation with constant coefficients.

Example 3.1.2: The gasdynamic equations (1.2.11) for a planar motion become

$$\rho_{,t} + (\rho u)_{,x} = 0, \ \ (\rho u)_{,t} + (p + \rho u^2)_{,x} = 0, \ \ (\rho E)_{,t} + (u(p + \rho E))_{,x} = 0, \tag{3.1.5}$$

with $p = p(\rho, e)$ as the equation of state; here $E = e + (u^2/2)$ is the total specific energy. We know that for smooth flows, the system (3.1.5) can be equivalently written in a nonconservative form (1.2.9), namely

$$\rho_{,t} + u\rho_{,x} + \rho u_{,x} = 0, \ \ u_{,t} + uu_{,x} + (1/\rho)p_{,x} = 0, \ \ S_{,t} + uS_{,x} = 0, \tag{3.1.6}$$

with $p = p(\rho, S)$ as the equation of state. The Jacobian of the system (3.1.6) has distinct eigenvalues.

$$\lambda_1 = u - a < \lambda_2 = u < \lambda_3 = u + a, \tag{3.1.7}$$

where $a = \sqrt{(\partial p/\partial \rho)_S}$ is the speed of sound; the associated eigenvectors can be taken as

$$\mathbf{r}_1 = (\rho, -a, 0)^{tr}, \mathbf{r}_2 = (\partial p/\partial s, 0, -a^2)^{tr}, \mathbf{r}_3 = (\rho, a, 0)^{tr}. \tag{3.1.8}$$

Thus, we have

$$\nabla \lambda_1 \cdot \mathbf{r}_1 = -(a + \rho \partial a/\partial \rho), \ \ \nabla \lambda_2 \cdot \mathbf{r}_2 = 0, \ \ \nabla \lambda_3 \cdot \mathbf{r}_3 = a + \rho \partial a/\partial \rho, \tag{3.1.9}$$

implying thereby that the first and third characteristic fields are genuinely nonlinear, while the second one is linearly degenerate.

3.2   Weak Solutions and Jump Condition

Because of the dependence of λ_i on the dependent variable $\mathbf{u}$, smooth solutions to the system (3.1.1), in general, do not exist globally in time, and one is bound to consider weak solutions.

The vector function $\mathbf{u}(x,t)$ is called a generalized or weak solution of (3.1.1) if it satisfies the integral form of the conservation law, i.e., if

$$\int_{x_1}^{x_2} \mathbf{u}(x,t_2)dx - \int_{x_1}^{x_2} \mathbf{u}(x,t_1)dx = \int_{t_1}^{t_2} (\mathbf{f}(\mathbf{u}(x_1,t)) - f(\mathbf{u}(x_2,t)))dt,$$

holds for all x_1, x_2, t_1 and t_2. An alternative formulation of the weak solution, by integrating (3.1.1) against smooth test functions with compact support, leads to the following definition.

Definition 3.2.1. A bounded measurable function $\mathbf{u}(x,t)$ with a given initial condition

$$\mathbf{u}(x,0) = \mathbf{u}_0(x), \tag{3.2.1}$$

is a weak solution of (3.1.1) if, and only if

$$\int_0^\infty \int_{-\infty}^\infty (\phi_{,t}\mathbf{u} + \phi_{,x}f(\mathbf{u}))dxdt + \int_{-\infty}^\infty u_0(x)\phi(0,x)dx = 0, \tag{3.2.2}$$

for every smooth function $\phi(x,t)$ with compact support in the set $\{(x,t)|\ x \in \mathbb{R}, t \geq 0\}$. A discontinuity in the weak solution satisfying the Rankine-Hugoniot jump condition

$$s[\mathbf{u}] = [f(\mathbf{u})], \tag{3.2.3}$$

where s is the speed of discontinuity $x = x(t)$, follows from (3.2.2) by choosing the test function to concentrate at the discontinuity (see Smoller [189]).

3.3 Entropy Conditions

As in the scalar case, weak solution of the Cauchy problem (3.1.1), (3.2.1) is not necessarily unique, and therefore we need to introduce some admissibility criterion, which enables us to choose the physically relevant solution among all the weak solutions.

3.3.1 Admissibility criterion I (Entropy pair)

Given a smooth convex entropy $\eta(\mathbf{u})$ for (3.1.1) with entropy flux $q(\mathbf{u})$, i.e.,

$$\nabla q = \nabla\eta(\nabla\mathbf{f}(\mathbf{u})), \tag{3.3.1}$$

the weak solution $\mathbf{u}(x,t)$ of (3.1.1), (3.2.1) is an entropy solution if it satisfies the inequality

$$(\eta(\mathbf{u}))_{,t} + (q(\mathbf{u}))_{,x} \leq 0, \tag{3.3.2}$$

in the distributional sense.

Indeed, it can be shown that a piecewise C^1 function $\mathbf{u}$, which is a weak solution of (3.1.1) and (3.2.1) satisfies (3.3.2) if, and only if, it satisfies the jump inequality

$$s[\eta(\mathbf{u})] - [q(\mathbf{u})] \leq 0, \tag{3.3.3}$$

where s is the speed of discontinuity; see, for details, Lax [102].

It may be noticed that (3.3.1) is a first order system of n equations for the scalar variables η and q; thus, for $n \geq 3$, the system is overdetermined and may not have solutions. However, in many practical problems, it is possible to find an entropy function, which has a physical meaning.

Example 3.3.1: Let us consider the following model for isentropic or polytropic gasdynamics, referred to as the p-system (in Lagrangian coordinates)

$$V_{,t} - u_{,x} = 0, \; u_{,t} + (p(V))_{,x} = 0, \; x \in \mathbb{R}, \; t > 0, \tag{3.3.4}$$

where V is the specific volume i.e., $V = 1/\rho$ with ρ as density, u the velocity and $p = p(V)$ the pressure, which is a known function of V. This system is strictly hyperbolic provided that we assume $p'(V) < 0$; in that case, the functions, $\eta(V, u) = (u^2/2) - \int_0^V p(s)ds$, and $q(V, u) = up(V)$ constitute an entropy pair (η, q) for the p-system (3.3.4). Note that the Hessian matrix of η, namely, $H \equiv (\partial^2\eta/\partial u_i \partial u_j)_{1\leq i,j\leq 2}$ is given by

$$H(V, u) = \begin{bmatrix} -p'(V) & 0 \\ 0 & 1 \end{bmatrix},$$

which is positive-definite and, thus, η is strictly convex.

3.3.2   Admissibility criterion II (Lax)

Another admissibility criterion due to Lax ([101]) is particularly useful as its application is not restricted to piecewise smooth solutions. For a strictly hyperbolic system (3.1.1), whose solution $\mathbf{u}(x, t)$ has a jump discontinuity, it can be stated as follows.

Let $\mathbf{u}_l$ and $\mathbf{u}_r$, respectively, be the values of $\mathbf{u}$ on the left and right sides of the discontinuity, moving with speed s, satisfying

$$s(\mathbf{u}_l - \mathbf{u}_r) = f(\mathbf{u}_l) - f(\mathbf{u}_r). \tag{3.3.5}$$

Then the discontinuity is admissible, provided that for some index $k, 1 \leq k \leq n$, the following inequalities hold if the kth characteristic field is genuinely nonlinear

$$\lambda_k(\mathbf{u}_l) > s > \lambda_k(\mathbf{u}_r), \quad \lambda_{k-1}(\mathbf{u}_l) < s < \lambda_{k+1}(\mathbf{u}_r). \tag{3.3.6}$$

The inequalities (3.3.6), which persist under small perturbations, i.e.,

$$|\mathbf{u_l} - \mathbf{u_r}| \ll 1,$$

can be written in the form

$$\lambda_{k-1}(\mathbf{u}_l) < s < \lambda_k(\mathbf{u}_l) \text{ and } \lambda_k(\mathbf{u_r}) < s < \lambda_{k+1}(\mathbf{u_r}),$$

implying thereby that $n - k + 1$ (respectively, k) characteristics with speeds $\lambda_k(\mathbf{u}_l), \lambda_{k+1}(\mathbf{u}_l), \ldots, \lambda_n(\mathbf{u}_l)$ (respectively, $\lambda_1(\mathbf{u_r}), \lambda_2(\mathbf{u_r}), \ldots, \lambda_k(\mathbf{u_r})$) impinge on the line of discontinuity from the left (respectively, right). Thus, at a point of discontinuity there are $n + 1$ incoming characteristics, leading to $n + 1$ scalar data; this, together with n jump conditions (3.3.5), suffices to determine uniquely the $2n + 1$ unknowns $\mathbf{u}_l, \mathbf{u}_r$ and s.

Remarks 3.3.1: If a discontinuity has less (respectively, more) than $n + 1$ characteristics impinging on it, it is called undercompressive (respectively, overcompressive). Such discontinuities occur in physical phenomena like magnetohydrodynamics (MHD), nonlinear elasticity, multiphase flows and combustion, where the governing system of PDEs is not strictly hyperbolic. For such systems, the concept of entropy is considerably more difficult, and is still an area of current research. It is conjectured that the entropy solutions are necessarily unique, but the conjecture is still wide open in the case of systems, even in one-dimension, when the system is nonconvex and the jump in the initial states is not small enough. It is believed that every entropy solution can be considered as a viscosity limit, but the criterion of viscous profiles is also inadequate to resolve the nonuniqueness of solutions (see, for instance, Azevedo et al. [8], Shearer [183], and Wu [213]).

For global existence of weak solutions to the initial value problem (3.1.1), (3.2.1), whose characteristic fields are all genuinely nonlinear and total variation of initial data is sufficiently small, the reader is referred to the fundamental paper of Glimm [62].

3.3.3 k-shock wave

A discontinuity satisfying (3.3.5) and (3.3.6) is called a k-shock wave, or simply a k-shock. In fact, it is assumed that $\lambda_0 = -\infty$ and $\lambda_{n+1} = \infty$, so that the conditions (3.3.6) assert that no two shocks of different families fit in a single inequality, and the characteristics in the kth family, just as in the scalar case, disappear into the shock as time advances, resulting in its stability. It may be remarked that for discontinuities of small amplitude (weak) shocks, the entropy condition (3.3.3) is equivalent to the Lax's entropy condition; for details concerning uniqueness and stability of entropy weak solutions, the reader is referred to Smoller [189], Dafermos [45], and Bressan [25].

3.3.4 Contact discontinuity

A discontinuity in a linearly degenerate field is called a contact discontinuity. If $\mathbf{u}_l$ and $\mathbf{u}_r$ are connected by a discontinuity in the kth characteristic field, which is linearly degenerate, then it can be shown that $\mathbf{u}_l$ and $\mathbf{u}_r$ lie on the same integral curve of $\mathbf{r}_k(\mathbf{u})$ so that

$$\lambda_k(\mathbf{u}_l) = s = \lambda_k(\mathbf{u}_r). \tag{3.3.7}$$

Equation (3.3.7) asserts that the characteristics in the kth family are parallel to the contact discontinuity propagating with speed s.

Thus, for systems with both genuinely nonlinear and linearly degenerate fields, the Lax's shock inequalities (3.3.6) should read

$$\lambda_k(\mathbf{u}_\ell) \geq s \geq \lambda_k(\mathbf{u}_r), \ \lambda_{k-1}(\mathbf{u}_\ell) < s < \lambda_{k+1}(\mathbf{u}_r). \tag{3.3.8}$$

3.4   Riemann Problem

The Riemann problem consists in finding weak solutions of (3.1.1) with initial data

$$\mathbf{u}(x,0) = \begin{cases} \mathbf{u}_\ell, & x < 0 \\ \mathbf{u}_r, & x > 0, \end{cases} \tag{3.4.1}$$

where $\mathbf{u}_\ell$ and $\mathbf{u}_r$ are constant vectors.

This problem occurs naturally in the shock-tube experiment of gasdynamics, where a long thin cylindrical tube has a gas in two different states, separated by a thin membrane. The gas is at rest on both sides of the membrane with constant pressures and densities on each side. The membrane is removed at time $t = 0$, and the problem is to determine the subsequent gas motion, which will be a solution of a Riemann problem. Here, we are concerned with the piecewise smooth solutions of (3.1.1), which is assumed to be strictly hyperbolic, with Riemann data (3.4.1). But before we explore these issues, we introduce certain basic notions, which will be used in subsequent discussion.

3.4.1   Simple waves

Simple waves are solutions of (3.1.1) of the form

$$\mathbf{u}(x,t) = \mathbf{v}(\xi(x,t)), \ x \in \mathbb{R}, \ t > 0. \tag{3.4.2}$$

Inserting (3.4.2) into (3.1.1), we find that $\mathbf{v}'(\xi)$, which denotes the derivative of $\mathbf{v}$ with respect to ξ, is an eigenvector of $\mathbf{A}$ for the eigenvalue $dx/dt = -\xi_{,t}/\xi_{,x}$, i.e., there exists an index $k, 1 \leq k \leq n$, such that

$$\xi_{,t} + \lambda_k(\mathbf{v}(\xi))\xi_{,x} = 0, \tag{3.4.3}$$

where $\mathbf{v}$ solves the ordinary differential equation

$$d\mathbf{v}(\alpha)/d\alpha = \mathbf{r}_k(\mathbf{v}(\alpha)), \ \alpha \in \mathbb{R}. \tag{3.4.4}$$

If (3.4.3) and (3.4.4) hold, we call the function defined by (3.4.2) a k-simple wave; equation (3.4.3) implies that the characteristics of the kth family, determined by the solution of $dx/dt = \lambda_k$, are straight lines along which ξ is constant. Thus, for each characteristic field $\lambda_k, 1 \leq k \leq n$, one can find a k-simple wave solution $\mathbf{u}$, which remains constant along the characteristics of the kth field.

3.4.2 Riemann invariants

Consider the system (3.1.1) with distinct eigenvalues $\lambda_k, 1 \leq k \leq n$, of the Jacobian matrix $\mathbf{A}$. Corresponding to these eigenvalues λ_k, there are n left eigenvectors $\mathbf{l}_k$ and n right eigenvectors $\mathbf{r}_k$, such that

$$\mathbf{l}_k \mathbf{A} = \lambda_k \mathbf{l}_k, \quad \mathbf{A} \mathbf{r}_k = \lambda_k \mathbf{r}_k \quad (k\text{-unsummed}). \tag{3.4.5}$$

Premultiplication of the system (3.1.1) by $\mathbf{l}_k$ gives

$$\mathbf{l}_k \cdot \frac{d\mathbf{u}}{dt} = 0, \quad \text{along} \quad \frac{dx}{dt} = \lambda_k. \tag{3.4.6}$$

Suppose that there exists a smooth scalar-valued function $\omega(\mathbf{u})$, which remains constant along a jth characteristic, then ω must satisfy

$$\nabla \omega \cdot \frac{d\mathbf{u}}{dt} = 0, \quad \text{along} \quad \frac{dx}{dt} = \lambda_j, \tag{3.4.7}$$

where ∇ is the gradient operator with respect to $(u_1, u_2, \ldots, u_n)$ space. Equation (3.4.7), in view of (3.4.6), implies that the vector $\nabla \omega$ is parallel to $\mathbf{l}_j$. Since $\mathbf{l}_k$ and $\mathbf{r}_k$ are bases of $\mathbb{R}^n$, and also $\mathbf{l}_j \cdot \mathbf{r}_k = 0$ for $j \neq k$, we obtain

$$\nabla \omega \cdot \mathbf{r}_k = 0. \tag{3.4.8}$$

If equation (3.4.8) has a solution for $\omega(\mathbf{u})$, it is called a k-Riemann invariant. By solving the first order differential equation (3.4.8) for ω, one may construct in the vicinity of any point $\mathbf{u}$, the $(n-1)$ k-Riemann invariants, whose gradients are linearly independent. Further, in view of (3.4.2) and (3.4.4), it immediately follows from (3.4.8) that a k-Riemann invariant is constant along the trajectories of the vector field $\mathbf{r}_k$. Indeed, it may be noticed that corresponding to each characteristic field λ_k there exist $n-1$ Riemann invariants, which remain constant in the kth simple wave of (3.1.1); this provides a connection between the simple wave solutions and the Riemann invariants. For example, one readily verifies that the functions $w_1(V, u) = u - \int^V \sqrt{-p'(y)}dy$ and $w_2(V, u) = u + \int^V \sqrt{-p'(y)}dy$ are, respectively, the 1- and 2-Riemann invariants of the system (3.3.4). Similarly, it may be verified that the three pairs of the functions $(S, u + \int(c/\rho)d\rho); (u, p); (S, u - \int(c/\rho)d\rho)$ are respectively, the 1-, 2- and 3-Riemann invariants of the system (3.1.6).

We now construct weak solutions of the Riemann problem (3.1.1), (3.4.1) assuming that (3.1.1) is strictly hyperbolic and $\mathbf{u}_l$ and $\mathbf{u}_r$ are sufficiently close.

3.4.3 Rarefaction waves

Given a fixed state $\mathbf{u}_o$, the kth rarefaction wave curve $R_k(\mathbf{u}_o)$ is defined to be the trajectory in $\mathbb{R}^n$ of the solution of (3.4.4), satisfying $\mathbf{v}(\alpha_o) = \mathbf{u}_o$; thus, $R_k(\mathbf{u}_o)$ is an integral curve of $\mathbf{r}_k(\mathbf{u})$ through a given state $\mathbf{u}_o$. Once the solution $\mathbf{v}$ of (3.4.4) is known, which exists at least locally, i.e., for α close to α_o, we look for ξ as a smooth solution of (3.4.3) or the scalar conservation law

$$\xi_{,t} + (f_k(\xi))_{,x} = 0 \tag{3.4.9}$$

with

$$f'_k(\alpha) = \lambda_k(\mathbf{v}(\alpha)). \tag{3.4.10}$$

Since

$$f''_k(\alpha) = \frac{d\lambda_k}{d\alpha} = \nabla\lambda_k(\mathbf{v}(\alpha)) \cdot \mathbf{r}_k(\mathbf{v}(\alpha)), \tag{3.4.11}$$

the flux function f_k in (3.4.9) is strictly convex (or concave), if the characteristic field $\lambda_k(\mathbf{u})$ is genuinely nonlinear. Thus, with the requirement (3.1.3) of genuine nonlinearity, (3.4.11) implies that $\lambda_k(\mathbf{u})$ is strictly monotone along the curve $R_k(\mathbf{u}_o)$, and we may divide $R_k(\mathbf{u}_o)$ into parts $R_k^+(\mathbf{u}_o)$ and $R_k^-(\mathbf{u}_o)$ such that $R_k(\mathbf{u}_o) = R_k^-(\mathbf{u}_o) \cup R_k^+(\mathbf{u}_o)$, where $R_k^+(\mathbf{u}_o) = \{\mathbf{u} \in R_k(\mathbf{u}_o) | \lambda_k(\mathbf{u}) \geq \lambda_k(\mathbf{u}_o)\}$ and $R_k^-(\mathbf{u}_o) = \{\mathbf{u} \in R_k(\mathbf{u}_o) | \lambda_k(\mathbf{u}) < \lambda_k(\mathbf{u}_o)\}$. Thus, for a state $\mathbf{u}_r \in R_k^+(\mathbf{u}_l)$, we can construct a k-rarefaction wave connecting $\mathbf{u}_l$ on the left and $\mathbf{u}_r$ on the right. Indeed, we choose ξ_l and $\xi_r \in \mathbb{R}$ with $\xi_l < \xi_r$, so that $\mathbf{u}_l = \mathbf{v}(\xi_\ell), \mathbf{u}_r = \mathbf{v}(\xi_r)$; further, since $\mathbf{u}_r \in R_k^+(\mathbf{u}_l)$, we have $\lambda_k(\mathbf{u}_r) > \lambda_k(\mathbf{u}_l)$ as $\lambda_k(\mathbf{u})$ is increasing along $R_k(\mathbf{u}_l)$ and subsequently from (3.4.11), we have $f'_k(\xi_r) > f'_k(\xi_l)$. The case $\xi_\ell > \xi_r$, which corresponds to the concave flux f_k, can be treated similarly. In Section 2.1.3, we have already studied the scalar Riemann problem for (3.4.9) with flux function strictly convex, whose unique weak solution is a continuous rarefaction wave connecting the states ξ_ℓ and ξ_r. Thus, it follows that the function

$$\mathbf{u}(x,t) = \begin{cases} \mathbf{u}_l, & x < t\lambda_k(\mathbf{u}_l) \\ \mathbf{v}(x/t), & t\lambda_k(\mathbf{u}_l) < x < t\lambda_k(\mathbf{u}_r) \\ \mathbf{u}_r, & x > t\lambda_k(\mathbf{u}_r), \end{cases} \tag{3.4.12}$$

is a continuous self-similar weak solution of (3.1.1) and (3.4.1). Solution (3.4.12) is called a k-centered simple wave or a k-rarefaction wave connecting the states $\mathbf{u}_l$ and $\mathbf{u}_r$.

3.4.4 Shock waves

Here, we consider the possibility that the states $\mathbf{u}_l$ and $\mathbf{u}_r$ may be joined by a shock discontinuity along which a weak solution $\mathbf{u}$ of (3.1.1) satisfies the R-H condition

$$s(\mathbf{u}_\ell - \mathbf{u}_r) = f(\mathbf{u}_\ell) - f(\mathbf{u}_r), \tag{3.4.13}$$

where s is the shock speed.

Now suppose, we fix the left state $\mathbf{u}_l$ and look for all the states $\mathbf{u}_r$, which can be connected to $\mathbf{u}_l$ by a shock satisfying (3.4.13), then (3.4.13) is equivalent to a system of n equations in $n+1$ unknowns $\mathbf{u}_r$ and s. This observation motivates one to study the Hugoniot curves $H(\mathbf{u}_o)$, through a fixed point $\mathbf{u}_o$, defined as

$$H(\mathbf{u}_o) = \{\mathbf{u} \in \mathbb{R}^n | f(\mathbf{u}) - f(\mathbf{u}_o) = \sigma(\mathbf{u} - \mathbf{u}_o)\},$$

for some $\sigma = \sigma(\mathbf{u}_o, \mathbf{u}) \in \mathbb{R}$. The jump condition (3.4.13) then implies that $\mathbf{u}_r \in H(\mathbf{u}_l)$ and $s = \sigma(\mathbf{u}_l, \mathbf{u}_r)$. Lax ([101], [102]) has shown that the existence of Hugoniot curves is a consequence of the implicit function theorem, i.e., in some neighborhood of $\mathbf{u}_o$, $H(\mathbf{u}_o)$ consists of the union of n smooth curves $H_k(\mathbf{u}_o), 1 \le k \le n$, with the properties

(i) The curve $H_k(\mathbf{u}_o)$ passes through $\mathbf{u}_o$ with tangent $\mathbf{r}_k(\mathbf{u}_o)$.

(ii) The Hugoniot curve $H_k(\mathbf{u}_o)$ and the rarefaction curve $R_k(\mathbf{u}_o)$ have second order contact at $\mathbf{u} = \mathbf{u}_o$.

(iii) The shock speed $\sigma(\mathbf{u}_o, \mathbf{u})$ tends to the characteristic speed $\lambda_k(\mathbf{u}_o)$ as $\mathbf{u} \to \mathbf{u}_o$. Moreover, the shock speed is the average of the characteristic speeds on both sides up to second order terms in $|\mathbf{u} - \mathbf{u}_o|$, i.e.,

$$\sigma(\mathbf{u}_o, \mathbf{u}) = \frac{1}{2}(\lambda_k(\mathbf{u}) + \lambda_k(\mathbf{u}_o)) + O(|\mathbf{u} - \mathbf{u}_o|^2) \text{ as } \mathbf{u} \to \mathbf{u}_o \text{ with } \mathbf{u} \in H_k(\mathbf{u}_o).$$

It may be noticed from the property (iii) above that $H_k(\mathbf{u}_o)$ can be partitioned in the same way as the rarefaction curve R_k, i.e., $H_k(\mathbf{u}_o) = H_k^-(\mathbf{u}_o) \bigcup H_k^+(\mathbf{u}_o)$, so that $\lambda_k(\mathbf{u}) \ge \lambda_k(\mathbf{u}_o)$ for $\mathbf{u} \in H_k^+(\mathbf{u}_o)$ and $\lambda_k(\mathbf{u}) < \lambda_k(\mathbf{u}_o)$ for the state $\mathbf{u} \in H_k^-(\mathbf{u}_o)$. Thus, if the system (3.1.1) is strictly hyperbolic and the kth characteristic field is genuinely nonlinear, then in a small neighborhood of a given state $\mathbf{u}_l$, the curve $\mathbf{u}_r \in R_k^+(\mathbf{u}_l)$ is a rarefaction wave, and $\mathbf{u}_r \in H_k^-(\mathbf{u}_l)$ is a k-shock satisfying the Lax entropy condition (3.3.6); however, for a linearly degenerate field, $H_k(\mathbf{u}_l) = R_k(\mathbf{u}_l)$ and $\sigma(\mathbf{u}_l, \mathbf{u}_r) = \lambda_k(\mathbf{u}_r) = \lambda_k(\mathbf{u}_l)$, i.e., the nearby states $\mathbf{u}_l$ and $\mathbf{u}_r$ are connected by a contact discontinuity of speed σ.

The foregoing results allow one to form the composite curves through $\mathbf{u}_l$

$$W_k(\mathbf{u}_l) = \begin{cases} R_k^+(\mathbf{u}_l) \bigcup H_k^-(\mathbf{u}_l), & k^{\text{th}} \text{ characteristic field genuinely nonlinear} \\ R_k(\mathbf{u}_l) = H_k(\mathbf{u}_l), & k^{\text{th}} \text{ characteristic field linearly degenerate}, \end{cases}$$
(3.4.14)

for each k, $1 \le k \le n$; the wave curves $W_k(\mathbf{u}_l)$ are used as building blocks towards the construction of a solution of the Riemann problem for (3.1.1) in the general case. For details, besides the Lax's fundamental paper [101], reader can consult Liu [111], Smoller [189], and Evans [54]; indeed, the following theorem, which we state without proof, gives us the sufficient conditions for the initial values such that the corresponding Riemann problem for a strictly hyperbolic system (3.1.1) has an admissible solution as a set of elementary waves, such as rarefaction waves, contact discontinuities and shock waves.

Theorem 3.4.1 *Suppose that each characteristic field is either genuinely nonlinear or linearly degenerate. Then the Riemann problem (3.1.1), (3.4.1) has a unique solution consisting of at most $(n+1)$ constant states separated by elementary waves provided that the states $\mathbf{u}_l$ and $\mathbf{u}_r$ are sufficiently closed.*

Remarks 3.4.1 The Lax's Theorem, referred to as above, has been generalized by Liu [110] to systems that are strictly hyperbolic, but neither convex nor linearly degenerate; the inequality $\sigma(\mathbf{u}, \mathbf{u}_l) > \sigma(\mathbf{u}_r, \mathbf{u}_l)$ for each $\mathbf{u} \in H_k(\mathbf{u}_l)$ and lying between $\mathbf{u}_l$ and $\mathbf{u}_r$, is referred to as the Liu's entropy criterion. However, there are examples which illustrate that Liu's entropy condition is no longer sufficiently discriminating to single out a unique solution for nonstrictly hyperbolic systems (see Dafermos [45]); the current status of the theory of Riemann problem for such systems is far from definitive with regard to both existence and uniqueness of solutions. Indeed, such systems often admit along with undercompressive and overcompressive shocks, a new type of singularity, called delta shock, which supports point masses; for further details regarding delta shocks; see Li et al. [107] and Zheng [215].

It is shown in [93] that there are examples of strictly hyperbolic and genuinely nonlinear systems for which the Riemann problem for a pair of constant states $\mathbf{u}_l$ and $\mathbf{u}_r$ has no solution if $|\mathbf{u}_l - \mathbf{u}_r|$ is large.

Example 3.4.1 (Polytropic ideal gas equations): We now return to the one-dimensional gasdynamic equations (1.2.11) for a planar ($m = 0$) motion, which is studied here both as motivation and an example with basic notions explicitly discussed. The system governing the one-dimensional planar motion of a polytropic ideal gas is

$$\rho_{,t} + (\rho u)_{,x} = 0, \quad (\rho u)_{,t} + (p + \rho u^2)_{,x} = 0, \quad (\rho E)_{,t} + (u(p + \rho E))_{,x} = 0, \quad (3.4.15)$$

with $E = (p/\rho)(\gamma - 1)^{-1} + (u^2/2)$ as the total specific energy, and $\gamma > 1$.

The system (3.4.15) admits three characteristic families corresponding to the eigenvalues $\lambda_1 = u - a < \lambda_2 = u < \lambda_3 = u + a$, where $a = (\gamma p/\rho)^{1/2}$ is the sound speed. The associated right eigenvectors can be chosen to be

$$\mathbf{r}_1 = (\rho, -a, \rho a^2)^{tr}, \quad \mathbf{r}_2 = (1, 0, 0)^{tr}, \quad \mathbf{r}_3 = (\rho, a, \rho a^2)^{tr}.$$

Thus, $\nabla \lambda_1 \cdot \mathbf{r}_1 = -(1 + \gamma)a/2$, $\nabla \lambda_2 \cdot \mathbf{r}_2 = 0$ and $\nabla \lambda_3 \cdot \mathbf{r}_3 = (1 + \gamma)a/2$, which show that the first and the third characteristic fields, $\lambda_1 = u - a$ and $\lambda_3 = u + a$, are genuinely nonlinear, while the second characteristic field, $\lambda_2 = u$, is linearly degenerate.

i) Simple waves and Riemann invariants

As noticed earlier in (3.4.8), any solution ω of the partial differential equation

$\nabla \omega \cdot \mathbf{r}_k = 0$, $1 \leq k \leq 3$, is a k-Riemann invariant associated with the k-th characteristic field of the system (3.4.15). It is easy to see that the pairs $(p\rho^{-\gamma}, u + 2a(\gamma - 1)^{-1})$, (u, p) and $(p\rho^{-\gamma}, u - 2a(\gamma - 1)^{-1})$ are respectively the 1-, 2- and the 3-Riemann invariants. Again, as noticed earlier that in a k-simple wave all k-Riemann invariants are constant, it follows that the flow described by

$$x = (u - a)t + g(u), \quad u + 2a/(\gamma - 1) = \text{const}, \qquad (3.4.16)$$

where the function $g(u)$ is arbitrary, is a 1-simple wave for the genuinely nonlinear characteristic field $\lambda_1 = u - a$. In fact, the functions $p\rho^{-\gamma}$ and $u + 2a/(\gamma - 1)$ are the constants of the flow; they make it possible to express the (nonconstant) unknowns of the system as a function of one of them. Equations (3.4.16) determine the velocity, and therefore all other quantities, as an implicit function of x and t. If we assume that there is a point in the simple wave region for which $u = 0$, as usually happens in practice, $(3.4.16)_2$ yields $a = a_o - ((\gamma - 1)/2)u$, where $a = a_o$ for $u = 0$; since $p\rho^{-\gamma} = p_o\rho_o^{-\gamma}$ in the wave region, where the subscript o refers to the point where the gas is at rest, we have

$$\rho = \rho_o \left(1 - \frac{(\gamma - 1)u}{2a_o} \right)^{2/\gamma - 1}, \quad p = p_o \left(1 - \frac{(\gamma - 1)u}{2a_o} \right)^{2\gamma/\gamma - 1}, \qquad (3.4.17)$$

and $x = t(-a_o + (\gamma + 1)(u/2)) + g(u)$, which may be written in the form

$$u = \phi \left(x - \left(-a_o + \frac{(\gamma + 1)u}{2} \right) t \right), \qquad (3.4.18)$$

with ϕ as another arbitrary function. Equations (3.4.17) and (3.4.18) express density, pressure and velocity as functions of x and t in the simple wave region. Since the velocities of different points in the wave profile are different, the profile changes its shape in the course of time. For expansion waves, the solution (3.4.17) – (3.4.18) is complete, but in compression parts of the disturbance, breaking occurs and shocks appear, and the wave ceases to be a simple wave; when shocks appear, we need to re-examine the arguments leading to $p\rho^{-\gamma} = \text{const}$ and (3.4.16).

We proceed in a similar manner for a 3-simple wave associated with the genuinely nonlinear characteristic field $\lambda_3 = u + a$. The second characteristic field, which is a 2-Riemann invariant, is linearly degenerate. In this case, there are no shocks or expansion waves as the characteristics of the 2-simple wave are parallel lines $x = ut + \text{const.}$; in fact, these are contact discontinuities with u and p as constants.

ii) Contact discontinuities and shocks

The R-H jump condition (3.2.3) for the system (3.4.15) can be written as

$$s[\rho] = [\rho u], \quad s[\rho u] = [p + \rho u^2],$$
$$s[\rho(e + \tfrac{1}{2}u^2)] = [(\rho(e + \tfrac{1}{2}u^2) + p)]. \qquad (3.4.19)$$

which can be rewritten in a more convenient form by introducing the relative velocity $v = u - s$ as

$$[m] = 0, \quad [p + mv] = 0, \quad m[a^2 + \frac{\gamma - 1}{2} v^2] = 0, \qquad (3.4.20)$$

where $m = \rho v$, the mass flux through the discontinuity curve. There are two cases according as $m = 0$ and $m \neq 0$.

For $m = 0$, the R-H conditions (3.4.19) admit a solution $s = u_l = u_r, p_l = p_r$ with an arbitrary jump discontinuity in the density. Such a discontinuity, for which

$$[p] = 0, \quad [u] = 0, \quad [\rho] \neq 0, \qquad (3.4.21)$$

is called a contact discontinuity in the sense of Lax (see equation (3.3.7)); this is never crossed by the gas particles. Since $s = u_l = u_r$, the characteristic curves of the second family, $dx/dt = \lambda_2 = u$, coincide on both sides with the discontinuity curve that appears in a linearly degenerate characteristic field $\lambda_2 = u$.

For $m \neq 0$, we have a shock discontinuity that may be a 1-shock or a 3-shock in the sense of Lax (see inequalities (3.3.6)). For a 1-shock, we have $s < u_l - a_l$ and $u_r - a_r < s < u_r$; since $v_r > 0$ and $v_l > a_l > 0$, we have $u_l, u_r > s$, and therefore it is a backward facing wave, which is crossed by the gas particles moving from left to right. Similarly, for a 3-shock $s > u_l, u_r$ and the gas particles cross form right to left; it is a forward facing wave. In both the shocks, we have $v_1^2 > a_1^2$ and $a_2^2 > v_2^2$, where the subscripts 1 and 2 refer to the states ahead of the shock and behind the shock, respectively; in fact for 1-shock the state ahead is on the left, whereas for a 3-shock, the state ahead is on the right side. Thus, for an admissible shock, the gas velocity relative to the shock front is supersonic at the front side and subsonic at the back side. Furthermore, since

$$\frac{2}{\gamma - 1} a_1^2 + a_1^2 < \frac{2}{\gamma - 1} a_1^2 + v_1^2 = \frac{2}{\gamma - 1} a_2^2 + v_2^2 < \frac{2}{\gamma - 1} a_2^2 + a_2^2, \qquad (3.4.22)$$

it follows that $a_1 < a_2$; here, we have used $a_1 < |v_1|$, $a_2 > |v_2|$ and the jump relation (3.4.20)$_3$. Also, since $a_1 < a_2$ equation (3.4.22) implies that $|v_1| > |v_2|$, and then $\rho_1 v_1 = \rho_2 v_2$ shows that $\rho_2 > \rho_1$. Subsequently, equation (3.4.20)$_2$ shows that $p_2 > p_1$. Thus, a shock satisfying the entropy condition is always compressive, i.e., $p_2 > p_1, \rho_2 > \rho_1, a_2 > a_1, |v_1| > |v_2|$. It may be noticed that for a 1-shock, both v_1 and v_2 are positive, whereas for a 3-shock, both v_1 and v_2 are negative.

iii) Shock Hugoniot

Some simple algebraic identities for a shock ($m \neq 0$) may be derived from the conditions (3.4.20); indeed, the elimination of m between (3.4.20)$_1$ and (3.4.20)$_2$ yields $v_2^2 - v_1^2 = -(p_2 - p_1)(V_2 + V_1)$, where $V = 1/\rho$ is the specific volume. This relation together with (3.4.20)$_3$ yields, on eliminating the velocity v, that

$$(V_2 - \mu^2 V_1)p_2 - (V_1 - \mu^2 V_2)p_1 = 0, \qquad (3.4.23)$$

where $\mu^2 = (\gamma - 1)/(\gamma + 1)$. The relation (3.4.23), which involves only p and V is called the Hugoniot relation. If we define the Hugoniot function with center (V_1, p_1) as

$$H(V, p) = (V - \mu^2 V_1)p - (V_1 - \mu^2 V)p_1. \qquad (3.4.24)$$

then the equation (3.4.23) may be written as $H(V_2, p_2) = 0$. The curve $H(V, p) = 0$, with (V_1, p_1) as center, is a rectangular hyperbola lying in the $V - p$ plane, and it represents all possible states which can be connected to the state (V_1, p_1) through a shock. It may be noticed that the part of the hyperbola for which $V > \mu^{-2} V_1$ has no physical meaning, as the pressure becomes negative; in fact, along the H-curve the values of V lie between $\mu^2 V_1$ and $\mu^{-2} V_1$, whereas the pressure varies between 0 and $+\infty$. As observed from the general theory in Sections 3.4.3 and 3.4.4 that the Hugoniot curve $H_k(\mathbf{u}_o)$ and the rarefaction curve $R_k(\mathbf{u}_o)$ have second order contact at $\mathbf{u} = \mathbf{u}_o$, it follows that 1- and 3-shock curves are projected in the $V - p$ plane onto the Hugoniot curve, while the 1- and 3-rarefaction curves are projected onto the isentropic curve $S(V_1, p_1) = S_o$ through the point (V_1, p_1). For further details, see Godlewski and Raviart [63].

iv) Weak and strong shocks

For a 3-shock, the flow quantities behind the shock can be expressed in terms of s and the state-ahead parameters as follows:

$$\frac{u_2 - u_1}{a_1} = \frac{2(M^2 - 1)}{(\gamma + 1)M}, \quad \frac{\rho_2}{\rho_1} = \frac{(\gamma + 1)M^2}{(\gamma - 1)M^2 + 2}, \quad \frac{p_2 - p_1}{p_1} = \frac{2\gamma(M^2 - 1)}{(\gamma + 1)},$$
$$(3.4.25)$$

where $M = (s - u_1)/a_1$ is the Mach number of a 3-shock relative to the flow ahead; it may be noticed that $M > 1$. It is often convenient to express the shock Mach number in terms of the shock-strength parameter $z = (p_2 - p_1)/p_1$ as $M = (1 + ((\gamma + 1)/2\gamma)z)^{1/2}$, so that the right-hand sides in the shock formulas (3.4.25) can be expressed in terms of z. This enables one to deduce certain important properties of shocks; for instance, the expression for the entropy of a polytropic gas, given by (1.2.10), shows that $d(S_2 - S_1)/dz > 0$, implying thereby that for the entropy S to increase across the shock, the shock must be compressive with $p_2 > p_1$.

For weak shocks ($M \to 1$, equivalently $z \to 0$), the relations (3.4.25) may be expressed as

$$M = 1 + \left(\frac{\gamma + 1}{4\gamma}\right)z + O(z^2), \quad \frac{u_2 - u_1}{a_1} = \frac{z}{\gamma} + O(z^2),$$
$$\frac{\rho_2 - \rho_1}{a_1} = \frac{1}{\gamma}z + O(z^2), \quad \frac{S_2 - S_1}{c_v} = \frac{\gamma^2 - 1}{12\gamma^2}z^3 + O(z^4), \qquad (3.4.26)$$

showing thereby that for shocks of weak or moderate strength, it is a reasonable approximation to neglect changes in the entropy and the Riemann invariants.

For strong shocks ($z \to \infty$), it is required that $s \gg u_1$ so that $M \sim z/a_1$ and $M \gg 1$. Thus, equations (3.4.25) may be approximated by

$$u_2 \sim (2/(\gamma+1))s, \quad \rho_2/\rho_1 \sim (\gamma+1)/(\gamma-1), \quad p_2 \sim (2/(\gamma+1))\rho_1 s^2. \quad (3.4.27)$$

It may be noticed that the density ratio across an infinitely strong shock remains only finite.

Example 3.4.2 (Isentropic system): Here, we consider the Riemann problem for (3.3.4) with initial conditions (3.4.1), where $\mathbf{u}_l = (V_l, u_l)^{tr}$ and $\mathbf{u}_r = (V_r, u_r)^{tr}$. The system (3.3.4) is of the form (3.1.1) by introducing $\mathbf{u} = (V, u)^{tr}$ and $\mathbf{f}(\mathbf{u}) = (-u, p(V))^{tr}$. We notice that the system (3.3.4) is strictly hyperbolic as its Jacobian matrix has two real distinct eigenvalues $\lambda_{1,2} = \mp\sqrt{-p'(V)}$ with the corresponding eigenvectors $\mathbf{r}_{1,2} = (\pm 1, \sqrt{-p'(V)})^{tr}$. Since $\nabla\lambda_k \cdot \mathbf{r}_k > 0$, $k = 1, 2$, it follows from (3.1.3) that the characteristic fields λ_1 and λ_2 are genuinely nonlinear. For the characteristic field λ_1, we find a rarefaction wave solution using (3.4.4) as $\mathbf{u} = (V(x,t), \int^{V(x,t)} \sqrt{-p'(y)}dy)^{tr}$, and call it the 1-rarefaction wave solution, where $V(x,t)$ is determined from (3.4.3) as $x/t = -\sqrt{-p'(V)}$; this gives us V as a function of x/t, using inverse function theorem. We now fix the state $\mathbf{u}_l$ and find states $\mathbf{u}_r$ which can be connected with $\mathbf{u}_l$ by 1-rarefaction wave. Since $p''(V) > 0$, the characteristic speed λ_1 is increasing in V, we must have $V_l < V_r$. Moreover, since $du/dV = \sqrt{-p'(V)}$, we have $u_r - u_l = \int_{V_l}^{V_r} \sqrt{-p'(y)}dy$, which shows that $u_r > u_l$. In other words, the possible initial states of the form (3.4.1), which generate a 1-rarefaction wave, lie on the curve

$$R_1 = \{(V, u) : u - u_l = \int_{V_l}^{V} \sqrt{-p'(y)}dy; V > V_l, u > u_l\}.$$

Similarly, the 2-rarefaction wave curve through the point $\mathbf{u}_l$ is given by

$$R_2 = \{(V, u) : u - u_l = \int_{V_l}^{V} \sqrt{-p'(y)}dy; V < V_l, u > u_l\}.$$

Next, the shock curves S_1 and S_2 through $\mathbf{u}_l$ are obtained from the R-H conditions

$$s(V - V_l) = -(u - u_l), \quad s(u - u_l) = p(V) - p(V_l), \quad (3.4.28)$$

where s is the shock speed. As the 1-shock wave and 2-shock wave satisfy the inequalities (3.3.6), i.e.,

$$s < \lambda_1(\mathbf{u}_l), \quad \lambda_1(\mathbf{u}_r) < s < \lambda_2(\mathbf{u}_r) \text{ and } s > \lambda_2(\mathbf{u}_r), \quad \lambda_1(\mathbf{u}_l) < s < \lambda_2(\mathbf{u}_l), \quad (3.4.29)$$

we find that $s < 0$ for 1-shock wave and $s > 0$ for 2-shock wave. Indeed, for

1-shock, we find that $-\sqrt{-p'(V_r)} < s < -\sqrt{-p'(V_l)}$, and for a 2-shock wave $\sqrt{-p'(V_r)} < s < \sqrt{-p'(V_l)}$; eliminating s from (3.4.28) and using the shock inequalities (3.4.29), the 1-shock curve S_1 and 2-shock curve S_2 are defined as

$$S_1 = \{(V,u) : u - u_l = -\sqrt{(V-V_l)(p(V_l) - p(V))}; V_l > V, u_l > u\},$$

$$S_2 = \{(V,u) : u - u_l = -\sqrt{(V-V_l)(p(V_l) - p(V))}; V_l < V, u_l < u\}.$$

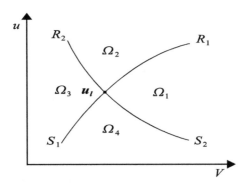

Figure 3.4.1: Rarefaction and shock curves.

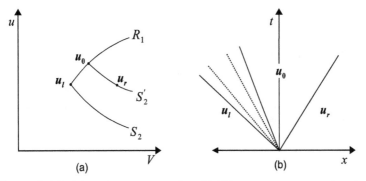

Figure 3.4.2: (a) $\mathbf{u}_r$ lies in region Ω_1; (b) Wave structure in region Ω_1.

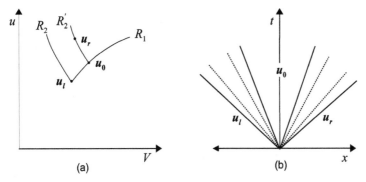

Figure 3.4.3: (a) $\mathbf{u}_r$ lies in region Ω_2; (b) Wave structure in region Ω_2.

We notice that the system under consideration has four elementary waves, namely the 1-rarefaction wave R_1, the 2-rarefaction wave R_2, the 1-shock wave S_1 and the 2-shock wave S_2. We have already seen that the Riemann problem (3.3.4), (3.4.1) admits a solution in the form of a centered rarefaction wave in the two cases $\mathbf{u}_r \in R_1$, $V_r > V_l$, or else $\mathbf{u}_r \in R_2$, $V_r < V_l$. On the other hand, we get a shock connecting these constant states provided that either $\mathbf{u}_r \in S_1$, $V_r < V_l$, or else $\mathbf{u}_r \in S_2$, $V_r > V_l$. We now fix $\mathbf{u}_l$ and superimpose the four curves R_1, R_2, S_1 and S_2 which divide a neighborhood of $\mathbf{u}_l$ into four regions $\Omega_i, i = 1, 2, 3, 4$ in the $V - u$ plane (see Fig.3.4.1); the structure of the general solution to the Riemann problem is now determined by the location of the state $\mathbf{u}_r$ with respect to the curves R_1, R_2 and S_1, S_2. If $\mathbf{u}_r$ lies in region Ω_1 between R_1 and S_2, then there is a point $\mathbf{u}_o \in R_1$ that lies on the curve S_2' of constant state that can be connected to $\mathbf{u}_r$ by a 2-shock (see Fig. 3.4.2(a)). Since $\mathbf{u}_o$ can be connected to $\mathbf{u}_l$ by the 1-rarefaction wave R_1 and to $\mathbf{u}_r$ by a 2-shock wave S_2', the resulting weak solution consists of a 1-rarefaction wave and a 2-shock wave (see Fig. 3.4.2(b)). If $\mathbf{u}_r \in \Omega_2$, then there is a point $\mathbf{u}_o \in R_1$ which lies on the curve R_2' of constant states that can be connected to $\mathbf{u}_r$ by a 2-rarefaction wave R_2'; thus the weak solution consists of two centered rarefaction waves (see Figs. 3.4.3 (a) and (b)). For $\mathbf{u}_r \in \Omega_3$, the weak solution consists of 1-shock S_1 and a 2-rarefaction wave, whereas for $\mathbf{u}_r \in \Omega_4$, the solution consists of two shocks. Thus, the Riemann problem for (3.3.4), (3.4.1) admits a weak solution consisting of three constant states connected by rarefaction waves or shocks.

Remarks : The existence of a linearly degenerate characteristic field, satisfying (3.1.4), allows for the possibility of an additional type of discontinuous weak solution, called a contact discontinuity, the speed of which satisfies (3.3.7), and which connects the nearby states $\mathbf{u}_l$ and $\mathbf{u}_r$, having the same Riemann invariant with respect to this linearly degenerate field; for details, see [189].

3.5 Shallow Water Equations

We consider the Riemann problem for a nonlinear hyperbolic system that arises in shallow water theory (see Kevorkian [92]). The method of analysis used here finds its origin in references [101], [189], and [149].

The system of equations, which governs the one-dimensional modified shallow water equations, can be written in conservation form as [89]

$$h_{,t} + (hu)_{,x} = 0,$$
$$(hu)_{,t} + (hu^2 + gHh + gh^2/2)_{,x} = 0, \quad t > 0, \ x \in \mathbb{R} \qquad (3.5.1)$$

where u is x-component of fluid velocity, h the water depth, g the acceleration due to gravity and $H = (k_0/g)$ the reduced factor characterizing advective transport of impulse with k_0 as a positive constant.

To carry out the characteristic analysis of (3.5.1), it is convenient to use the primitive variables $U = (h, u)^{tr}$, rather than the vector of conserved variables. Then for smooth solutions, system (3.5.1) is equivalent to

$$U_{,t} + AU_{,x} = 0, \qquad (3.5.2)$$

where A is 2×2 matrix with elements $A_{11} = A_{22} = u$, $A_{21} = c^2/h$ and $A_{12} = h$; here, $c = \sqrt{g(h+H)}$ is the speed of propagation of the surface disturbance. The eigenvalues of A are $\lambda_1 = u - c$ and $\lambda_2 = u + c$. Thus, the system (3.5.2) is strictly hyperbolic when $c > 0$ and, therefore, admits discontinuities and piecewise continuous solutions, which are called bores and dilatation waves, respectively. Let $\vec{r}_1 = (h, -c)^{tr}$ and $\vec{r}_2 = (h, c)^{tr}$ be the right eigenvectors corresponding to the eigenvalues λ_1 and λ_2, respectively. For the characteristic field λ_1, we have $\nabla\lambda_1 \cdot \vec{r}_1 = (-gh - 2c^2)/2c$ which is different from zero, and therefore the first characteristic field is genuinely nonlinear. Similarly, the second characteristic field λ_2 is also genuinely nonlinear. The waves associated with $\vec{r}_1$ and $\vec{r}_2$ characteristic fields will be either bores or dilatation waves. Because the characteristic fields are genuinely nonlinear, we can expect to solve the Riemann problem for (3.5.1) with bores and dilatation waves.

3.5.1   Bores

Let h_l, $u_l = u(h_l)$ and h, $u = u(h)$ denote, respectively, the left and the right-hand states of either a bore or a dilatation wave. Here, we compute bore curves for the hyperbolic system (3.5.1). We fix, once and for all, a state (h_l, u_l) and compute the state (h, u) such that there exists a speed σ satisfying the Rankine-Hugoniot jump conditions

$$\sigma[h] = [hu], \qquad (3.5.3)$$
$$\sigma[hu] = [hu^2 + gHh + gh^2/2], \qquad (3.5.4)$$

where $[.]$ denotes the jump across a discontinuity curve $x = x(t)$ and $\sigma = dx/dt$ is the bore speed.

Lemma 3.5.1 *Let B_1 and B_2, respectively denote 1-bore and 2-bore associated with λ_1 and λ_2 characteristic fields. Let the states U_l and U satisfy the Rankine-Hugoniot jump conditions (3.5.3) and (3.5.4). Then the bore curves satisfy*

$$u = u_l \mp (h - h_l)\phi(h, h_l), \qquad (3.5.5)$$

where $\phi(h, h_l) = \sqrt{\frac{g(h+h_l+2H)}{2hh_l}}$; indeed, on B_1, we have $\frac{du}{dh} < 0$ and $\frac{d^2u}{dh^2} > 0$, whilst on B_2 we have $\frac{du}{dh} > 0$ and $\frac{d^2u}{dh^2} < 0$.

Proof: The σ-elimination of (3.5.3) and (3.5.4) yields (3.5.5) and then differentiating (3.5.5) with respect to h, we obtain

$$\frac{du}{dh} = \mp\left(\phi(h, h_l) + (h - h_l)\frac{d\phi(h, h_l)}{dh}\right). \tag{3.5.6}$$

It is easy to show using (3.5.6) that $\frac{du}{dh} < 0$ on B_1, and $\frac{du}{dh} > 0$ on B_2. Differentiating (3.5.6) with respect to h, we get

$$\frac{d^2u}{dh^2} = \mp\left(2\frac{d\phi(h, h_l)}{dh} + (h - h_l)\frac{d^2\phi(h, h_l)}{dh^2}\right). \tag{3.5.7}$$

Since $2\frac{d\phi(h,h_l)}{dh} + (h - h_l)\frac{d^2\phi(h,h_l)}{dh^2} = -\frac{g^2(h_l+2H)(4hh_l+(h_l+2H)(h+3h_l))}{16h^4h_l^2\phi(h,h_l)^3}$, which is negative for all values of h, we obtain, in view of (3.5.7), that $\frac{d^2u}{dh^2} > 0$ for 1-bore, and $\frac{d^2u}{dh^2} < 0$ for 2-bore. $\qquad\square$

We now show that the bores satisfy the Lax stability conditions.

Lemma 3.5.2 *Across 1-bore (respectively, 2-bore), $h > h_l$ and $u < u_l$ (respectively, $h < h_l$ and $u < u_l$) if, and only if, the Lax conditions hold, i.e., 1-bore satisfies*

$$\sigma < \lambda_1(U_l), \quad \lambda_1(U) < \sigma < \lambda_2(U), \tag{3.5.8}$$

while the 2-bore satisfies

$$\lambda_1(U_l) < \sigma < \lambda_2(U_l), \quad \lambda_2(U) < \sigma. \tag{3.5.9}$$

Proof: First let us consider 1-bore and prove $\sigma < \lambda_1(U_l)$. On 1-bore $h_l < h$, implying thereby that $h_l < (h_l + h)/2$, which means that

$$c_l = \sqrt{g(h_l + H)} < \sqrt{\frac{g(h + h_l + 2H)}{2}}. \tag{3.5.10}$$

Also, since $h > \sqrt{hh_l}$, it follows from (3.5.10) that $c_l < h\phi(h, h_l)$, which implies that $-c_l > -h\phi(h, h_l)$; in view of equation (3.5.5) for 1-bore, we get $-c_l > \frac{h(u-u_l)}{h-h_l}$, implying thereby that

$$\sigma = \frac{hu - h_l u_l}{h - h_l} < u_l - c_l = \lambda_1(U_l). \tag{3.5.11}$$

Next, because $h_l < h$ on 1-bore, we have $(h + h_l + 2H)/2 < h + H$, which implies that $\sqrt{\frac{g(h+h_l+2H)}{2}} < c$, or equivalently

$$-c < -\sqrt{\frac{g(h + h_l + 2H)}{2}}. \tag{3.5.12}$$

Also, because $h_l < \sqrt{hh_l}$, which implies that $h_l\phi(h, h_l) < \sqrt{\frac{g(h+h_l+2H)}{2}}$, we have

$$-\sqrt{\frac{g(h + h_l + 2H)}{2}} < -h_l\phi(h, h_l). \qquad (3.5.13)$$

In view of (3.5.12) and (3.5.13), we get $-c < -h_l\phi(h, h_l)$; since for 1-bore, $u - u_l = -(h - h_l)\phi(h, h_l)$, it follows that $-c < \frac{h_l(u-u_l)}{h-h_l}$, and hence

$$u - c = \lambda_1(U) < \frac{hu - h_l u_l}{u - u_l}. \qquad (3.5.14)$$

Also, from (3.5.12) and (3.5.13), we obtain $-h_l\phi(h, h_l) < c$, which in view of (3.5.5) yields $\frac{h_l(u-u_l)}{h-h_l} < c$, and hence

$$\frac{hu - h_l u_l}{u - u_l} = \sigma < u + c = \lambda_2(U). \qquad (3.5.15)$$

Therefore, 1-bore satisfies Lax conditions; proof for 2-bore follows on similar lines. Conversely, we assume for 1-bore that the left- and right-hand states satisfy Lax conditions (3.5.8), and show that $h > h_l$ and $u < u_l$. Let us define $v = \sigma - u$; then since $\sigma < \lambda_1(U_l)$, it follows that $v_l < -c_l$, implying thereby that $v_l < 0$, and hence $\sigma < u_l$. Similarly, by using second condition $\lambda_1(U) < \sigma < \lambda_2(U)$, we get $u - c < \sigma < u + c$, which implies that $-c < v < c$, showing thereby that $|v| < c$. From (3.5.3), we have $hv = h_l v_l$. Since h and h_l are positive, both v and v_l must have the same sign; further, since $v_l < 0$, we have $v < 0$. For 1-bore, the fluid velocity on both sides of the bore is greater than the bore velocity, and therefore the particles cross the bore from left to right.

Since for 1-bore we have $v_l^2 > c_l^2$ and $v^2 < c^2$, equation (3.5.4), namely, $h_l v_l^2 + gHh_l + gh_l^2/2 = hv^2 + gHh + gh^2/2$ implies that $h_l c_l^2 + gHh_l + gh_l^2/2 < h_l v_l^2 + gHh_l + gh_l^2/2 = hv^2 + gHh + gh^2/2 < hc^2 + gHh + gh^2/2$, showing thereby that $\frac{3}{2}h_l^2 + 2Hh_l < \frac{3}{2}h^2 + 2Hh$; thus, it follows that $h_l < h$. Moreover, from (3.5.3), we have $v = \frac{h_l v_l}{h}$; since $h_l < h$, one infers that $u - u_l = v_l - v = v_l - \frac{h_l v_l}{h} = v_l(1 - \frac{h_l}{h})$, which is negative, and hence $u < u_l$. The corresponding results for 2-bore can be proved in a similar way, and we shall not reproduce the details. $\qquad\qquad\square$

3.5.2 Dilatation waves

Here, we construct the dilatation wave curves and recall that an n-dilatation wave ($n = 1, 2$), connecting the states U_l and U_r, is a solution to (3.5.2) of the form

$$U(x, t) = \begin{cases} U_l, & \frac{x}{t} \leq \lambda_n(U_l) \\ U(\frac{x}{t}), & \lambda_n(U_l) \leq \frac{x}{t} \leq \lambda_n(U_r) \\ U_r, & \frac{x}{t} \geq \lambda_n(U_r), \end{cases} \qquad (3.5.16)$$

where $U(\eta)$, $\eta = x/t$, is a solution to the system of ordinary differential equations $(A - \eta I)(\dot{h}, \dot{u})^{tr} = 0$; here, I is 2×2 identity matrix and an overhead dot denotes differentiation with respect to the variable η. If $(\dot{h}, \dot{u})^{tr} = (0, 0)$, then h and u are constant; but, as we are interested in nonconstant solutions, we consider $(\dot{h}, \dot{u})^{tr} \neq (0, 0)$ and then it follows that $(\dot{h}, \dot{u})^{tr}$ is an eigenvector of the matrix A corresponding to the eigenvalue η. Since the matrix A has two real and distinct eigenvalues λ_1 and λ_2, there are two families of dilatation waves, D_1 and D_2 which denote, respectively, 1-dilatation waves and 2-dilatation waves.

First we consider 1-dilatation waves. Since, $(A - \lambda_1 I)(\dot{h}, \dot{u})^{tr} = 0$ with $\lambda_1 = u - c$, we have, $c\dot{h} + h\dot{u} = 0$, implying thereby that

$$\Pi_1 \equiv u + \psi(h) = constant, \tag{3.5.17}$$

where $\psi(h) = 2\sqrt{g(h + H)} + \sqrt{gH} \ln\{\frac{\sqrt{g(h+H)} - \sqrt{gH}}{\sqrt{g(h+H)} + \sqrt{gH}}\}$. Equation (3.5.17) represents D_1 curves with Π_1 as the 1-Riemann invariant. Similarly, 2-dilatation wave curves are given by

$$\Pi_2 \equiv u - \psi(h) = constant, \tag{3.5.18}$$

where Π_2 is the 2-Riemann invariant; indeed, the integral curves of the vector fields $\vec{r}_1$ and $\vec{r}_2$ are nothing but the level sets of the Riemann invariants Π_1 and Π_2, respectively.

Theorem 3.5.1 *On D_1 (respectively, D_2), the Riemann invariant Π_1 (respectively, Π_2) is constant.*

Proof: Let U be an n-dilatation wave of the form (3.5.16), and let Π be an n-Riemann invariant; here $n = 1, 2$. Since, U is continuous and Π is assumed to be smooth, the function $\Pi : (x, t) \to \Pi(U)$ is continuous for $t > 0$. Obviously, $\Pi(U)$ is constant for $x/t \leq \lambda_n(U_l)$ and $x/t \geq \lambda_n(U_r)$.

Further, since $\eta = x/t$, we have

$$\frac{d\Pi(U)}{d\eta} = \nabla\Pi(U).\dot{U}. \tag{3.5.19}$$

As $\dot{U}$ is parallel to $\vec{r}_n$, the right-hand side of (3.5.19) is zero, and this proves the theorem. □

Theorem 3.5.2 *The D_1 curve is convex and monotonic decreasing, while D_2 curve is concave and monotonic increasing.*

Proof: The 1-dilatation curve is given by

$$u = u_l + \psi(h_l) - \psi(h), \qquad h \leq h_l \tag{3.5.20}$$

which on differentiation with respect to h, yields $du/dh = -c/h < 0$, and subsequently,

$$\frac{d^2 u}{dh^2} = \frac{c^2 + gH}{2ch^2}. \tag{3.5.21}$$

Since c, h, g, H are positive, it follows from (3.5.21) that $\frac{d^2 u}{dh^2} > 0$ and, therefore, u is convex with respect to h for 1-dilatation waves. In a similar way, we can prove for 2-dilatation waves.   □

Lemma 3.5.3 *Across 1-dilatation waves (respectively, 2-dilatation waves), $h \leq h_l$ and $u_l \leq u$ (respectively, $h \geq h_l$ and $u \geq u_l$) if, and only if, the characteristic speed increases from left- to right-hand state.*

Proof: Since $\frac{dc}{dh} = (g/2c) > 0$, c is an increasing function of h; this implies that for 1-dilatation waves, $c(h) \leq c(h_l)$ or equivalently $-c_l \leq -c$. The inequalities $u_l \leq u$ and $-c_l \leq -c$ imply that $\lambda_1(U_l) \leq \lambda_1(U)$. In a similar way, we can prove $\lambda_2(U_l) \leq \lambda_2(U)$ for 2-dilatation waves.
Conversely, for 1-dilatation waves, since $\lambda_1(U_l) \leq \lambda_1(U)$, we have

$$c - c_l \leq u - u_l. \tag{3.5.22}$$

Further, because in 1-dilatation wave region Π_1 is constant, we have $u - u_l = \psi(h_l) - \psi(h)$, which in view of (3.5.22) yields $c - c_l \leq \psi(h_l) - \psi(h)$, implying thereby that $h \leq h_l$, and $u - u_l = \psi(h_l) - \psi(h) \geq 0$. Hence, $h \leq h_l$ and $u \geq u_l$. In the same way, one can prove that for 2-dilatation waves $h \geq h_l$ and $u \geq u_l$.   □

3.5.3   The Riemann problem

In what follows, we consider the Riemann problem for the system (3.5.1) with piecewise constant initial data consisting of just two constant states, which in terms of primitive variables are $U_l = (h_l, u_l)^{tr}$ to the left of $x = x_0$, and $U_r = (h_r, u_r)^{tr}$ to the right of $x = x_0$, separated by a discontinuity at $x = x_0$, i.e.,

$$U(x, t_0) = \begin{cases} U_l, & x < x_0 \\ U_r, & x > x_0. \end{cases} \tag{3.5.23}$$

We solve this problem in the class of functions consisting of constant states, separated by either bores or dilatation waves. The solution of the Riemann problem consists of at most three constant states (including U_l and U_r), which are separated either by a bore or a dilatation wave.

Theorem 3.5.3 *The bore and dilatation wave curves for 1-family, i.e., B_1 and D_1 (respectively 2-family, i.e., B_2 and D_2) have the second order contact at U_l.*

Proof: In order to prove that B_1 and D_1 have the second order contact at U_l, we have to show that B_1 and D_1 curves at $h = h_l$, up to second derivatives, are equal. The equation for 1-dilatation wave is given by (3.5.20), and from (3.5.18) we obtain

$$u|_{h=h_l} = u_l, \quad \frac{du}{dh}\Big|_{h=h_l} = -\frac{c_l}{h_l}, \quad \left(\frac{d^2u}{dh^2}\right)\Big|_{h=h_l} = \frac{c_l^2 + gH}{2c_lh_l^2}. \qquad (3.5.24)$$

The equation for 1-bore is given in (3.5.5) and from (3.5.6) and (3.5.7), we get

$$u|_{h=h_l} = u_l, \quad \frac{du}{dh}\Big|_{h=h_l} = -\frac{c_l}{h_l}, \quad \left(\frac{d^2u}{dh^2}\right)\Big|_{h=h_l} = \frac{c_l^2 + gH}{2c_lh_l^2}. \qquad (3.5.25)$$

Thus u, $\frac{du}{dh}$ and $\frac{d^2u}{dh^2}$ at $h = h_l$ have the same value for 1-bore and 1-dilatation wave curve. Therefore, B_1 and D_1 have the second order contact at U_l. Proof for 2-family follows on similar lines. $\square$

When U_r is sufficiently close to U_l, the existence and uniqueness of the solution of Riemann problem for system (3.5.1) in the class of elementary waves follow from the general theorem of Lax, which applies to any system of conservation laws that is strictly hyperbolic and genuinely nonlinear in each characteristic field (see Lax [100] and Godlewski and Raviart [63]). We now show that the solution of the Riemann problem for system (3.5.1) exists for any arbitrary initial data.

We consider the physical variables as a coordinate system; let us draw the curves B_1, B_2, D_1 and D_2, from (3.5.5), (3.5.17) and (3.5.18), respectively, for $g = 1$ and $H = 0.1$, in the (h, u)-plane as shown in Figure 3.5.1; these curves divide the (h, u)-plane into four disjoint open regions I, II, III and IV for a given left state U_l. Indeed, we fix U_l and allow U_r to vary; if U_r lies on any of the above four curves, then we have seen how to solve the problem. We assume that U_r belongs to one of the four open regions I, II, III and IV as shown in Figure 3.5.1.

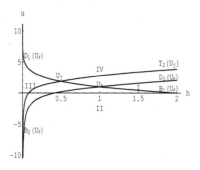

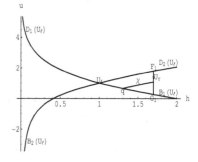

Figure 3.5.1:   Wave curves in the (h, u)-plane

Figure 3.5.2: U_r is in region I

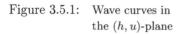

Following [189], we define, for $\hat{U} \in \mathbb{R}^2$, $B_n(\hat{U}) = \{(h, u) : (h, u) \in B_n(\hat{U})\}$, $D_n(\hat{U}) = \{(h, u) : (h, u) \in D_n(\hat{U})\}$, and $T_n(\hat{U}) = B_n(\hat{U}) \cup D_n(\hat{U}), n = 1, 2$. For fixed $U_l \in \mathbb{R}^2$, we consider the family of curves $\mathcal{S} = \{T_2(\hat{U}) : \hat{U} \in T_1(U_l)\}$. As the (h, u) plane is covered univalently by the family of curves $\mathcal{S}$, i.e., through each point U_r, there passes exactly one curve $T_2(\hat{U})$ of $\mathcal{S}$ and the solution to the Riemann problem is given as follows; we connect $\hat{U}$ to U_l on the right by a 1-wave (either bore or dilatation wave), and then we connect U_r to $\hat{U}$ on the right by a 2-wave (either B_2 or D_2). Indeed, depending on the position of U_r we have different wave configurations.

Theorem 3.5.4 *For any* $U_l = (h_l, u_l)$, $U_r = (h_r, u_r)$ *with* $h_l, h_r > 0$, *the Riemann problem is solvable.*

Proof: We allow $U_r \in \mathbb{R}^2$ to vary so that it may lie in region I, II, III or IV. If $U_r \in I$, then draw a vertical line $h = h_r$ as shown in Figure 3.5.2, which meets D_2 and B_1 uniquely at $F_1 = (h_1, u_1)$ and $G_1 = (h_2, u_2)$, respectively. We notice that the sub-family of curves in $\mathcal{S}$, consisting of the set $\{T_2(\hat{U}) \equiv T_2(\hat{h}, \hat{u}) : h_l \leq \hat{h} \leq h_r\}$ induces a continuous mapping $q \to \chi(q)$ from the arc $U_l G_1$ to line segment $G_1 F_1$; indeed, the region I is covered by curves in $\mathcal{S}$. So, let us suppose that (h_m, u_m) is the point which is mapped to U_r. Then

$$u = u_l - (h_m - h_l)\phi(h_m, h_l) - \psi(h_m) + \psi(h_r), \tag{3.5.26}$$

which on differentiation yields $\frac{du}{dh_m} = -(h_m - h_1)\frac{d\phi(h_m, h_l)}{dh_m} - \phi(h_m, h_l) - \frac{c_m}{h_m} < 0$, implying thereby that (h_m, u_m) is unique. Similarly, we can prove uniqueness if U_r is in region II, III or IV.

Thus if $U_r \in I$, then the solution to Riemann problem consists of 1-bore and a 2-dilatation wave connecting U_l to U_r. Suppose U_r is in region II; then the solution consists of bores B_1 and B_2 joining U_l to U_r. Let $U_r \in III$; then the solution of the Riemann problem is obtained by joining U_l to U_r by D_1 followed by B_2. Let U_r lie in region IV, then the solution consists of 1-dilatation wave and 2-dilatation wave. Thus the set $\{T_2(\hat{U}) : \hat{U} \in T_1(U_l)\}$ covers the entire half space $h > 0$ in a 1-1 fashion, and the Riemann problem is solvable for any arbitrary initial data.

Since 1-dilatation wave curve and 2-bore curve defined by (3.5.17) and (3.5.5), respectively, diverge to ∞ and $-\infty$, as $h \to 0$, and 2-dilatation wave curve and 1-bore curve defined by (3.5.18) and (3.5.5), respectively, diverge to ∞ and $-\infty$, as $h \to \infty$, we can find the solution to the Riemann problem for arbitrary U_r; this means that in contrast to the p-system, the vacuum state does not occur in this case. $\square$

3.5.4 Numerical solution

For a given left state U_l and a right state U_r, Karelsky and Petrosyan [89] have discussed how to find the unknown state U_m analytically for all the possible cases. Following Toro [200], we give below a numerical scheme to find the unknown state U_m, and discuss the influence of H on the unknown state U_m in the (x, t)-plane.

Case a: For $h_l < h_m$ and $h_r \geq h_m$, we eliminate u_m from (3.5.5) and (3.5.18) to obtain

$$u_r - u_l + \psi(h_m) - \psi(h_r) + (h_m - h_l)\phi(h_m, h_l) = 0. \qquad (3.5.27)$$

Case b: For $h_l \geq h_m$ and $h_r \geq h_m$, eliminating u_m from (3.5.17) and (3.5.18), we get

$$u_r - u_l - \psi(h_l) - \psi(h_r) + 2\psi(h_m) = 0. \qquad (3.5.28)$$

Case c: For $h_l \geq h_m$ and $h_r < h_m$, eliminating u_m from (3.5.17) and (3.5.5), we get

$$u_r - u_l + \psi(h_m) - \psi(h_l) - (h_r - h_m)\phi(h_r, h_m) = 0. \qquad (3.5.29)$$

Case d: For $h_l < h_m$ and $h_r < h_m$, we obtain from (3.5.5) that

$$u_r - u_l + (h_m - h_l)\phi(h_m, h_l) - (h_r - h_m)\phi(h_r, h_m) = 0. \qquad (3.5.30)$$

Thus, for all the four possible wave patterns (3.5.27) – (3.5.30), we obtain a single nonlinear equation

$$f_r(h_m, U_r) + f_l(h_m, U_l) + u_r - u_l = 0, \qquad (3.5.31)$$

where

$$f_l(\rho_m, U_l) = \begin{cases} (h_m - h_l)\phi(h_m, h_l), & \text{if } h_m > h_l, \\ \psi(h_m) - \psi(h_l), & \text{if } h_m \leq h_l, \end{cases} \qquad (3.5.32)$$

and

$$f_r(\rho_m, U_r) = \begin{cases} (h_m - h_r)\phi(h_m, h_r), & \text{if } h_m > h_r, \\ \psi(h_m) - \psi(h_r), & \text{if } h_m \leq h_r. \end{cases} \qquad (3.5.33)$$

We solve (3.5.31) for h_m by using Newton-Raphson iterative procedure with a stop criterion when the relative error is less than 10^{-8}; the initial guess for h_m is taken to be the average value of h_l and h_r. Once h_m is known, the solution for the x-component of fluid velocity u_m can be obtained from (3.5.5) or (3.5.17) (respectively, from (3.5.5) or (3.5.18)) depending on whether the 1-wave (respectively, 2-wave) is a bore or a dilatation wave. In case of dilatation waves, we have to find the solution inside the wave region. For 1-dilatation wave, the slope of the characteristic from $(0, 0)$ to (x, t) is

$$\frac{dx}{dt} = \frac{x}{t} = u - c, \qquad (3.5.34)$$

where the fluid velocity u and the speed of propagation c of surface disturbance are functions of the unknown h.

Since Π_1 is constant in 1-dilatation wave region, we have

$$u = u_l + \psi(h_l) - \psi(h), \qquad (3.5.35)$$

which in view of (3.5.34), yields

$$u_l + \psi(h_l) - \psi(h) - \frac{x}{t} - c = 0. \qquad (3.5.36)$$

Equation (3.5.36) is solved for h, using Newton-Raphson method, and then u is found from (3.5.35). In a similar way, we find the solution inside the 2-dilatation wave.

When $h_l < h_m$ and $h_r < h_m$, the solution of the Riemann problem consists of 1-bore and 2-bore; indeed, for Test 1 (see Table 1), the solution profiles are shown in Figure 3.5.3 at time $t = 0.1$. When $h_l \geq h_m$ and $h_r \geq h_m$, the solution consists of a 1-dilatation wave and a 2-dilatation wave, and the solution profiles for Test 2, at time $t = 0.25$, are shown in Figure 3.5.4. When $h_l < h_m$ and $h_r \geq h_m$, the solution consists of a 1-bore and a 2-dilatation wave, and the solution profiles for Test 3, at time $t = 0.17$, are shown in Figure 3.5.5. Similarly, when $h_l \geq h_m$ and $h_r < h_m$, the solution of the Riemann problem consists of a 1-dilatation wave and a 2-bore, and the solution profiles for Test 4, at time $t = 0.35$, are shown in Figure 3.5.6. Table 2 shows how for a fixed U_l and U_r (namely, Test 4), the intermediate state U_m (unknown state) is influenced by the presence of advective transport of impulse (H); indeed, an increase in H makes the dilatation wave weaker and bore stronger, see Figure 3.5.6.

Table 1 : Four Riemann problem tests

Test	h_l	u_l	h_m	u_m	h_r	u_r	Result
1	1.0	1.0	2.13116	0.0	1.0	−1.0	$B_1 B_2$
2	1.0	−0.5	0.65516	0.0	1.0	0.5	$D_1 D_2$
3	0.8	1.1	1.32917	0.54619	1.7	0.9	$B_1 D_2$
4	3.0	0.0	2.20694	0.54306	1.5	0.0	$D_1 B_2$

Table 2 : Intermediate state influenced by the parameter H

$H = 0$		$H = 0.1$		$H = 0.2$	
h_m	u_m	h_m	u_m	h_m	u_m
2.18076	0.51062	2.20694	0.54306	2.20732	0.55705

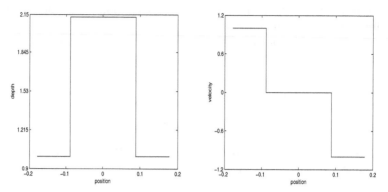

Figure 3.5.3: Exact solution for depth and velocity at $t = 0.1$

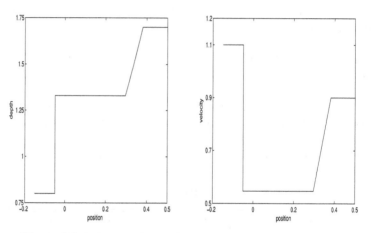

Figure 3.5.4: Exact solution for depth and velocity at $t = 0.17$

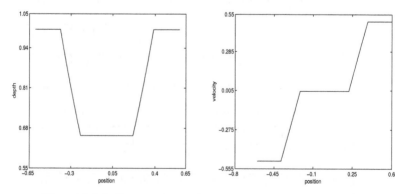

Figure 3.5.5: Exact solution for depth and velocity at $t = 0.25$

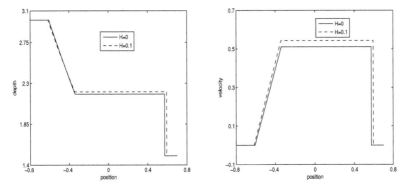

Figure 3.5.6: Exact solution for depth and velocity at $t = 0.35$

3.5.5   Interaction of elementary waves

The interaction of elementary waves, obtained from the Riemann problem (3.5.1), (3.5.23), gives rise to new emerging elementary waves. We define the initial function, with two jump discontinuities at x_1 and x_2, as follows.

$$U(x, t_0) = \begin{cases} U_l, & -\infty < x \le x_1 \\ U_*, & x_1 < x \le x_2 \\ U_r, & x_2 < x < \infty, \end{cases} \tag{3.5.37}$$

with an appropriate choice of U_* and U_r in terms of U_l and arbitrary x_1 and $x_2 \in \mathbb{R}$. With the above initial data, we have two Riemann problems locally. An elementary wave of the first Riemann problem may interact with an elementary wave of the second Riemann problem, and a new Riemann problem is formed at the time of interaction. For one dimensional Euler equations, a discussion of the interaction of elementary waves may be found in Courant and Friedrichs [41], Chorin [35], and Chang and Hsiao [31]; in Section 3.5.7, we include a discussion of the interaction of shock and rarefaction waves belonging to the same family; the interaction of waves belonging to the same family is referred to as overtaking. The notation, $D_2 B_1 \rightarrow B_1 D_2$, used in the sequel, means that a 2-dilatation wave D_2 of the first Riemann problem (connecting U_l to U_*) interacts with the 1-bore of the second Riemann problem (connecting U_* to U_r), and leads to a new Riemann problem (connecting U_l to U_r via U_m), the solution of which consists of the bore B_1 and a 2-dilatation wave D_2 (i.e., $B_1 D_2$). The possible interactions of elementary waves belonging to different families, referred to as collision, consist of $B_2 B_1$, $B_2 D_1$, $D_2 D_1$ and $D_2 B_1$, while the elementary wave interactions belonging to the same family consist of $B_2 B_2$, $B_1 B_1$, $D_1 B_1$, $B_1 D_1$, $B_2 D_2$ and $D_2 B_2$.

3.5.6 Interaction of elementary waves from different families

i) Collision of two bores $(B_2 B_1)$:

We consider that U_l is connected to U_* by a 2-bore of the first Riemann problem and U_* is connected to U_r by a 1-bore of the second Riemann problem. In other words, for a given U_l, we choose U_* and U_r in such a way that $h_* < h_l$, $u_* = u_l + (h_* - h_l)\phi(h_*, h_l)$ and $h_* < h_r$, $u_r = u_* - (h_r - h_*)\phi(h_r, h_*)$. Since the speed of the 2-bore of the first Riemann problem is greater than the speed of the 1-bore of the second Riemann problem, B_2 overtakes B_1. In order to show that for any arbitrary state U_l, the state U_r lies in the region II (see Figure 3.5.1), it is sufficient to prove that $(h - h_*)\phi(h, h_*) - (h - h_l)\phi(h, h_l) + (h_l - h_*)\phi(h_l, h_*) > 0$ for $h_* < h_l$ and $h_* < h$.

Since ϕ is a decreasing function with respect to the second argument, and $h_* < h_l$, we have $\phi(h, h_*) > \phi(h, h_l)$, which implies that

$$(h - h_l)\phi(h, h_l) - (h_l - h_*)\phi(h_l, h_*) < (h - h_l)\phi(h, h_l) < (h - h_*)\phi(h, h_*).$$

Hence $(h - h_*)\phi(h, h_*) - (h - h_l)\phi(h, h_l) + (h_l - h_*)\phi(h_l, h_*) > 0$, i.e., the curve $B_1(U_*)$ lies below the curve $B_1(U_l)$, and therefore U_r lies in the region II. Thus, in view of the results presented in the preceding section, it follows that the outcome of the interaction is $B_2 B_1 \rightarrow B_1 B_2$; the computed results illustrate this case in Figure 3.5.7.

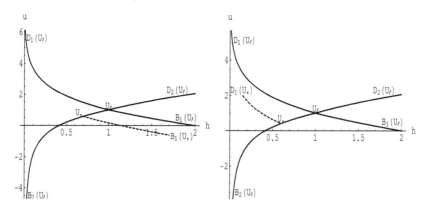

Figure 3.5.7: Collision $B_2 B_1$ Figure 3.5.8: Collision $B_2 D_1$

ii) Collision of a bore and dilatation wave $(B_2 D_1)$:

Here $U_* \in B_2(U_l)$ and $U_r \in D_1(U_*)$. That is, for a given U_l, we choose U_* and U_r such that $h_* < h_l$, $u_* = u_l + (h_* - h_l)\phi(h_*, h_l)$ and $h_r \leq h_*$, $u_r = u_* + \psi(h_*) - \psi(h_r)$. Since 2-bore has a greater velocity than 1-dilatation wave, it follows that B_2 overtakes D_1. Moreover, since for any given U_l

$$\psi(h_l) - \psi(h_*) - (h_* - h_l)\phi(h_*, h_l) > 0 \text{ for } h < h_* < h_l,$$

it follows that the curve $D_1(U_*)$ lies below the curve $D_1(U_l)$; hence U_r lies in the region III, and subsequently $B_2D_1 \rightarrow D_1B_2$. The computed results illustrate this case in Figure 3.5.8.

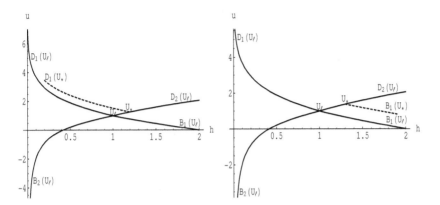

Figure 3.5.9: Collision D_2D_1 Figure 3.5.10: Collision D_2B_1

iii) Collision of two dilatation waves (D_2D_1):

We consider $U_* \in D_2(U_l)$ and $U_r \in D_1(U_*)$. In other words, for a given U_l, we choose U_* and U_r such that $h_l \leq h_*$, $u_* = u_l - \psi(h_l) + \psi(h_*)$ and $h_r \leq h_*$, $u_r = u_* + \psi(h_*) - \psi(h_r)$. Since the trailing end of the 2-dilatation wave has velocity (bounded above) greater than that of the 1-dilatation wave, the interaction will take place in a finite time. Further, since $h_l < h_*$ and ψ is an increasing function of h, we have $\psi(h_l) < \psi(h_*)$, and therefore the curve $D_1(U_*)$ lies above the curve $D_1(U_l)$; hence U_r lies in the region IV and the outcome of the interaction is $D_2D_1 \rightarrow D_1D_2$. The computed results illustrate this case in Figure 3.5.9.

iv) Collision of a dilatation wave and a bore (D_2B_1):

Here $U_* \in D_2(U_l)$ and $U_r \in B_1(U_*)$, i.e., for a given U_l, we choose U_* and U_r such that $h_l \leq h_*$, $u_* = u_l - \psi(h_l) + \psi(h_*)$ and $h_* < h_r$, $u_r = u_* - (h_* - h_r)\phi(h_*, h_r)$. Since the 1-bore speed of the second Riemann problem is less than the speed of trailing end of the 2-dilatation wave of the first Riemann problem, the bore B_1 penetrates D_2 in a finite time. For any given U_l, we show that $U_r \in I$; for this, it is enough to show that

$$\psi(h_*) - \psi(h_l) + (h - h_l)\phi(h, h_l) - (h - h_*)\phi(h_*, h) > 0. \qquad (3.5.38)$$

Since $\psi(h)$ is an increasing function of h, we have $\psi(h_*) > \psi(h_l)$ for $h_l < h_*$; hence, the inequality (3.5.38) follows, implying thereby that the curve $B_1(U_*)$ lies above the curve $B_1(U_l)$, and U_r lies in the region I. Thus the outcome of the interaction is $D_2B_1 \rightarrow B_1D_2$; the computed results illustrate this case in Figure 3.5.10.

3.5.7   Interaction of elementary waves from the same family

i) A 2-bore overtakes another 2-bore (B_2B_2):

We consider the situation in which U_l is connected to U_* by a 2-bore of the first Riemann problem and U_* is connected to U_r by a 2-bore of the second Riemann problem. In other words, for a given left state U_l, the intermediate state U_* and the right state U_r are chosen such that $h_* < h_l$ and $u_* = u_l + (h_* - h_l)\phi(h_*, h_l)$ with Lax stability conditions

$$\lambda_1(U_l) < \sigma_2(U_l, U_*) < \lambda_2(U_l), \qquad \lambda_2(U_*) < \sigma_2(U_l, U_*), \quad (3.5.39)$$

and $h_r < h_*$, $u_r = u_* + (h_r - h_*)\phi(h_r, h_*)$ with

$$\lambda_1(U_*) < \sigma_2(U_*, U_r) < \lambda_2(U_*), \qquad \lambda_2(U_r) < \sigma_2(U_*, U_r), \quad (3.5.40)$$

where $\sigma_2(U_l, U_*)$ is the speed of the bore connecting U_l to U_*, and similarly $\sigma_2(U_*, U_r)$ is the speed of the bore connecting U_* to U_r. From (3.5.39) and (3.5.40), we obtain $\sigma_2(U_*, U_r) < \sigma_2(U_l, U_*)$, i.e., the 2-bore of second Riemann problem overtakes the 2-bore of the first Riemann problem after a finite time, and gives rise to a new Riemann problem with data U_l and U_r. In order to solve this problem, we must determine the region in which U_r lies with respect to U_l. We claim that U_r lies in region III so that the solution of the new Riemann problem consists of D_1 and B_2. In other words, to prove our claim, we need to show that $B_2(U_*)$ lies entirely in the region III; to show this we are required to prove that for $h < h_* < h_l$,

$$(h_l - h)\phi(h, h_l) - (h_l - h_*)\phi(h_l, h_*) - (h_* - h)\phi(h, h_*) > 0. \quad (3.5.41)$$

Let us define a new function

$$f_1(h) = (h_l - h)\phi(h, h_l) - (h_l - h_*)\phi(h_l, h_*) - (h_* - h)\phi(h, h_*),$$

so that $f_1(h_*) = 0$, and differentiate $f_1(h)$ with respect to h to obtain

$$\frac{df_1}{dh} = (h_l - h)\frac{d\phi(h, h_l)}{dh} - \phi(h, h_l) - \left((h_* - h)\frac{d\phi(h, h_*)}{dh} - \phi(h, h_*)\right). \quad (3.5.42)$$

Now we define

$$f_2(h, h_l) = (h_l - h)\frac{d\phi(h, h_l)}{dh} - \phi(h, h_l). \quad (3.5.43)$$

Since $f_2(h, h_l)$ is a decreasing function with respect to the second variable h_l for $h < h_l$, we have $f_2(h, h_l) < f_2(h, h_*)$, and it follows from (3.5.42) that $f_1(h)$ is a decreasing function of h, implying thereby that $f_1(h) > f_1(h_*) = 0$. Hence, $B_2B_2 \to D_1B_2$; the computed results illustrate this situation in Figure 3.5.11.

ii) A 1-bore overtakes another 1-bore $(B_1 B_1)$:

The analytical proof that U_r lies in the region I, so that $B_1 B_1 \rightarrow B_1 D_2$, is similar to the previous case.

iii) A 1-dilatation wave overtakes a 1-bore $(D_1 B_1)$:

In this case, U_l is connected to U_* by a 1-dilatation wave of the first Riemann problem and U_* is connected to U_r by a 1-bore of the second Riemann problem. That is, for a given U_l, we choose U_* and U_r in such a way that $h_* \leq h_l$, $u_* = u_l + \psi(h_l) - \psi(h_*)$ and $h_* < h_r$, $u_r = u_* - (h_r - h_*)\phi(h_r, h_*)$.

First we show that $D_1(U_l)$ lies above the curve $B_1(U_*)$ for $h_* < h \leq h_l$; in other words, for $h_* < h \leq h_l$

$$\psi(h_*) - \psi(h) + (h - h_*)\phi(h, h_*) > 0. \qquad (3.5.44)$$

Let us define $f_3(h) = \psi(h_*) - \psi(h) + (h - h_*)\phi(h, h_*)$, so that $f_3(h_*) = 0$. We claim that $f_3'(h) > 0$. Let us assume on the contrary that $f_3'(h) \leq 0$, which implies that $(h - h_*)\frac{d\phi(h, h_*)}{dh} + \phi(h, h_*) \leq \frac{c}{h}$; squaring both sides and simplifying, we obtain

$$(h - h_*)^2((h_* + 2H)^2 + 4h(h + h_* + 2H)) \leq 0, \qquad (3.5.45)$$

which is a contradiction, because the left-hand side of the inequality (3.5.45) is strictly positive. Thus $f_3'(h) > 0$, implying thereby that $f_3(h) > f_3(h_*) = 0$; hence $D_1(U_l)$ lies above the curve $B_1(U_*)$ for $h_* < h \leq h_l$. Next, we prove that $B_1(U_l)$ lies above the curve $B_1(U_*)$ for $h_l < h$; for this, it is sufficient to prove that

$$\psi(h_*) - \psi(h_l) + (h - h_*)\phi(h, h_*) - (h - h_l)\phi(h, h_l) > 0, \ \forall \ h_l < h. \ (3.5.46)$$

Let us define $f_4(h) = \psi(h_*) - \psi(h_l) + (h - h_*)\phi(h, h_*) - (h - h_l)\phi(h, h_l)$, so that $f_4(h_l) = f_3(h_l) > 0$. It may be noticed that

$$f_4'(h) = \phi(h, h_*) + (h - h_*)\frac{d\phi(h, h_*)}{dh} - \phi(h, h_l) - (h - h_l)\frac{d\phi(h, h_l)}{dh} > 0,$$

because $\phi(h, h_*) + (h - h_*)\frac{d\phi(h, h_*)}{dh}$ is a decreasing function with respect to h_* for $h_* < h$; thus, $f_4'(h) > 0$ for $h_* < h_l < h$, and hence $\psi(h_*) - \psi(h_l) + (h - h_*)\phi(h, h_*) - (h - h_l)\phi(h, h_l) > 0$.

Lastly, we show that $B_2(U_l)$ and $B_1(U_*)$ intersect at some point $(\tilde{h}_1, \tilde{u}_1)$, where $h_* < \tilde{h}_1 < h_l$. To prove this, we define a new function

$$f_5(h) = \psi(h_l) - \psi(h_*) - (h - h_*)\phi(h, h_*) - (h - h_l)\phi(h, h_l) \text{ for } h_* \leq h \leq h_l.$$

Since $f_5(h_l) = -f_3(h_l) < 0$ and $f_5(h_*) > 0$, by virtue of monotonicity and the intermediate value property, there exists a unique $\tilde{h}_1$, between h_* and h_l, such that $f_5(\tilde{h}_1) = 0$. Thus, the intersection of $B_2(U_l)$ and $B_1(U_*)$ is uniquely determined; the computed results are shown in Figure 3.5.12. Thus, depending on the value of h_r we distinguish three cases:

a) When $h_r < \tilde{h}_1$, $U_r \in III$ and the outcome of the interaction is $D_1 B_1 \to D_1 B_2$; indeed, the 1-bore is weak compared to the 1-dilatation wave.

b) When $h_r = \tilde{h}_1$, U_r lies on $B_2(U_l)$ and the outcome of the interaction is $D_1 B_1 \to B_2$; thus, when two waves of first family interact, they annihilate each other, and give rise to a wave of the second family.

c) When $h_r > \tilde{h}_1$, $U_r \in II$ and the outcome of the interaction is $D_1 B_1 \to B_1 B_2$; this means that the 1-bore of the second Riemann problem, which is strong compared to the 1-dilatation wave of the first Riemann problem, overtakes the trailing end of the 1-dilatation wave, producing a reflected bore $B_2(U_m, U_r)$ that connects a new constant state U_m on the left to the known state U_r on the right. The transmitted wave, after interaction, is the 1-bore that joins U_l on the left to the state U_m on the right.

iv) A 1-bore overtakes a 1-dilatation wave ($B_1 D_1$):
Here $U_* \in B_1(U_l)$ and $U_r \in D_1(U_*)$. That is, for a given U_l, we choose U_* and U_r such that $h_l < h_*$, $u_* = u_l - (h_* - h_l)\phi(h_*, h_l)$ and $h_r \leq h_*$, $u_r = u_* + \psi(h_*) - \psi(h_r)$. In the (x, t) plane, the speed $\lambda_1(U_*)$ of the trailing end of the 1-dilatation wave is less than the velocity $\sigma_1(U_l, U_*)$, and therefore the 1-dilatation wave from right overtakes the 1-bore from left after a finite time.

First we show that $B_1(U_l)$ lies above the curve $D_1(U_*)$ for $h_l < h < h_*$; for this we need to show $\psi(h) - \psi(h_*) - (h - h_l)\phi(h, h_l) + (h_* - h_l)\phi(h_*, h_l) > 0$ for $h_l < h < h_*$. To prove this, we define a new function

$$f_6(h) = \psi(h) - \psi(h_*) - (h - h_l)\phi(h, h_l) + (h_* - h_l)\phi(h_*, h_l) \text{ for } h_l < h < h_*.$$

Then, one can show that $f_6'(h) = \frac{c}{h} - \phi(h, h_l) - (h - h_l)\frac{d\phi(h, h_l)}{dh} < 0$ for $h_l < h < h_*$, implying thereby that $f_6(h) > f_6(h_*) = 0$.

Next, we show that $D_1(U_*)$ lies below the curve $D_1(U_l)$ for $h \leq h_l < h_*$, i.e., $\psi(h_l) - \psi(h_*) + (h_* - h_l)\phi(h_*, h_l) > 0$ for $h \leq h_l < h_*$. Since the left-hand side of this inequality, for $h \leq h_l < h_*$, turns out to be $f_6(h_l)$, which has already been shown to be positive, the conclusion follows.

Lastly, we show that $B_2(U_l)$ and $D_1(U_*)$ intersect uniquely at some point, say, $(\tilde{h}_2, \tilde{u}_2)$; for this, it is enough to show that the equation

$$\psi(h) - \psi(h_*) + (h - h_l)\phi(h, h_l) + (h_* - h_l)\phi(h_*, h_l) = 0,$$

has a unique root $\tilde{h}_2$ such that $\tilde{h}_2 < h_l$. To establish this, we define a new function $f_7(h) = \psi(h) - \psi(h_*) + (h - h_l)\phi(h, h_l) + (h_* - h_l)\phi(h_*, h_l)$; since $f_7(h_l) > 0$, and $f_7(h)$ takes negative values as h is close to zero, in view of monotonicity and the intermediate value property, it follows that the curves $D_1(U_*)$ and $B_2(U_l)$ intersect uniquely; the computed results are shown in Figure 3.5.13. Here, again, we distinguish three cases depending on the value of h_r:

a) When $h_r > \tilde{h}_2$, $U_r \in II$ and the outcome of the interaction is $B_1 D_1 \rightarrow B_1 B_2$, i.e., the 1-bore is sufficiently strong compared to the 1-dilatation wave which, after interaction, produces a new elementary wave.

b) When $h_r = \tilde{h}_2$, $U_r \in B_2(U_l)$ and the outcome of the interaction is $B_1 D_1 \rightarrow B_2$, i.e., the interaction of elementary waves of the first family gives rise to a new elementary wave of the second family.

c) When $h_r < \tilde{h}_2$, $U_r \in III$ and the outcome of the interaction is $B_1 D_1 \rightarrow D_1 B_2$.

v) A 2-bore overtakes a 2-dilatation wave $(B_2 D_2)$:

The $B_2 D_2$ interaction takes place when $U_* \in B_2(U_l)$ and $U_r \in D_2(U_*)$. In other words, for a given U_l, we choose U_* and U_r in such a way that $h_* < h_l$, $u_* = u_l + (h_* - h_l)\phi(h_*, h_l)$ and $h_* \le h_r$, $u_r = u_* - \psi(h_*) + \psi(h_r)$.

First we show that for $h_* < h < h_l$, $B_2(U_l)$ lies above $D_2(U_*)$, i.e.,

$$\psi(h_*) - \psi(h) + (h - h_l)\phi(h, h_l) + (h_l - h_*)\phi(h_*, h_l) > 0, \ \forall \ h \in (h_*, h_l]. \ (3.5.47)$$

To prove this, we define a new function $f_8(h) = \psi(h_*) - \psi(h) + (h - h_l)\phi(h, h_l)$ $+(h_l - h_*)\phi(h_*, h_l)$ so that $f_8(h_*) = 0$. Since $f_8'(h) > 0$, we have $f_8(h) > f_8(h_*)$; it follows that $f_8(h) > 0$, implying thereby that $B_2(U_l)$ lies above $D_2(U_*)$.

Next, we show that the curve $D_2(U_l)$ lies above the curve $D_2(U_*)$ for $h_* < h_l \le h$; for this it is enough to prove $\psi(h_*) - \psi(h_l) + (h_l - h_*)\phi(h_*, h_l) > 0$ for $h_* < h_l \le h$. We notice that the left-hand side of this inequality is $f_8(h_l)$, which has already been shown to be positive, and hence the curve $B_2(U_l)$ lies above the curve $D_2(U_*)$ for $h_* < h_l \le h$.

Lastly, we show that $D_2(U_*)$ and $B_1(U_l)$ intersect uniquely, say, at $(\tilde{h}_3, \tilde{u}_3)$ for $h_* < h_l < \tilde{h}_3$. Now, we define $f_9(h) = \psi(h) - \psi(h_*) + (h - h_l)\phi(h, h_l) + (h_* - h_l)\phi(h_*, h_l)$ for $h_* < h_l \le h$ so that $f_9(h_l) < 0$, and we can choose a constant $K > 0$ such that $f_9(h) > 0$ for all $h > K$. Then there exists an $\tilde{h}_3$ such that $f_9(\tilde{h}_3) = 0$. Thus, $B_2(U_*)$ and $B_1(U_l)$ intersect uniquely at $(\tilde{h}_3, \tilde{u}_3)$, as $D_2(U_*)$ and $B_1(U_l)$ are monotone; the computed results are shown in Figure 3.5.14. Here, again, the following cases arise:

a) When $h_r < \tilde{h}_3$, $U_r \in II$ and the outcome of the interaction is $B_2 D_2 \rightarrow B_1 B_2$; this means that the strength of D_2 is small compared to the elementary wave B_2, and B_2 annihilates D_2 in a finite time. The strength of reflected B_1 wave is small compared to the incident waves B_2 and D_2.

b) When $h_r = \tilde{h}_3$, $U_r \in B_1(U_l)$ and the outcome of the interaction is $B_2 D_2 \rightarrow B_1$, showing thereby that the reflected shock B_1 is weak compared to the incident waves D_2 and B_2.

c) When $h_r > \tilde{h}_3$, $U_r \in I$ and the outcome of the interaction is $B_2 D_2 \rightarrow B_1 D_2$, implying thereby that D_2 is stronger than B_2.

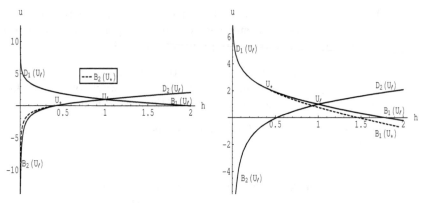

Figure 3.5.11: B_2 overtakes B_2 Figure 3.5.12: B_1 overtakes D_1

Here $U_* \in D_2(U_l)$ and $U_r \in B_2(U_*)$. Thus, for a given U_l, we choose U_* and U_r such that $h_l \leq h_*$, $u_* = u_l - \psi(h_l) + \psi(h_*)$ and $h_r < h_*$, $u_r = u_* + (h_r - h_*)\phi(h_r, h_*)$.

Now, we show that $D_2(U_l)$ lies above $B_2(U_*)$ for $h_l \leq h < h_*$, i.e.,

$$\psi(h) - \psi(h_*) - (h - h_*)\phi(h, h_*) > 0, \quad \forall\, h_l \leq h < h_*. \qquad (3.5.48)$$

To prove this, we define a new function $f_{10}(h) = \psi(h) - \psi(h_*) - (h - h_*)\phi(h, h_*)$ for $h_l \leq h \leq h_*$ so that $f_{10}(h_*) = 0$; this, in view of the expressions for $c(h)$ and $\phi(h, h_*)$, yields $\frac{df_{10}(h)}{dh} = \frac{-[(h-h_*)^2((h_*+2H)^2+4h(h+h_*+2H))]}{4h^2 h_* \phi(h,h_*)} < 0$, implying thereby that $f_{10}(h) > f_{10}(h_*) = 0$. Hence, the result.

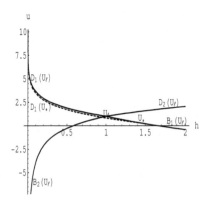

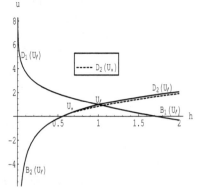

Figure 3.5.13: D_1 overtakes B_1 Figure 3.5.14: D_2 overtakes B_2

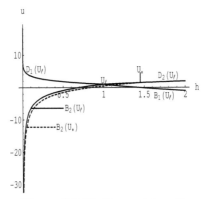

Figure 3.5.15: B_2 overtakes D_2

vi) A 2-dilatation wave overtakes a 2-bore (D_2B_2):

Next, we show that $B_2(U_l)$ lies above the curve $B_2(U_*)$ for $h < h_l < h_*$; for this it is sufficient to prove $\psi(h_l) - \psi(h_*) - (h - h_*)\phi(h, h_*) + (h - h_l)\phi(h, h_l) > 0$ for $h < h_l < h_*$. In order to prove this inequality, we define a new function $f_{11}(h) = \psi(h_l) - \psi(h_*) - (h - h_*)\phi(h, h_*) + (h - h_l)\phi(h, h_l)$ for $h \leq h_l < h_*$ so that $f_{11}(h_l) = f_{10}(h_l) > 0$. Since $\phi(h, h_l) + (h - h_l)\phi(h, h_l)$ is an increasing function with respect to the second argument for $h_l > h$, it follows that $\frac{df_{11}(h)}{dh} < 0$, implying thereby that $f_{11}(h) > f_{11}(h_l) > 0$.

Lastly, we show that $B_1(U_l)$ and $B_2(U_*)$ intersect uniquely at a point, say, $(\tilde{h}_4, \tilde{u}_4)$, where $h_l < \tilde{h}_4 < h_*$. The proof for this follows on similar lines as discussed earlier; the computed results are shown in Figure 3.5.15. Here, also, we encounter the following possibilities:

a) When $h_r > \tilde{h}_4$, $U_r \in I$ and the outcome of the interaction is $D_2B_2 \to B_1D_2$; this means that D_2 is strong compared to the elementary wave B_2, and the strength of reflected B_1 is small compared to the incident waves B_2 and D_2.

b) When $h_r = \tilde{h}_4$, $U_r \in B_1(U_l)$ and the outcome of the interaction is $D_2B_2 \to B_1$.

c) When $h_r < \tilde{h}_4$, $U_r \in II$ and the outcome of the interaction is $D_2B_2 \to B_1B_2$, implying thereby that the elementary wave B_2 is strong compared to D_2.

Chapter 4

Evolution of Weak Waves in Hyperbolic Systems

We have already noticed in Chapter 2 that the solutions of conservation law (1.1.1) with $m = 1$, which are piecewise smooth, admit discontinuities across certain smooth curves in the (x, t) plane. In this chapter, we study systems (1.1.1) in 1- and 3-space dimensions. We consider only piecewise smooth solutions so that there exists a smooth orientable surface $\sum : \phi(\mathbf{x}, t) = 0$, with outward space-time normal $(\phi_{,t}, \phi_{,1}, \phi_{,2}, \phi_{,3})$, across which $\mathbf{u}$ suffers a jump discontinuity satisfying the R-H condition

$$G[\mathbf{u}] = [f_{,j}]n_j,$$

and outside of which $\mathbf{u}$ is a C^1 function; here $G = -\phi_{,t}/(\phi_{,j}\phi_{,j})^{1/2}$ is the propagation speed of $\sum$ in the direction $n_j = \phi_{,j}/(\phi_{,i}\phi_{,i})^{1/2}$, $1 \leq i, j \leq 3$. Explicit formulae for the discontinuities in the first and higher order derivatives of $\mathbf{u}$ across $\sum$, which are of the nature of compatibility conditions for the existence of discontinuities, have been derived by Thomas [196] as well as Truesdell and Toupin [201]. A surface $\sum$ across which the field variable $\mathbf{u}$ or its derivative is discontinuous is called the singular surface or a wave-front; it is only such a wave that is being studied here. Indeed, the use of compatibility conditions that hold on $\sum$ enable us to obtain some interesting results in the general theory of surfaces of discontinuity in continuum mechanics.

4.1 Waves and Compatibility Conditions

A wave may be conceived as a moving surface across which some of the field variables or their derivatives, describing the material medium, undergo certain kinds of discontinuities that are carried along by the surface. The discontinuities across the surface are found to be interrelated. The relations connecting the field variables, or their derivatives on the two sides of the discontinuity surface are usually referred to as compatibility conditions, and they arise as a direct consequence of the dynamical conditions which determine the behavior of the material medium. The first set of compatibility conditions, known as the R-H jump conditions, is essentially a consequence of the conservation laws that hold across the discontinuity surface. The relations which connect the first order derivatives of the field variables on the two sides of the discontinuity surface with its speed of propagation are known as compatibility conditions of the first order. In a similar manner, the compatibility conditions of second and higher order, which relate respectively to the second and higher order derivatives, follow from the assumption of a smooth wave front.

Let $\phi(\mathbf{x}, t) = 0$ or $x_i = x_i(y^1, y^2, t)$ represent the moving surface $\sum$; here $y^\alpha (\alpha = 1, 2)$ are the curvilinear coordinates of the surface $\sum(t)$ which is moving with speed G. Our convention is that $G > 0$, which corresponds to a singular surface propagating in the direction of $\mathbf{n}$. Let z be a field variable which suffers a jump discontinuity across $\sum(t)$, defined as $z = z_2 - z_1$, where z_1 and z_2 are the values of z immediately ahead of and behind the wave front $\sum(t)$, respectively. The compatibility conditions can now be stated as (see Thomas [196])

$$
\begin{aligned}
[z_{,i}] &= d^{(1)} n_i + g^{\alpha\beta} [z]_{,\alpha} x_{i,\beta}; \quad [z_{,t}] = -d^{(1)} G + \frac{\delta}{\delta t} [z], \\
[z_{,ij}] &= d^{(2)} n_i n_j + g^{\alpha\beta} (d^{(1)}_{,\alpha} + g^{\sigma\tau} b_{\alpha\sigma} [z]_{,\tau})(n_i x_{j,\beta} + n_j x_{i,\beta}) \\
&\quad + g^{\alpha\beta} g^{\sigma\tau} ([z]_{,\alpha\sigma} - d^{(1)} b_{\alpha\sigma}) x_{i,\beta} x_{j,\tau}, \\
\left[\frac{\partial^2 z}{\partial x_i \partial t} \right] &= \left(-d^{(2)} G + \frac{\delta}{\delta t} d^{(1)} - g^{\alpha\beta} [z]_{,\alpha} x_{k,\beta} \frac{\delta n_k}{\delta t} \right) n_i + g^{\alpha\beta} \left[\frac{\partial z}{\partial t} \right]_{,\alpha} x_{i,\beta}, \\
\left[\frac{\partial^2 z}{\partial t^2} \right] &= d^{(2)} G^2 - G \frac{\delta}{\delta t} d^{(1)} + G g^{\alpha\beta} [z]_{,\alpha} x_{i,\beta} \frac{\delta n_i}{\delta t} + \frac{\delta}{\delta t} \left[\frac{\partial z}{\partial t} \right],
\end{aligned}
\tag{4.1.1}
$$

where $d^{(1)} = [z_{,i}] n_i$, $d^{(2)} = [z_{,ij}] n_i n_j$, and $\delta/\delta t = \partial/\partial t + G n_i \partial/\partial x_i$ denotes the time derivative as apparent to an observer moving with the wave front along the normal $\mathbf{n}$; the quantities $g_{\alpha\beta} = x_{i,\alpha} x_{i,\beta}$ and $b_{\alpha\beta}$ are respectively the covariant components of the first and second fundamental surface tensors of the wave front $\sum(t)$.

Here, summation convention on repeated indices applies; the Latin indices (i, j, k) have the range 1,2,3 and the Greek indices $(\alpha, \beta, \sigma, \tau)$ the range $1, 2$, and a comma followed by a Greek index, say α, denotes partial derivative with respect to y^α. Quantities $\phi_{i,\alpha}$ are the components of a covariant vector, and $x_{i,\alpha}$ is the tangent vector to the surface $\sum(t)$ having the direction of the corresponding y^α curve. We also recall the following relations which we shall be using in our subsequent discussion

$$
\begin{aligned}
&n_i x_{i,\alpha} = 0; \quad x_{i,\alpha\beta} = b_{\alpha\beta} n_i; \quad n_{i,\beta} = -g^{\alpha\sigma} b_{\alpha\beta} x_{i,\sigma}, \\
&\delta n_i / \delta t = -g^{\alpha\beta} G_{,\alpha} x_{i,\beta}; \quad g^{\alpha\beta} b_{\alpha\beta} = b^\beta_\beta = 2\Omega,
\end{aligned}
\tag{4.1.2}
$$

where Ω and $g^{\alpha\beta}$ are respectively the mean curvature and the contravariant components of the metric tensor of the surface $\sum$; the Gaussian curvature of $\sum$ is given by $K = det(b^\alpha_\alpha)$.

Equations (4.1.1) are called the geometrical and kinematical conditions of compatibility of the first and second order; these conditions are iterated to yield higher order compatibility conditions for the derivatives of z (see

Nariboli [130], Grinfeld [65], Estrada and Kanwal [53], and Shugaev and Shtemenko [184]).

The singularity $\sum(t)$ is called a wave of order N if all the derivatives of z of order less than N are continuous across $\sum(t)$, while at least an Nth order derivative of z suffers a jump discontinuity across $\sum(t)$. When $N = 0$, the singular surface is called a shock wave, while for $N = 1$, it is called a weak discontinuity wave or a first order discontinuity, or an acceleration wave; a wave of order $N \geq 2$ is called a mild discontinuity (see Varley and Cumberbatch [205], Coleman et al. [39], and Ting [198]).

The analysis of acceleration waves and mild discontinuities for a nonlinear system proceeds on similar lines.

The above compatibility conditions, when applied to the system of partial differential equations, which describe the variation of field variables in space and time, yield in succession the normal speed of propagation and then the transport equation that governs the variation of wave amplitude. When the speed of propagation G is independent of $\mathbf{x}$ and $\mathbf{n}$, the propagation is called isotropic and homogeneous, and the transport equation for the variation of wave amplitude can be easily integrated along the normal trajectories of $\sum(t)$. This is the simplest case, and it corresponds to the physical situation when the wave propagates into a medium which is in a uniform state at rest, or when the medium ahead is nonuniform but the problem is mathematically one-dimensional, such as planar, cylindrically or spherically symmetric waves (see, e.g., Chen [32] and the references therein, Singh and Sharma [186], Pai et al. [139], Schmitt [160], and Sharma and Radha [169]). The wave propagation is called anisotropic only if the speed G depends on the normal to the surface $\sum(t)$. If the wave is propagating into a nonuniform medium then the speed G depends on the spatial coordinates also, and the propagation phenomenon is designated as anisotropic and nonhomogeneous. In this case, the transport equation for the variation of the wave amplitude along the normal trajectories involves the surface derivative terms, which pose a real difficulty in obtaining its solution; but if we transform the transport equation to a differential equation along bicharacteristic curves of the governing system of partial differential equations, this difficulty disappears. In order to proceed further, we need to observe certain relations involving the bicharacteristic curves or rays on $\sum(t)$.

4.1.1 Bicharacteristic curves or rays

As for the wave front $\sum : \phi(\mathbf{x}, t) = 0$, the δ-time derivative vanishes, its normal speed of propagation G, which in general is a function of $\mathbf{x}$ and $\mathbf{n}$, satisfies the first order partial differential equation:

$$H \equiv G(\phi_{,i}\phi_{,i})^{1/2} + \phi_{,t} = 0. \tag{4.1.3}$$

The characteristic curves of (4.1.3), known as bicharacteristic rays, are the curves in space-time whose parametric representation is obtained by solving the ODEs:

$$\frac{dx_i}{dt} = \frac{\partial H}{\partial \phi_{,i}} \Big/ \frac{\partial H}{\partial \phi_{,t}}, \quad \frac{d}{dt}(\phi_{,i}) = -\frac{\partial H}{\partial x_i} \Big/ \frac{\partial H}{\partial \phi_{,t}},$$

which can be rewritten as

$$\frac{dx_i}{dt} = Gn_i + (\delta_{ij} - n_i n_j)\frac{\partial G}{\partial n_j}, \quad \frac{dn_i}{dt} = (n_i n_j - \delta_{ij})\frac{\partial G}{\partial x_j}, \tag{4.1.4}$$

where $n_i = \phi_{,i}/(\phi_{,j}\phi_{,j})^{1/2}$. In equation (4.1.4), the vector $dx_i/dt \equiv V_i$ is identified as ray velocity vector; equation $(4.1.4)_2$ describes the time variation of the normal n_i as we move along the rays.

Thus the time rate of change of the field variable z, as apparent to an observer moving with the wave front in the ray direction with velocity V_i, is given by $dz/dt = \partial z/\partial t + V_i z_{,i}$; equation $(4.1.4)_1$ shows that the component of V_i along normal to the surface $\sum$ is just G, while its component tangential to $\sum$ is given by $V_\alpha = V_i x_{i,\alpha} = x_{i,\alpha}\frac{\partial G}{\partial n_i}$ and, thus, the ray derivative is related to the δ-time derivative as

$$\frac{dz}{dt} = \frac{\delta z}{\delta t} + V^\alpha z_{,\alpha}, \tag{4.1.5}$$

where $V^\alpha = g^{\alpha\beta}V_\beta$. Equation (4.1.5), which will be used in our subsequent analysis, shows that the rays coincide with the normal trajectories only if G is independent of n_i.

4.1.2   Transport equations for first order discontinuities

We consider the quasilinear system (1.1.2) for m unknowns $u_i = u_i(x_1, x_2, x_3, t), 1 \leq i \leq m$,

$$\mathbf{u}_{,t} + \mathbf{A}_J \mathbf{u}_{,J} = \mathbf{g}, \quad J = 1, 2, 3, \tag{4.1.6}$$

where $\mathbf{A}_J$ and $\mathbf{g}$ are known functions of $\mathbf{u}$, and (x_1, x_2, x_3) are the Cartesian coordinates of a point in Euclidean space.

For the system (4.1.6), it is possible to consider a particular class of solutions that characterizes the first order discontinuity waves or in the language of continuum mechanics, the acceleration waves. We consider the case that (4.1.6) admits constant solution $\mathbf{u} = \mathbf{u}_o$, and we require that $\mathbf{g}(\mathbf{u}_o) = 0$. Let $\sum(t) : \phi(\mathbf{x}, t) = 0$ be the moving surface (wave front) that separates the space into two subspaces; ahead of the wave front, we have the known uniform field $\mathbf{u}_o$ and behind the wave front, unknown perturbed field $\mathbf{u}(\mathbf{x}, t)$. Across $\phi = 0$, the field $\mathbf{u}$ is continuous (i.e., $[\mathbf{u}] = 0$) but its derivatives are allowed to have simple jump discontinuities; the compatibility conditions can now be stated as follows

$$[\mathbf{u}_{,I}] = \boldsymbol{\pi}^{(1)}n_I; \quad [\mathbf{u}_{,t}] = -\boldsymbol{\pi}^{(1)}G,$$

$$[\mathbf{u}_{,IJ}] = \boldsymbol{\pi}^{(2)}n_In_J + g^{\alpha\beta}(n_Ix_{J,\beta} + n_Jx_{I,\beta})\boldsymbol{\pi}^{(1)}_{,\alpha} - \boldsymbol{\pi}^{(1)}b_{\alpha\sigma}g^{\alpha\beta}g^{\sigma\tau}x_{I,\beta}x_{J,\tau},$$

$$\left[\frac{\partial^2 \mathbf{u}}{\partial x_I \partial t}\right] = -\left(\boldsymbol{\pi}^{(2)}G + \delta_t\boldsymbol{\pi}^{(1)}\right)n_I - (\boldsymbol{\pi}^{(1)}G)_{,\alpha}g^{\alpha\beta}x_{I,\beta},$$

$$\left[\frac{\partial^2 \mathbf{u}}{\partial t^2}\right] = \boldsymbol{\pi}^{(2)}G^2 - 2G\delta_t\boldsymbol{\pi}^{(1)} - \boldsymbol{\pi}^{(1)}\delta_tG, \tag{4.1.7}$$

$$[\mathbf{u}_{,IJK}] = \boldsymbol{\pi}^{(3)}n_In_Jn_K + g^{\alpha\beta}(\boldsymbol{\pi}^{(2)}_{,\alpha} + 2g^{\gamma\delta}b_{\delta\alpha}\boldsymbol{\pi}^{(1)}_{,\gamma})P^{IJK}_\beta$$
$$+ g^{\alpha\beta}g^{\gamma\delta}\left(\boldsymbol{\pi}^{(1)}_{,\gamma\alpha} - b_{\gamma\alpha}\boldsymbol{\pi}^{(2)} - g^{\sigma\tau}b_{\alpha\sigma}b_{\gamma\tau}\boldsymbol{\pi}^{(1)}\right)Q^{IJK}_{\beta\delta}$$
$$- g^{\alpha\beta}g^{\gamma\delta}g^{\sigma\tau}b_{\alpha\sigma}\boldsymbol{\pi}^{(1)}_{,\gamma}M^{IJK}_{\beta\tau\delta} + \boldsymbol{\pi}^{(1)}_{,\gamma}g^{\alpha\beta}g^{\gamma\delta}_{,\alpha}(n_Ix_{J,\beta} + n_Jx_{I,\beta})x_{K,\delta}$$
$$- \boldsymbol{\pi}^{(1)}g^{\alpha\beta}g^{\sigma\tau}(g^{\gamma\delta}_{,\sigma}b_{\gamma\alpha} + g^{\gamma\delta}b_{\gamma\alpha,\sigma})x_{I,\beta}x_{J,\tau}x_{K,\delta},$$

$$\left[\frac{\partial^3 \mathbf{u}}{\partial x_I \partial x_J \partial t}\right] = g^{\alpha\beta}(n_Ix_{J,\beta} + n_Jx_{I,\beta})\left(-\boldsymbol{\pi}^{(2)}G + \delta_t\boldsymbol{\pi}^{(1)}\right)_{,\alpha}$$
$$+ g^{\alpha\beta}g^{\sigma\tau}(n_Ix_{J,\beta} + n_Jx_{I,\beta})b_{\alpha\sigma}(-\boldsymbol{\pi}^{(1)}G)_{,\tau}$$
$$+ \left(-\boldsymbol{\pi}^{(3)}G + \delta_t\boldsymbol{\pi}^{(2)} + 2g^{\alpha\beta}\boldsymbol{\pi}^{(1)}_{,\alpha}G_{,\beta}\right)n_In_J$$
$$+ \left\{\left(\boldsymbol{\pi}^{(2)}G - \delta_t\boldsymbol{\pi}^{(1)}\right)b_{\alpha\sigma} - (\boldsymbol{\pi}^{(1)}G)_{,\alpha\sigma}\right\}g^{\alpha\beta}g^{\sigma\tau}x_{I,\beta}x_{J,\tau},$$

where I, J, K take values 1,2,3 and $\delta_t \equiv \delta/\delta t$; (n_1, n_2, n_3) denotes unit normal to the wave front $\sum$, $\boldsymbol{\pi}^{(1)} = [\mathbf{u}_{,I}]n_I$, $\boldsymbol{\pi}^{(2)} = [\mathbf{u}_{,IJ}]n_In_J$, $\boldsymbol{\pi}^{(3)} = [\mathbf{u}_{,IJK}]n_In_Jn_K$, and

$$P^{IJK}_\beta = n_In_Jx_{K,\beta} + n_Jn_Kx_{I,\beta} + n_Kn_Ix_{J,\beta},$$
$$Q^{IJK}_{\beta\delta} = n_Ix_{J,\beta}x_{K,\delta} + n_Jx_{K,\delta}x_{I,\beta} + n_Kx_{I,\beta}x_{J,\delta},$$
$$M^{IJK}_{\beta\tau\delta} = x_{I,\beta}x_{J,\tau}x_{K,\delta} + x_{J,\delta}x_{K,\tau}x_{I,\beta} + x_{K,\beta}x_{I,\delta}x_{J,\tau}.$$

Taking a jump in equation (4.1.6) across the singularity surface $\sum$, and using the compatibility conditions $(4.1.7)_1$ and $(4.1.7)_2$ together with the continuity of $\mathbf{u}$, we obtain

$$(A^\circ_j n_J - G\mathbf{I})\boldsymbol{\pi}^{(1)} = 0, \tag{4.1.8}$$

where the superscript-$\circ$ denotes evaluation just ahead of the wave-front at $\mathbf{u} = \mathbf{u}_o$, and $\mathbf{I}$ is the $m \times m$ unit matrix. If $\boldsymbol{\pi}^{(1)} \neq 0$, its coefficient matrix in (4.1.8) must be singular, i.e.,

$$\det(A^\circ_j n_J - G\mathbf{I}) = 0, \tag{4.1.9}$$

which implies that the singularity surface $\phi = 0$ must be one of the characteristic surfaces for the system (4.1.6); in other words, discontinuities in the first derivatives of $\mathbf{u}$ can occur only on characteristic surfaces. The characteristic equation (4.1.9) is also called the eikonal equation. Hyperbolicity of (4.1.6) ensures that there is at least one possible speed of propagation G; strict hyperbolicity ensures that there are m possible speeds. If the characteristic root G of $\mathbf{A}_I^o n_I$ has multiplicity $r(\leq m)$, then there exist r linearly independent right eigenvectors $\mathbf{R}_q^o, 1 \leq q \leq r$, of $\mathbf{A}_I^o n_I$ such that

$$\boldsymbol{\pi}^{(1)} = \sigma_q^{(1)} \mathbf{R}_q^o, \tag{4.1.10}$$

where $\sigma_q^{(1)}$ are scalars, which remain undetermined at this stage. The evolutionary behavior of first order discontinuity waves can be studied merely by determining the transport equations for $\sigma_1^{(1)}, \sigma_2^{(1)}, \ldots, \sigma_r^{(1)}$.

The characteristics of the eikonal equation (4.1.9) are called bicharacteristic curves or rays of the system (4.1.6); if the multiplicity of the root G does not depend on $\mathbf{n}$, these curves are given by (see Varley and Cumberbatch [205])

$$\left(\mathbf{L}_\ell^o \cdot \mathbf{R}_p^o\right) \frac{dx_I}{dt} = \mathbf{L}_\ell^o \mathbf{A}_I^o \mathbf{R}_p^o, \quad \ell, p = 1, 2, \ldots r$$

where $\mathbf{L}_\ell^o$ are r linearly independent left eigenvectors of $\mathbf{A}_I^o n_I$ associated with the wave speed G so that

$$\mathbf{L}_\ell^o (\mathbf{A}_I^o n_I - G\mathbf{I}) = 0. \tag{4.1.11}$$

Next, we differentiate the equation (4.1.6) partially with respect to x_K, take jumps across $\sum$, multiply the resulting equations by n_K, and then make use of equations $(4.1.7)_1$, $(4.1.7)_3$ and $(4.1.7)_4$ together with the identity $[\mathbf{U} \cdot \mathbf{V}] = [\mathbf{U}] \cdot \mathbf{V}^o + \mathbf{U}^o \cdot [\mathbf{V}] + [\mathbf{U}] \cdot [\mathbf{V}]$ to obtain

$$
\begin{aligned}
-(\mathbf{A}_J^o n_J - G\mathbf{I})\boldsymbol{\pi}^{(2)} = & \ \delta_t \boldsymbol{\pi}^{(1)} + g^{\alpha\beta} x_{J,\beta} \mathbf{A}_J^o \boldsymbol{\pi}_{,\alpha}^{(1)} \\
& + \boldsymbol{\pi}^{(1)} \cdot (\nabla \mathbf{A}_J)^o \boldsymbol{\pi}^{(1)} n_J - \boldsymbol{\pi}^{(1)} \cdot (\nabla \mathbf{g})^o,
\end{aligned}
\tag{4.1.12}
$$

where ∇ denotes the gradient operator with respect to the field variables $u_1, u_2, \ldots, u_m$.

Equation (4.1.12), on pre-multiplying by the left eigenvectors $\mathbf{L}_p^o$ and using (4.1.5), (4.1.10) and (4.1.11), yields

$$\alpha_{pq} \frac{d\sigma_q^{(1)}}{dt} + \mu_{pq}^{(1)} \sigma_q^{(1)} + \nu_{pq\ell}^{(1)} \sigma_q^{(1)} \sigma_\ell^{(1)} = 0, \tag{4.1.13}$$

where

$$
\begin{aligned}
\alpha_{pq} &= \mathbf{L}_p^o \mathbf{R}_q^o, \quad \nu_{pq\ell}^{(1)} = \mathbf{L}_p^o (\mathbf{R}_q^o \cdot (\nabla \mathbf{A}_J)^o n_J) \mathbf{R}_\ell^o, \\
\mu_{pq}^{(1)} &= \mathbf{L}_p^o \left(\delta_t \mathbf{R}_q^o + \mathbf{A}_J^o \mathbf{R}_{q,\alpha}^o g^{\alpha\beta} x_{J,\beta} - \mathbf{R}_q^o \cdot (\nabla \mathbf{g})^o \right),
\end{aligned}
$$

and d/dt is the ray derivative defined in (4.1.5).

Equations (4.1.13), which constitute a system of r coupled nonlinear ordinary differential equations for the variation of $\sigma_q^{(1)}$ at an acceleration wavefront along bicharacteristic curves, are analogous to the transport equations derived by Varley and Cumberbatch [205], Ting [198], Boillat and Ruggeri [21], and Estrada and Kanwal [53]. For $r = 1$, equation (4.1.13) reduces to a single Bernoulli equation of the form

$$\frac{d\sigma}{dt} + \mu(t)\sigma = \beta(t)\sigma^2, \quad t > 0 \tag{4.1.14}$$

where σ is the wave amplitude, and the coefficients μ and β are, in general, functions of t which are known if $\mathbf{u}_o$ is known in the region into which the front is spreading.

4.1.3 Transport equations for higher order discontinuities

Transport equations for the discontinuities in the second order derivatives of the field variables at an acceleration front can be obtained as an extension of the preceding analysis. For this, we consider the enlarged system of equations comprising (4.1.6) together with the equations that result when (4.1.6) is differentiated twice with respect to the space coordinates $\mathbf{x}$, i.e.,

$$\frac{\partial}{\partial t}\mathbf{u}_{,JK} + \mathbf{u}_{,JK} \cdot (\nabla\mathbf{A}_I)\mathbf{u}_{,I} + 2\mathbf{u}_{,J} \cdot (\nabla\mathbf{A}_I)\mathbf{u}_{,IK} + \mathbf{u}_{,J}\mathbf{u}_{,K}(\nabla\nabla\mathbf{A}_I)\mathbf{u}_{,I}$$
$$+\mathbf{A}_I\mathbf{u}_{,IJK} = \mathbf{u}_{,JK} \cdot (\nabla\mathbf{g}) + \mathbf{u}_{,J}\mathbf{u}_{,K}(\nabla\nabla\mathbf{g}), \tag{4.1.15}$$

where

$$\mathbf{u}_{,JK} \cdot (\nabla\mathbf{A}_I) = \frac{\partial^2 u_i}{\partial x_J \partial x_K}\frac{\partial \mathbf{A}_I}{\partial u_i}, \quad \mathbf{u}_{,J}\mathbf{u}_{,K} \cdot (\nabla\nabla\mathbf{A}_I) = \frac{\partial u_i}{\partial x_J}\frac{\partial u_j}{\partial x_K}\frac{\partial^2 \mathbf{A}_I}{\partial u_i \partial u_j}.$$

Next, we take a jump in (4.1.15) across the acceleration front $\sum$, multiply the resulting equation by $n_J n_k$, and use the compatibility conditions $(4.1.7)_1$, $(4.1.7)_3$, $(4.1.7)_6$ and $(4.1.7)_7$ to obtain

$$
\begin{aligned}
-(\mathbf{A}_J^o n_J - G\mathbf{I})\boldsymbol{\pi}^{(3)} = {} & \delta_t\boldsymbol{\pi}^{(2)} + g^{\alpha\beta}x_{I,\beta}\mathbf{A}_{I,\alpha}^o\boldsymbol{\pi}^{(2)} + \boldsymbol{\pi}^{(2)} \cdot (\nabla\mathbf{A}_I)^o\boldsymbol{\pi}^{(1)}n_I \\
& +2\boldsymbol{\pi}^{(1)} \cdot (\nabla\mathbf{A}_I)^o\boldsymbol{\pi}^{(2)}n_I - \boldsymbol{\pi}^{(2)} \cdot (\nabla\mathbf{g})^o \\
& +2\boldsymbol{\pi}^{(1)} \cdot (\nabla\mathbf{A}_I)^o g^{\alpha\beta}\boldsymbol{\pi}_{,\alpha}^{(1)}x_{I,\beta} - 2g^{\alpha\beta}\boldsymbol{\pi}_{,\alpha}^{(1)}G_{,\beta} \\
& +2g^{\alpha\beta}g^{\gamma\delta}b_{\delta\alpha}\mathbf{A}_I^o\boldsymbol{\pi}_{,\gamma}^{(1)}x_{I,\beta} \\
& +\boldsymbol{\pi}^{(1)}\boldsymbol{\pi}^{(1)} \cdot (\nabla\nabla\mathbf{A}_I)^o\boldsymbol{\pi}^{(1)}n_I - \boldsymbol{\pi}^{(1)}\boldsymbol{\pi}^{(1)} \cdot (\nabla\nabla\mathbf{g})^o,
\end{aligned}
\tag{4.1.16}
$$

where use has been made of $(4.1.2)_1$ and $(4.1.2)_4$. We notice from (4.1.12) and (4.1.16) that

$$\boldsymbol{\pi}^{(2)} = \sigma_q^{(2)}\mathbf{R}_q^o + \mathbf{e}^{(1)}, \quad \boldsymbol{\pi}^{(3)} = \sigma_q^{(3)}\mathbf{R}_q^o + \mathbf{e}^{(2)}, \tag{4.1.17}$$

where $\mathbf{e}^{(1)}$ (respectively, $\mathbf{e}^{(2)}$) is a particular solution depending on $\boldsymbol{\pi}^{(1)}$ and its derivatives (respectively, on $\boldsymbol{\pi}^{(1)}$, $\boldsymbol{\pi}^{(2)}$ and their derivatives) as well as on the geometrical and kinematical characteristics of the discontinuity surface.

Equation (4.1.16), on pre-multiplying by the left eigenvector $\mathbf{L}_p^o$, and using (4.1.4), (4.1.10), (4.1.11) and (4.1.17)$_1$ yields the following transport equations for $\sigma_q^{(2)}$.

$$\alpha_{pq}\frac{d\sigma_q^{(2)}}{dt} + \mu_{pq}^{(2)}\sigma_q^{(2)} + \nu_p^{(2)} = 0, \tag{4.1.18}$$

where $\mu_{pq}^{(2)}$ and $\nu_p^{(2)}$, which depend on $\sigma_q^{(1)}$ and its derivatives as well as on the geometric and kinematic characteristics of the discontinuity surface, are defined as

$$
\begin{aligned}
\mu_{pq}^{(2)} &= \mathbf{L}_p^o\left\{\delta_t\mathbf{R}_q^o + \boldsymbol{A}_I^o\mathbf{R}_{q,\alpha}^o g^{\alpha\beta}x_{I,\beta} + \mathbf{R}_q^o\cdot\nabla\boldsymbol{A}_I)^o n_I\mathbf{R}_\ell^o\sigma_\ell^{(1)}\right.\\
&\quad \left.+2\mathbf{R}_\ell^o\cdot(\nabla\boldsymbol{A}_I)^o n_I\mathbf{R}_q^o\sigma_\ell^{(1)} - \mathbf{R}_q^o\cdot(\nabla\mathbf{g})^o\right\},\\
\nu_p^{(2)} &= \mathbf{L}_p^o\left\{2\mathbf{R}_q^o\cdot(\nabla\boldsymbol{A}^I)^o g^{\alpha\beta}(\mathbf{R}_\ell^o\sigma_{\ell,\alpha}^{(1)} + R_{\ell,\alpha}^o\sigma_\ell^{(1)})\sigma_q^{(1)}x_{I,\beta} - 2g^{\alpha\beta}(\mathbf{R}_q^o\sigma_{q,\alpha}^{(1)}\right.\\
&\quad +\mathbf{R}_{q,\alpha}^o\sigma_q^{(1)})G_{,\beta} + 2g^{\alpha\beta}g^{\gamma\delta}b_{\delta\alpha}\boldsymbol{A}_I^o(\mathbf{R}_q^o\sigma_{q,\alpha}^{(1)} + \mathbf{R}_{q,\alpha}^o\sigma_q^{(1)})x_{I,\beta}\\
&\quad +\mathbf{R}_q^o\mathbf{R}_\ell^o\cdot(\nabla\nabla\boldsymbol{A}_I)^o n_I\mathbf{R}_m^o\sigma_\ell^{(1)}\sigma_\ell^{(1)}\sigma_m^{(1)} - \mathbf{R}_q^o\mathbf{R}_\ell^o\cdot(\nabla\nabla\mathbf{g})^o\sigma_q^{(1)}\sigma_\ell^{(1)}\\
&\quad +\delta_t\mathbf{e}^{(1)} + g^{\alpha\beta}x_{I,\beta}\boldsymbol{A}_I^o\mathbf{e}_{,\alpha}^{(1)} + \mathbf{e}^{(1)}\cdot(\nabla\boldsymbol{A}_I)^o\sigma_q^{(1)}\mathbf{R}_q^o n_I\\
&\quad \left.+2\mathbf{R}_q^o\cdot(\nabla\boldsymbol{A}_I)^o\mathbf{e}^{(1)}\sigma_q^{(1)}n_I - e^{(1)}\cdot(\nabla\mathbf{g})^o\right\}.
\end{aligned}
$$

It may be noticed that unlike (4.1.13), the system of equations (4.1.18) is linear along the rays.

Transport equations for the discontinuities in higher order derivatives at an acceleration front can be obtained following the same procedure; indeed, it can be shown that for $n \geq 2$

$$\boldsymbol{\pi}^{(n)} = \sigma_q^{(n)}\mathbf{R}_q^o + \mathbf{e}^{(n-1)},$$

$$\alpha_{pq}\frac{d\sigma_q^{(n)}}{dt} + \mu_{pq}^{(n)}\sigma_q^{(n)} + \nu_p^{(n)} = 0, \tag{4.1.19}$$

where $\mathbf{e}^{(n-1)}, \mu_{pq}^{(n)}$ and $\nu_p^{(n)}$ depend on $\sigma_q^{(1)}, \sigma_q^{(2)}\ldots, \sigma_q^{(n-1)}$ and their derivatives as well as on the geometric and kinematic characteristics of the surface $\sum$. It may be remarked that the transport equations for $\sigma_q^{(1)}$ are nonlinear, whereas the transport equations for $\sigma_q^{(n)}, n \geq 2$ are linear.

4.1.4 Transport equations for mild discontinuities

For second order waves $(N = 2)$, $\mathbf{u}$ and its first order derivatives are continuous, while discontinuities in the second and higher order derivatives are permitted, and the compatibility conditions now become

$$[\mathbf{u}_{,IJ}] = \pi^{(2)}n_In_J; \quad \left[\frac{\partial}{\partial t}\mathbf{u}_{,I}\right] = -\pi^{(2)}Gn_I,$$

$$[\mathbf{u}_{,IJK}] = \pi^{(3)}n_In_Jn_K + g^{\alpha\beta}\pi^{(2)}_{,\alpha}P^{IJK}_\beta - g^{\alpha\beta}g^{\gamma\delta}b_{\gamma\alpha}\pi^{(2)}Q^{IJK}_{\beta\delta};$$

$$[\mathbf{u}_{,IJKL}] = \pi^{(4)}n_In_Jn_Kn_L + g^{\alpha\beta}(\pi^{(3)}_{,\alpha} + 3g^{\gamma\delta}b_{\delta\alpha}\pi^{(2)}_{,\gamma})n_In_Jn_Kx_{L,\beta}$$

$$+g^{\alpha\beta}\left\{(\pi^{(3)}n_L + g^{\gamma\delta}\pi^{(2)}_{,\gamma}x_{L,\delta})_{,\alpha} + 2g^{\gamma\delta}b_{\delta\alpha}(\pi^{(2)}n_L)_{,\gamma}\right\}P^{IJK}_\beta$$

$$+g^{\alpha\beta}g^{\gamma\delta}\left\{(\pi^{(2)}n_L)_{,\gamma\alpha} - b_{\gamma\alpha}\left(\pi^{(3)}n_L + g^{\sigma\tau}\pi^{(2)}_{,\sigma}x_{L,\tau}\right)\right.$$

$$\left.-g^{\sigma\tau}b_{\alpha\sigma}b_{\gamma\tau}\pi^{(2)}n_L\right\}Q^{IJK}_{\beta\delta} - g^{\alpha\beta}g^{\gamma\delta}g^{\sigma\tau}b_{\alpha\sigma}(\pi^{(2)}n_L)_{,\gamma}M^{IJK}_{\beta\tau\delta}$$

$$+(\pi^{(2)}n_L)_{,\gamma}g^{\alpha\beta}g^{\gamma\delta}_{,\alpha}(n_Ix_{J,\beta} + n_Jx_{I,\beta})x_{K,\delta}$$

$$-\pi^{(2)}n_Lg^{\alpha\beta}g^{\sigma\tau}(g^{\gamma\delta}_{,\sigma}b_{\gamma\alpha} + g^{\gamma\delta}b_{\gamma\alpha,\sigma})x_{I,\beta}x_{J,\tau}x_{K,\delta},$$

$$\left[\frac{\partial}{\partial t}\mathbf{u}_{,IJK}\right] = \left(-G\pi^{(4)} + \delta_t\pi^{(3)} + 3g^{\alpha\beta}\pi^{(2)}_{,\alpha}G_{,\beta}\right)n_In_Jn_k$$

$$+g^{\alpha\beta}\left\{(-\pi^{(3)}G + \delta_t\pi^{(2)})_{,\alpha} + 2g^{\gamma\delta}b_{\delta\alpha}(-\pi^{(2)}G)_{,\gamma}\right\}P^{IJK}_\beta$$

$$+g^{\alpha\beta}g^{\gamma\delta}\left\{(-\pi^{(2)}G)_{,\gamma\alpha} - b_{\gamma,\alpha}\left(-\pi^{(3)}G + \delta_t\pi^{(2)}\right)\right.$$

$$\left.-g^{\sigma\tau}b_{\alpha\sigma}(-\pi^{(2)}G)\right\}Q^{IJK}_{\beta\delta} + g^{\alpha\beta}g^{\gamma\delta}g^{\sigma\tau}b_{\alpha\sigma}(\pi^{(2)}G)_{,\gamma}M^{IJK}_{\beta\tau\delta}$$

$$-(\pi^{(2)}G)_{,\gamma}g^{\alpha\beta}g^{\gamma\delta}_{,\alpha}(n_Ix_{J,\beta} + n_Jx_{I,\beta})x_{K,\delta}$$

$$+\pi^{(2)}Gg^{\alpha\beta}g^{\sigma\tau}\left(g^{\gamma\delta}_{,\sigma}b_{\gamma\alpha} + g^{\gamma\delta}b_{\gamma\alpha,\sigma}\right)x_{I,\beta}x_{J,\tau}x_{K,\delta}.$$

For such waves, equations, $(4.1.17)_1$, and $(4.1.18)$ apply with $\mathbf{e}^{(1)} = \mathbf{0}$, $\nu^{(2)}_p = 0$, showing that the transport equations for $\sigma^{(2)}_q$ are linear and homogeneous; however, the variations of $\sigma^{(n)}_q$ with $n > 2$ are governed by ODEs which are linear and nonhomogeneous. For instance, following the procedure used in the preceding subsection, the transport equations for $\sigma^{(3)}_q$ are

$$\alpha_{pq}\frac{d\sigma^{(3)}_q}{dt} + \mu^{(3)}_{pq}\sigma^{(3)}_q + \nu^{(3)}_p = 0,$$

where $\mu^{(3)}_{pq} = \mathbf{L}^o_p(\delta_t\mathbf{R}^o_q + g^{\alpha\beta}x_{I,\beta}\mathbf{A}^o_I\mathbf{R}^o_{q,\alpha} - \mathbf{R}^o_q \cdot (\nabla\mathbf{g})^o)$, and

$$\nu^{(3)}_p = \mathbf{L}^o_p\left\{\delta_t\mathbf{e}^{(2)} + g^{\alpha\beta}\mathbf{A}^o_I\mathbf{e}^{(2)}_{,\alpha}x_{I,\beta} + 3g^{\alpha\beta}(\sigma^{(2)}_q\mathbf{R}^o_q)_{,\alpha}G_{,\beta}\right.$$

$$+3\mathbf{R}^o_q \cdot (\nabla\mathbf{A}_I)^o\mathbf{R}^o_l\sigma^{(2)}_q\sigma^{(2)}_\ell n_I - \mathbf{e}^{(2)} \cdot (\nabla\mathbf{g})^o$$

$$\left.+3g^{\alpha\beta}g^{\gamma\delta}b_{\beta\gamma}\mathbf{A}^o_I(\sigma^{(2)}_q\mathbf{R}^o_q)_{,\alpha}x_{I,\delta}\right\},$$

with $\mathbf{e}^{(2)}$ depending on $\pi^{(2)}$ and its derivatives, as well as on the geometrical and kinematical characteristics of the discontinuity surface. It may be remarked that for a wave of order $N \geq 2$, the transport equations for $\sigma_q^{(N)}$ are linear and homogeneous, whereas the transport equations for $\sigma_q^{(n)}, n > N$ are linear and nonhomogeneous.

4.2 Evolutionary Behavior of Acceleration Waves

In the study of acceleration waves in material media, and elsewhere, a Bernoulli or Riccati type equation of the following form (see equation (4.1.14) is frequently encountered,

$$\frac{ds}{dt} = -\mu(t)s + \beta(t)s^2, t \geq 0; \tag{4.2.1}$$

where $s(t)$ is the wave amplitude, $\mu(t)$ depends upon the condition (state) of the medium ahead of the wave and the geometry of the wave surface, and $\beta(t)$ depends on the elastic response of the material (see, e.g., McCarthy [120], Chen [32], Ruggeri [158], and Lou and Ruggeri [113]). The present work is concerned with the general behavior of the wave amplitude $s(t)$, and especially with its limiting behavior; that is with the conditions under which it ultimately damps out $(s(t) \to 0)$, attains a stable wave from $(s(t) \to \text{const.})$, or develops into a shock wave in a finite time $\tilde{t}(|s(t)| \to \infty$ as $t \to \tilde{t})$.

When

$$\beta(t) = 0 \text{ for all } t, \tag{4.2.2}$$

which corresponds to a linear mechanical response of the material, equation (4.2.1) reduces to a linear equation. Such a situation arises in a number of physical problems. For example, shear waves in nonlinear viscoelastic fluids in equilibrium (see Coleman and Gurtin [38]), homothermal and frozen homentropic transverse waves in laminated bodies (see Bowen and Wang [24]) and waves in unstrained isotropic materials (see Scott [162]). In all these references, it has been shown that the corresponding wave amplitude $s(t)$ satisfies the differential equation (4.2.1) with $\beta(t) = 0$. On the other hand, in the most general case known so far, the function $\beta(t)$ is restricted to be of constant sign, so that

$$\text{either } \beta(t) < 0 \text{ for all } t, \text{ or }, \beta(t) > 0 \text{ for all } t, \tag{4.2.3}$$

and the results of Bailey and Chen [9], Bowen and Chen [23], and Wright [211], all of which are subject to these conditions, have been reported as theorems 3.2.1 to 3.2.6 in Chen [32]; see also Menon and Sharma ([122], [175]). Notice also that here the function is not allowed to vanish anywhere.

However, in an inhomogeneous medium interspersed with layers showing a linear response corresponding to (4.2.2) it is more reasonable to assume that the function $\beta(t)$ does not change sign, so that

$$\text{either } \beta(t) \leq 0 \text{ for all } t, \text{ or }, \beta(t) \geq 0 \text{ for all } t. \qquad (4.2.4)$$

The condition (4.2.4), which allows the function $\beta(t)$ to vanish over the parts of the range, includes both conditions (4.2.2) and (4.2.3) as special cases.

Here, we discuss the behavior of $s(t)$ under the condition (4.2.4); the present account is largely based on our paper [123]. Our results lead to some improvements over the known results even in the special case (4.2.3). In the following subsections, the function β is assumed to satisfy condition (4.2.4). Our theorems contain all the known results and, in some instances, when specialized to the case of an existing theorem, they provide a sharper result; it is shown that even in the simpler case, where β satisfies (4.2.3), our results are stronger and more extensive. Finally, we offer some comments on an alternative but equivalent method of analyzing the behavior of wave amplitudes and the development of shock solutions in quasi-linear hyperbolic systems due to Jeffrey [80].

4.2.1 Local behavior

In this subsection, which will be needed at the end of subsection 4.2.3, we restate Theorem 3.2.1 due to Chen [32]. There are no restrictions on the functions μ and β at present.

Define the function λ by

$$\lambda(t) = \mu(t)/\beta(t), \qquad (4.2.5)$$

which is defined only for those t for which $\beta(t) \neq 0$. Equation (4.2.1) then becomes

$$\begin{aligned} \frac{ds}{dt} &= \beta(t)s(t)[s(t) - \lambda(t)], \text{ if } \beta(t) \neq 0; \\ &= -\mu(t)s(t), \text{ if } \beta(t) = 0. \end{aligned} \qquad (4.2.6)$$

The following result is trivial.

Proposition 4.2.1 *Consider equation (1.1)*

(i) At any instant, if $\beta(t)s(t) < 0$, then

$$\frac{ds}{dt} > 0 \text{ if } s(t) < \lambda(t); \quad \frac{ds}{dt} < 0 \text{ if } s(t) > \lambda(t).$$

(ii) At any instant, if $\beta(t)s(t) \neq 0$, then

$$\frac{ds}{dt} = 0, \text{ if, and only if, } s(t) = \lambda(t).$$

(iii) At any instant, if $\beta(t)s(t) > 0$, then

$$\frac{ds}{dt} < 0 \text{ if } s(t) < \lambda(t); \quad \frac{ds}{dt} > 0 \text{ if } s(t) > \lambda(t).$$

Thus we may say at the instants where $\beta(t)s(t) < 0$, the function s tends to approach the function λ.

In Theorem 3.2.1 of [32], the above statements (i) and (ii) take the form "(i) if either $\beta(t) > 0$ and $s(t) < \lambda(t)$ or $\beta(t) < 0$ and $s(t) > \lambda(t)$, then $ds(t)/dt < 0$," which is not true because it is possible that $s(t) = 0$; "(ii) if $s(t) \neq 0, s(t) = \lambda(t)$ if, and only if $ds(t)/dt = 0$," which is again not true because $\beta(t)$ could be zero, in that case $ds/dt = -\mu(t)s(t)$.

4.2.2   Global behavior: The main results

The integral $F(t) = \int_0^t \mu(\tau)d\tau$ appears in the solution $s(t)$ below in (4.2.8). We shall assume that either $F(t)$ exists for all $t \in [0, \infty)$, or $F(t)$ becomes infinite at some finite $t^* > 0$. When $F(t)$ exists for all t (i.e., $\mu(t)$ is integrable over every finite sub-interval of $[0, \infty)$), then from the well known definition of an integral over $[0, \infty)$, it is possible that $F(t)$ may converge, diverge to $\pm\infty$, or not tend to a limit at all as $t \to \infty$. Further, we shall assume that

$$\text{either} \quad \beta(t) \geq 0 \text{ for all } t \text{ (i.e., } \operatorname{sgn}\beta = +1),$$
$$\text{or} \quad \beta(t) \leq 0 \text{ for all } t \text{ (i.e., } \operatorname{sgn}\beta = -1), \tag{4.2.7}$$

in the range $[0, \infty)$ or $[0, t^*)$ according to the condition satisfied by $F(t)$ above. The integral of $\beta(t) \exp(-F(t))$ appearing in (4.2.8) then exists if the integral of β exists, because $\exp(-F(t)) \geq 0$. The condition (4.2.7) may be expressed in words by asserting that "the function β does not change sign." Notice that if the integral of β vanishes over any interval, the solution of equation (4.2.1) reduces to that of the equation $ds/dt = -\mu s$ over the interval. In [9] and [32], the function β is not allowed to vanish.

The solution of (4.2.1) is easily shown to be ([32])

$$s(t) = \frac{\exp\left(-\int_0^t \mu(\tau)d\tau\right)}{(1/s(0)) - \int_0^t \beta(\tau)\exp\left(-\int_0^\tau \mu(s)ds\right)d\tau}, \quad t \geq 0. \tag{4.2.8}$$

Since $\operatorname{sgn}\beta$ is either $+1$ or -1 from (4.2.7), it follows immediately, that

$$s(t) = (\operatorname{sgn} s(0))\exp(-F(t))/|s(0)|^{-1} + G(t), \tag{4.2.9}$$

if $\operatorname{sgn} s(0) = -\operatorname{sgn}\beta$; and

$$s(t) = (\operatorname{sgn} s(0))\exp(-F(t))/|s(0)|^{-1} - G(t), \tag{4.2.10}$$

if $\operatorname{sgn} s(0) = \operatorname{sgn}\beta$, where

$$F(t) = \int_0^t \mu(\tau)d\tau, \quad G(t) = \int_0^t |\beta(\tau)|\exp(-F(\tau))d\tau. \tag{4.2.11}$$

We now state the main results.

Theorem 4.2.1 *Consider equation (4.2.1) with $s(t) \neq 0$. Suppose that μ and β are integrable on every finite subinterval of $[0, \infty)$, the function β does not change sign (see (4.2.7)) on $[0, \infty)$, and $\operatorname{sgn} s(0) = -\operatorname{sgn} \beta$. Let F, G be as in (4.2.11). Then*

(a) $s(t)$ is continuous, nonzero and of constant sign on $[0, \infty)$;

(b) $|s(t)|$ is bounded above if F is bounded below (i.e., $\liminf\limits_{t\to\infty} F(t) > -\infty$);

(c) $s(t)$ is bounded away from zero if F is bounded above (i.e., $\limsup\limits_{t\to\infty} F(t) < \infty$) and $G(\infty) < \infty$;

(d) the limiting behavior of s is as follows:

  (i) $\liminf\limits_{t\to\infty} s(t) = 0$ if $\limsup\limits_{t\to\infty} F(t) = \infty$.

  (ii) Suppose that $F(t)$ tends to a finite or infinite limit $F(\infty)$ as $t\to\infty$, then

$$\lim_{t\to\infty} s(t) = \begin{cases} 0, \text{ if } F(\infty) = +\infty; \\ \infty, \text{ if } F(\infty) = -\infty \text{ and } G(\infty) < \infty; \\ \lim\limits_{t\to\infty} |\mu/\beta|, \text{ if } F(\infty) = -\infty \text{ and } G(\infty) = \infty \text{ and } |\mu/\beta| \\ \quad \text{tends to an infinite or finite limit as } t\to\infty; \\ \{|1/s(0)| + G(\infty)\}^{-1} \exp(-F(\infty)), \text{ if } |F(\infty)| < \infty. \end{cases}$$

  (iii) Suppose that $F(t)$ does not tend to a limit as $t\to\infty$, due to the oscillation of $F(t)$ as $t\to\infty$. If $G(\infty) < \infty$, then $s(t)$ oscillates at $t\to\infty$. Moreover $|s(t)|$ is bounded away from zero (respectively, bounded above) if, and only if, $F(t)$ is bounded above (respectively, bounded below). If $G(\infty) = \infty$ and F is bounded below, then $\lim_{t\to\infty} s(t) = 0$.

Theorem 4.2.2 *Consider equation (4.2.1), with $s(0) \neq 0$. Suppose that μ and β are integrable on every finite sub-interval of $[0, \infty)$; the function β does not change sign on $[0, \infty)$ and $\operatorname{sgn} s(0) = \operatorname{sgn} \beta$. Let F, G be as in (4.2.11). Define the quantity α by*

$$\alpha = (G(\infty))^{-1} = \lim_{t\to\infty} (G(t))^{-1}. \tag{4.2.12}$$

The quantity α exists and $\alpha \geq 0$. Moreover $\alpha = 0$ if, and only if, $G(\infty) = \infty$.

Case 1. Suppose that either $|s(0)| > \alpha$, or, $|s(0)| = \alpha$ and $\int_t^\infty \beta(\tau)d\tau = 0$ for some $t < \infty$, then there exists a finite time $\tilde{t}$ such that

$$\int_0^{\tilde{t}} \beta(t) \cdot \exp(-F(t)dt = |1/s(0)|, \quad 0 < \tilde{t} < \infty, \tag{4.2.13}$$

and

$$\lim_{t \to \tilde{t}} |s(t)| = \infty.$$

Case 2. Suppose that $|s(0)| < \alpha$. If $F(t)$ tends to a finite or infinite limit $F(\infty)$ as $t \to \infty$, then

$$\lim_{t \to \infty} s(t) = \begin{cases} 0, & \text{if } F(\infty) = \infty, \\ \infty, & \text{if } F(\infty) = -\infty, \\ \left(\frac{1}{|s(0)|} - \frac{1}{\alpha}\right)^{-1} F(\infty), & \text{if } |F(\infty)| < \infty. \end{cases}$$

If $F(t)$ oscillates as $t \to \infty$, then $s(t)$ also oscillates as $t \to \infty$, and $|s(t)|$ is bounded away from zero (respectively, bounded above) if, and only if, F is bounded above (respectively, bounded below).

Case 3. Suppose that $|s(0)| = \alpha$ and $\int_t^\infty \beta(\tau) \neq 0$ for all finite t. Then

(i) $\lim_{t \to \infty} |s(t)| = \infty$ if F is bounded above;

(ii) If $F(t)$ tends to a finite or infinite limit $F(\infty)$ as $t \to \infty$, then

$$\lim_{t \to \infty} |s(t)| = \begin{cases} \infty, \text{ if } F(\infty) < \infty, \\ \lim_{t \to \infty} |\mu/\beta|, \text{ if } F(\infty) = \infty \text{ and } |\mu/\beta| \text{ tends} \\ \text{to a finite or infinite limit as } t \to \infty; \end{cases}$$

(iii) If $F(t)$ oscillates and is bounded above as $t \to \infty$, then $\lim_{t \to \infty} |s(t)| = \infty$.

Moreover, over the interval $[0, \tilde{t})$ in case 1, and over $[0, \infty)$ in cases 2, 3, the function $s(t)$ is continuous and nonzero.

Theorem 4.2.3 *Consider equation (4.2.1) with $s(0) \neq 0$. Let there exist a finite $t^* > 0$ such that μ and β are integrable on every sub-interval of $[0, t^*)$; the function β does not change sign on $[0, t^*)$ and $F(t^*) = +\infty$ or $-\infty$, where F, G are as in (4.2.11). Then $s(t)$ is continuous and nonzero over $[0, t^*)$.*
Define the quantity α by $\alpha = \{G(t^*)\}^{-1} = \lim_{t \to t^*} \{G(t)\}^{-1}$; the quantity α exists and $\alpha \geq 0$.

Case 1. Suppose that $\operatorname{sgn} s(0) = -\operatorname{sgn} \beta$. Then

$$\lim_{t \to t^*} s(t) = \begin{cases} 0, \text{ if } F(t^*) = \infty; \\ \infty, \text{ if } F(t^*) = -\infty \text{ and } \alpha > 0; \\ \lim_{t \to t^*} |\mu/\beta|, \text{ if } F(t^*) = -\infty \text{ and } \alpha = 0, \text{ and } |\mu/\beta| \text{ tends} \\ \text{to a finite or infinite limit as } t \to t^*. \end{cases}$$

Case 2. Suppose that $\operatorname{sgn} s(0) = \operatorname{sgn} \beta$.

(i) If $|s(0)| < \alpha$, then $\lim_{t \to t^*} |s(t)| = \begin{cases} 0, & \text{if } F(t^*) = \infty; \\ \infty, & \text{if } F(t^*) = -\infty. \end{cases}$

(ii) If either $|s(0)| > \alpha$, or, $|s(0)| = \alpha$ and $\int_t^{t^*} \beta(\tau)d\tau = 0$ for some $t < t^*$, then there exists a time $\tilde{t}$ as in case 1 of Theorem 4.2.3, with $0 < \tilde{t} < t^*$ and $\lim_{t \to \tilde{t}} |s(t)| = \infty$.

(iii) If $|s(0)| = \alpha$ and $\int_t^{t^*} \beta(\tau)d\tau \neq 0$ for all $t < t^*$, then $\tilde{t} = t^*$ and

$$\lim_{t \to t^*} |s(t)| = \begin{cases} \infty, & \text{if } F(t^*) = -\infty, \\ \lim_{t \to t^*} |\mu/\beta|, & \text{if } F(t^*) = +\infty \text{ and } |\mu/\beta| \text{ tends to a} \\ & \text{finite or infinite limit as } t \to t^*. \end{cases}$$

Moreover, over the interval $[0, t^*)$ in cases 1, 2(i), 2(iii), and over $[0, \tilde{t})$ in case 2 (ii), the function $s(t)$ is continuous and nonzero.

4.2.3   Proofs of the main results

In the following, the interval $[0, T)$ will stand for $[0, \infty)$ for Theorems 4.2.1 and 4.2.2, and for $[0, t^*)$ for Theorem 4.2.3. Since μ and β are both integrable over all finite subintervals of $[0, T)$, the functions F and G of (4.2.11) are well-defined and continuous over $[0, T)$. The following Lemma is obvious since a bounded monotone function over $[0, T)$ must have a limit as $t \to T$.

Lemma 4.2.1 *If μ and β are integrable on all finite sub-intervals of $[0, T)$, and β does not change sign on $[0, T)$, then*

(a) *the function $G(t)$ of (4.2.11) is continuous, nonnegative and monotonically nondecreasing over $[0, T)$, and $G(0) = 0$. Moreover, $G(t)$ tends to a finite or infinite limit $G(t)$ as $t \to T$.*

Therefore,

$$\alpha = G(T)^{-1} \tag{4.2.14}$$

exists and $\alpha \geq 0$. Also, $\alpha = 0$ if, and only if, $G(T) = \infty$.

(b) *The function $\exp(-F(t))$ is continuous and strictly positive over $[0, T)$, and its limit as $t \to T$ exists if, and only if, F tends to a (finite or infinite) limit as $t \to T$. Moreover, $\exp(-F(t))$ is bounded above (respectively, bounded away from zero) if, and only if, F is bounded below (respectively, bounded above) on $[0, T)$.*

Notice that, if $T = \infty$, the condition "F is bounded below" is equivalent to the condition " $\liminf_{t \to \infty} F(t) > -\infty$," because F is continuous on $[0, \infty)$. Similar remarks hold for the other conditions on $F(t)$.

Proof: The proofs of all our theorems are based on Lemma 4.2.1.

First, consider Theorem 4.2.1 (i.e., $T = \infty$) and case 1 of Theorem 4.2.3 (i.e. $T = t^*$), so that the equation (4.2.9) holds. The denominator of the right-hand side of (4.2.9) is bounded away from zero; in fact it exceeds unity, from Lemma 4.2.1. It is also bounded above if $G(\infty) < \infty$. The statements (a), (b), (c), d(i), d(iii) of Theorem 4.2.1, the first two statements in case 1 of Theorem 4.2.3, and the last statement of Theorem 4.2.3 are all easy consequences of Lemma 4.2.1. Consider now the statement d(ii) of Theorem 4.2.1 and the last statement in case 1 of Theorem 4.2.3. The only statement requiring a proof is the third one, where both the denominator and the numerator on the right-hand side of (4.2.9) tend to ∞ as $t\to\infty$. In this situation we can use l'Hopital's rule, so that

$$
\begin{aligned}
\lim_{t\to T} s(t) &= (\operatorname{sgn} s(0)) \lim_{t\to T} \left(\frac{\frac{d}{dt}(\exp(-F(t)))}{\frac{d}{dt}G(t)} \right) \\
&= (\operatorname{sgn} s(0)) \lim_{t\to T} (-\mu/|\beta|), \qquad (4.2.15)
\end{aligned}
$$

provided the last limit exists. Hence, if $|\mu/\beta|$ tends to a finite or infinite limit as $t\to T$, then

$$
\lim_{t\to T} |s(t)| = \lim_{t\to T} |\mu/\beta|,
$$

by taking the absolute values of both sides of (4.2.15).

Consider now the situation when $\operatorname{sgn} s(0) = \operatorname{sgn} \beta$, so that the equation (4.2.10) holds. A possible complication is introduced because the denominator on the right-hand side of (4.2.10) may vanish at some time $\tilde{t}$. If $\tilde{t}$ exists, then $t > 0$ because $G(0) = 0$. Since G is monotonic nondecreasing and continuous, there are three possibilities for the denominator $(|1/s(0)| - G(t))$:

(a) $G(T) < |1/s(0)|$; in this case the denominator is bounded away from zero over $[0, T)$, and we have the situation similar to Theorem 4.2.1;

(b) $G(T) > |1/s(0)|$; in this case there must exist a $\tilde{t} < T$, where $G(\tilde{t}) = |1/s(0)|$, so that (4.2.13) is satisfied;

(c) $G(T) = |1/s(0)|$; in this case, either $G(t) = |1/s(0)|$ for some $t < T$ which is case (b) above or, $G(t) < |1/s(0)|$ for all $t \in [0, T)$.

Since the numerator $\exp(-F(t))$ is nonzero over $[0, T)$, $G(t) = G(T)$ implies that $\int_t^T \beta(\tau)d\tau = 0$, and the limits of $s(t)$ as $t\to\tilde{t}$ in case (b), and as $t\to T$ in cases (a) and (c), are easily obtained, except when both the numerator and denominator tend to zero. In this exceptional case, therefore $F(t)\to +\infty, G(t)\to|1/s(0)|$ as $t\to T$, and $G(t) < |1/s(0)|$ for $t < T$, and we may apply l'Hopital's rule again to obtain from (4.2.9),

$$
\begin{aligned}
\lim_{t\to T} s(t) &= (\operatorname{sgn} s(0)) \lim_{t\to T} \left(\frac{\frac{d}{dt}(\exp(-F(t)))}{\frac{d}{dt}G(t)} \right) \\
&= (\operatorname{sgn} s(0)) \lim_{t\to T} (-\mu/-|\beta|), \qquad (4.2.16)
\end{aligned}
$$

provided the last limit exists. Therefore, as before, if $|\mu/\beta|$ tends to a finite or infinite limit as $t \to T$, we have the required limit of $|s(t)|$, by taking absolute values on both sides of (4.2.16).

This completes the proof of Theorem 4.2.2, by taking $t = \infty$, and case 2 of Theorem 4.2.3, by taking $T = t^*$.

We may remark that there is no confusion with signs in (4.2.15) and (4.2.16). If $(\mu/|\beta|) \to 0$ as $t \to T$, there is no problem because 0 can be regarded as having either sign, while if $\mu/|\beta|$ tends to a finite nonzero or infinite limit, then μ is also of constant sign for all sufficiently large t as $t \to T$. Therefore, if $\lim_{t \to T} F(t) = -\infty$ as in (4.2.15) and $\mu|\beta|$ tends to a finite nonzero or infinite limit, μ is negative as $t \to T$, i.e., $-\mu = |\mu|$. Similarly, in (4.2.16), μ is positive for t sufficiently close to T.

4.2.4 Some special cases

We first derive a few simple consequences of Theorem 4.2.1.

Corollary 4.2.1 *Consider equation (4.2.1). Let μ and β be integrable on every finite sub-interval of $[0, \infty)$, and let the function β not change sign on $[0, \infty)$. Suppose that $s_1(t), s_2(t)$ are two solutions of (4.2.1) with $s_1(0) \neq 0$, $s_2(0) \neq 0$; and $\operatorname{sgn} s_1(0) = \operatorname{sgn} s_2(0) = -\operatorname{sgn} \beta$. Let P denote any one of the following properties of a continuous function $s(t)$ defined on $[0, \infty)$:*
(a) $|s(t)|$ is bounded on $[0, \infty)$; (b) $|s(t)|$ is bounded away from zero on $[0, \infty)$; (c) $\lim_{t \to \infty} s(t) = 0$; (d) $\lim_{t \to \infty} |s(t)| = \infty$; (e) $\liminf_{t \to \infty} s(t) = 0$;
(f) $s(t)$ tends to a finite nonzero limit as $t \to \infty$; (g) $s(t)$ does not tend to a limit as $t \to \infty$.

If s_1 satisfies the property P, then s_2 also satisfies the Property P. Moreover,

(i) if $\lim_{t \to \infty} s_1(t) = 0$, then $\lim_{t \to \infty} s_2(t) = 0$, so that $\lim_{t \to \infty} |s_1(t) - s_2(t)| = 0$;

(ii) if $\liminf_{t \to \infty} s_1(t) = 0$, then $\liminf_{t \to \infty} s_2(t) = 0$, and $\liminf_{t \to \infty} |s_1(t) - s_2(t)| = 0$.

Proof: From Theorem 4.2.1, the limiting behavior of $s(t)$ is determined by μ, β and the sign of $s(0)$, in all cases except when both $F(\infty)$ and $G(\infty)$ are finite, i.e., in the last line of the case d(ii). Even in this exceptional case, $s(t)$ tends to a finite nonzero limit as $t \to \infty$, so that $s(t)$ is bounded and bounded away from zero over $[0, \infty)$. It follows that the property P is satisfied by both or neither of $s_1(t)$ and $s_2(t)$. This also proves statement (i). The situation (ii) arises when $\limsup_{t \to \infty} F(t) = \infty$, and

$$s_1(t) - s_2(t) = \left(\left(\frac{1}{|s_1(0)|} + G(t) \right)^{-1} - \left(\frac{1}{|s_2(0)|} + G(t) \right)^{-1} \right) \exp(-F(t)),$$

where $G(t) \geq 0$, so that (ii) is also proved. This concludes the proof of the corollary.   □

Consider now the situation in Theorem 4.2.1, where $\operatorname{sgn} s(0) = -\operatorname{sgn}\beta$, and suppose that $\liminf_{t\to\infty} |\beta(t)| \neq 0$. From the proposition in Subsection 4.2.1, the function s will tend to approach the function $\lambda = \mu/\beta$ as $t\to\infty$. Since s is of constant sign and never 0 or ∞ on $[0,\infty)$, we therefore know that $s\to 0$, if $\operatorname{sgn} s(0) = -\operatorname{sgn}\lambda(t)$ for all sufficiently large t, whereas $s(t)$ approaches $\lambda(t)$ if $\operatorname{sgn} s(0) = \operatorname{sgn}\lambda(t)$ for all sufficiently large t. Notice that since $\operatorname{sgn} s(0) = -\operatorname{sgn}\beta$, for any instant t, if $\operatorname{sgn} s(0) = -\operatorname{sgn}\lambda(t)$ then $\mu(t) > 0$, and if $\operatorname{sgn} s(0) = \operatorname{sgn}\lambda(t)$ then $\mu(t) < 0$. We are now in a position to investigate more closely the approach of the function s to the function λ. The following possibilities arise.

Case 1. Suppose that, for all sufficiently large t, $\operatorname{sgn}\lambda(t) = -\operatorname{sgn} s(0)$. Then there is a t_0 such that $\mu(t) \geq 0$ for all $t \geq t_0$. Thus $F(t)$ is monotonically nondecreasing when $t \geq t_0$. From Lemma 4.2.1, the integral $G(t)$ in (4.2.11) is nondecreasing and either converges, or diverges to $+\infty$, on $[0,\infty)$. If $G(\infty) = \infty$, it follows that $\lim_{t\to\infty} s(t) = 0$ because the numerator $\exp(-F(t))$ in (4.2.9) is bounded above. If $G(\infty) < \infty$, it follows that $\exp(-F(t))\to 0$, because $\liminf_{t\to\infty} |\beta| \neq 0$, and the integral of $\beta\exp(-F)$ converges over $[0,\infty)$, and $\exp(-F)$ is monotone over (t_0,∞). Thus, from (4.2.9), in this case also $\lim_{t\to\infty} s(t) = 0$.

Case 2. Suppose that $\lim_{t\to\infty} \lambda(t) = 0$. Since $\liminf_{t\to\infty} |\beta(t)| \neq 0$, for any sufficiently small $\epsilon > 0$, there exists t_0 such that

$$|\lambda(t)| < \epsilon/2 \text{ and } |\beta(t)| > \epsilon/2 \text{ for all } t \geq t_0.$$

We shall prove that $|s(t)| \leq \epsilon$ for all sufficiently large t. First we show that if $|s(t')| \leq \epsilon$ for some $t' \geq t_0$, then $|s(t)| \leq \epsilon$ for all $t \geq t'$. Since $\operatorname{sgn} s(t) = -\operatorname{sgn}\beta$ for all t, case (i) of Proposition 4.2.1 in Subsection 4.2.1 is applicable and the function s tends to approach the function λ, while $|\lambda(t)| < \epsilon/2$ for all $t \geq t'$. Therefore, the only possibility remaining to be examined is that $|s(t)| > \epsilon$ for all $t \geq t_0$: in this case we find from (4.2.6) that for $t \geq t_0$,

$$\left|\frac{d}{dt}\right| = |\beta(t)| \; |s(t)| \; |s(t) - \lambda(t)| \geq (\epsilon/2)(\epsilon)(\epsilon/2),$$

which is bounded away from zero, while the function s tends to approach the function λ where $|\lambda| < \epsilon/2$. Hence in a finite time $t' > t_0$ we will have $|s(t')| \leq \epsilon$. We have therefore proved that $|s(t)|\to 0$.

Case 3. Suppose that $\lambda(t)$ tends to a finite nonzero or infinite limit L as $t\to\infty$, where $\operatorname{sgn} s(0) = \operatorname{sgn} L$. Then for all sufficiently large t, say $t \geq t_1$, we have $\operatorname{sgn}\lambda(t) = \operatorname{sgn} L$, that is, $\mu(t) < 0$. Therefore $\exp(-F(t))$ is monotonically increasing over $[t_1,\infty)$. Since $\mu/\beta\to L \neq 0$ as $t\to\infty$, and $\liminf_{t\to\infty} |\beta| \neq 0$, it follows from (4.2.11) that $F(t) \to -\infty$ and $G(t) \to \infty$ as $t \to \infty$. Therefore, as in (4.2.15),

$$\lim_{t \to \infty} s(t) = \operatorname{sgn} s(0) \lim_{t \to \infty} \left[-\frac{\mu}{|\beta|} \right] = \lim_{t \to \infty} \left(\frac{\mu}{\beta} \right),$$

as a consequence of applying l'Hopital's rule.

Case 4. Suppose that $s(0) > 0$ and λ is bounded above, say by $\lambda(t) \leq K$ for all t. Then the argument used in case 2 shows that $s(t) \leq K + \epsilon$ for all sufficiently large t, i.e., $s(t)$ is bounded above. A similar result holds when $s(0) < 0$.

The following result has thus been established.

Corollary 4.2.2 *Consider equation (4.2.1), with $s(0) \neq 0$. Suppose that μ and β are integrable on $[0, \infty)$, the function β does not change sign on $[0, \infty)$, $\operatorname{sgn} s(0) = -\operatorname{sgn} \beta$, and $\liminf_{t \to \infty} |\beta(t)| \neq 0$.*

(a) *If, for all sufficiently large t, $\operatorname{sgn} \lambda(t) = -\operatorname{sgn} s(0)$, then $\lim_{t \to \infty} s(t) = 0$.*

(b) *If λ tends to a finite or infinite limit L as $t \to \infty$ and $\operatorname{sgn} s(0) = \operatorname{sgn} L$, then $\lim_{t \to \infty} s(t) = L$.*

(c)   (i) *If $s(0) > 0$ and λ is bounded above, then s is bounded above.*

     (ii) *If $s(0) < 0$ and λ is bounded below, then s is bounded below.*

Remarks 4.2.1: We explain briefly the relationship of our theorems to Theorems 3.2.1 through 3.2.6 of Chen [32]. The remarks (ii) – (v) refer to the special case where the function β satisfies the assumption (4.2.3).

(i) The proposition 4.2.1 of Subsection 4.2.1 is the corrected version of Theorem 3.2.1 of [32].

(ii) The Corollary 4.2.2 in Subsection 4.2.4 above is the modified version of Theorem 3.2.2 of [32]; the case (a) is ignored in [32]. The proof of case (b) is from [32].

(iii) The Corollary 4.2.1 in Subsection 4.2.4 corresponds to Theorem 3.2.3 of [32]; however, we have included more possibilities than are mentioned in [32].

(iv) Theorem 4.2.2 in Subsection 4.2.2 corresponds to Theorem 3.2.4 of [32]; however, the case 3 is ignored in [32].

(v) Our Theorem 4.2.3 of Subsection 4.2.2 is a modified version of Theorems 3.2.5 and 3.2.6 of [32] where the possibility $F(t^*) = +\infty$ and the case 2(iii) were ignored. There is a slip in the last step of the proof of

Theorem 3.2.6 of [32]; it is asserted there that if $F(t) \to -\infty$ as $t \to t^*$, i.e. $\mu(t) \to -\infty$ as $t \to t^*$, then $|\mu(t)/\beta(t)| \to \infty$ as $t \to t^*$. This is obviously not true, because $|\beta(t)|$ could be unbounded above (e.g., $\beta = \mu^2$) so that $|\mu/\beta|$ could tend to a finite limit.

4.3   Interaction of Shock Waves with Weak Discontinuities

The problem of the interaction of an acoustic wave with a shock has been studied by Swan and Fowles [193] as well as Van Moorhem and George [203] among several others. A detailed study directed towards gaining a better understanding of the wave interaction problem within the context of hyperbolic systems has been carried out by Jeffrey [80]. Brun [26] also developed a method by which to study the interaction problem within a general framework, the application of which to elasticity and magnetofluiddynamics has been carried out by Morro ([126], [127]). A further contribution to the study of wave interactions, which enables the evaluation of the reflected and transmitted amplitudes when a weak discontinuity wave encounters a shock wave, may be found in the paper by Boillat and Ruggeri [22]. The application of this work to interaction with a contact shock and a spherical blast wave has been carried out by Ruggeri [157] and Virgopia and Ferraioli [208], respectively.

It is a known fact that a shock undergoes an acceleration jump as a consequence of an interaction with a weak wave [98]; this fact has been accounted for in the work of Brun [26] and Boillat and Ruggeri [22]. The present discussion, which is largely based on [148], takes into account this physical fact, and shows that the general theory of wave interaction which originated from the work of Jeffrey leads to the results obtained by Brun and Boillat.

4.3.1   Evolution law for the amplitudes of C^1 discontinuities

The Bernoulli equation examined in the previous sections arises when the evolution law for weak discontinuities is derived directly from the governing system of equations, as in the references already cited or, for example, as in [81]. An alternative approach developed by Jeffrey [80] starts from a general first order quasilinear hyperbolic system

$$\mathbf{u}_{,t} + \mathbf{A}(\mathbf{u}, x, t)\mathbf{u}_{,x} + \mathbf{b}(\mathbf{u}, x, t) = 0, \qquad (4.3.1)$$

where the vector $\mathbf{u}$ has n components and $\mathbf{A}$ is an $n \times n$ square matrix with real eigenvalues and a full linearly independent set of eigenvectors.

The analysis proceeds by introducing a discontinuity in a derivative of

u, and then determining the evolution law governing its development as it propagates along a wavefront. Of central importance to this approach is a change of variable which is introduced from x and t to semi-characteristic coordinates t' and ϕ, thereby leading to a linear system of equations governing the evolution of the discontinuity in the derivative of **u** across the wavefront.

The method leads both to the determination of the time and place of shock formation when this occurs on the wavefront, and also to conditions which ensure that no shock will ever evolve (the exceptional condition) on the wavefront. Also arising from this approach is the identification of a bifurcation-type phenomenon in the event that eigenvalues change their multiplicity.

However, the system of linear equations arising from this method appears formally to be quite different from the Bernoulli equation which is found when the equation governing the amplitude of the weak discontinuity is derived directly. This problem has been considered by Boillat and Ruggeri [21] who showed them to be equivalent. In the process they also established how the nonlinear Bernoulli evolution equation and the linear system of equations are related by a discontinuous mapping involving the semi-characteristic coordinates.

Since system (4.3.1) is assumed to be hyperbolic, in general the matrix **A** possesses p distinct real eigenvalues $\lambda^{(i)}, i = 1, 2, \ldots, p$ (assumed to be ordered so that $\lambda^{(p)} < \lambda^{(p-1)} < \cdots < \lambda^{(1)}$) with multiplicities m_i such that

$$\sum_{i=1}^{p} m_i = n,$$ together with n linearly independent left (respectively; right) eigenvectors $\mathbf{L}^{(i,k)}$ (respectively; $\mathbf{R}^{(i,k)}$), $k = 1, 2, \ldots, m_i$ corresponding to the eigenvalues $\lambda^{(i)}$.

Let $\phi_i(x, t) = 0$ be the equation of the ith characteristic curve which is determined as the solution of $dx/dt = \lambda^{(i)}$ passing through (x_0, t_0). After introducing the characteristic coordinates (ϕ_i, t), if $\mathbf{\Pi}_i$ and X_i are the jumps in the derivatives of **u** and x with respect to ϕ_i across the ith characteristic curve, the transport equations for a C^1 discontinuity vector $\mathbf{\Pi}_i$ across the ith characteristic curve are given by the following system of ordinary differential equations (here we recall the main results of Jeffrey [80] which will be necessary in what follows)

$$\mathbf{L}^{(j,k)}\left(\mathbf{\Pi}_i - \mathbf{u}_{0,x}X_i\right) = 0, \quad \text{for } j = 1, 2, \ldots, i-1, i+1, \ldots, p \tag{4.3.2}$$
$$\text{and } k = 1, 2, \ldots, m_j$$

$$\frac{dX_i}{dt} = \left(\nabla\lambda^{(i)}\right)_0 \mathbf{\Pi}_i + \lambda^{(i)}_{0,x}X_i, \tag{4.3.3}$$

$$\left\{\left(\nabla\mathbf{L}^{(i,k)}\right)_0 \mathbf{\Pi}_i\right\}^{tr} \frac{d\mathbf{u}_0}{dt} + \mathbf{L}^{(i,k)}_{0,x}\frac{d\mathbf{u}_0}{dt}X_i + \mathbf{L}^{(i,k)}_0\frac{d\mathbf{\Pi}_i}{dt}$$

$$+ \left(\nabla\left(\mathbf{L}^{(i,k)}\mathbf{b}\right)\right)_0 \mathbf{\Pi}_i + \left(\mathbf{L}^{(i,k)}\mathbf{b}\right)_{0,x}X_i = 0, \quad \text{for } k = 1, 2, \ldots, m_i, \tag{4.3.4}$$

where the index i is fixed, the subscript 0 refers to the solution ahead of the ith characteristic curve, ∇ denotes the gradient operator with respect to the

n elements of $\mathbf{u}$, and d/dt denotes material time derivative following the ith characteristic curve. The set of $n + 1$ equations $(4.3.2) - (4.3.4)$ determine the vector $\mathbf{\Pi}_i$ and the scalar X_i, using the initial conditions $\mathbf{\Pi}_i(t_0) = \mathbf{\Pi}_{i0}$ and $X_i(t_0) = 0$. It should be noted that

$$\mathbf{\Pi}_i = \mathbf{\Lambda}_i \left(X_i + (x_{,\phi_i})_0 \right) + \mathbf{u}_{0,x} X_i, \tag{4.3.5}$$

where $\mathbf{\Lambda}_i$ is the jump in $\mathbf{u}_{,x}$ across the ith characteristic curve. In view of $(4.3.2)$ and $(4.3.5)$, it follows that there exist functions $\alpha_k^{(i)}$ such that

$$\mathbf{\Lambda}_i = \sum_{k=1}^{m_i} \alpha_k^{(i)}(t) \mathbf{R}_0^{(i,k)}. \tag{4.3.6}$$

The transport equations for C^1 discontinuities across the ith characteristic, in terms of the original independent variables x and t, are obtained by using $(4.3.3)$ and $(4.3.5)$ in $(4.3.4)$ in the following form

$$\mathbf{L}_0^{(i,k)} \frac{d\mathbf{\Lambda}_i}{dt} + \mathbf{L}_0^{i,k} \left(\mathbf{u}_{0,x} + \mathbf{\Lambda}_i \right) \left(\nabla \lambda^{(i)} \right)_0 \mathbf{\Lambda}_i + \left\{ \left(\nabla \mathbf{L}^{(i,k)} \right)_0 \mathbf{\Lambda}_i \right\}^{tr} \frac{d\mathbf{u}_0}{dt}$$
$$+ \left(\mathbf{L}_0^{(i,k)} \mathbf{\Lambda}_i \right) \left(\left(\nabla \lambda^{(i)} \right)_0 \mathbf{u}_{0,x} + \lambda_{0,x}^{(i)} \right) + \left(\nabla \left(\mathbf{L}^{(i,k)} \mathbf{b} \right) \right)_0 \mathbf{\Lambda}_i = 0. \tag{4.3.7}$$

When the expression for $\mathbf{\Lambda}_i$ given in $(4.3.6)$ is substituted into $(4.3.7)$, following the summation convention, we arrive at the Bernoulli type equations

$$a_j^{(i,k)} \frac{d\alpha_j^{(i)}}{dt} + b_{jp}^{(i,k)} \alpha_j^{(i)} \alpha_p^{(i)} + C_j^{(i,k)} \alpha_j^{(i)} = 0, \quad i, k - \text{unsummed} \tag{4.3.8}$$

where

$$a_j^{(i,k)} = \mathbf{L}_0^{(i,k)} \mathbf{R}_0^{(i,j)},$$

$$b_{jp}^{(i,k)} = \mathbf{L}_0^{(i,k)} \mathbf{R}_0^{(i,j)} \left(\nabla \lambda^{(i)} \right)_0 \mathbf{R}_0^{(i,p)},$$

$$C_j^{(i,k)} = \mathbf{L}_0^{(i,k)} \frac{d\mathbf{R}_0^{(i,j)}}{dt} + \left(\nabla \mathbf{R}^{(i,j)} \right)_0 \frac{d\mathbf{u}_0}{dt} + \mathbf{L}_0^{(i,k)} \mathbf{u}_{0,x} \left(\nabla \lambda^{(i)} \right)_0 \mathbf{R}_0^{(i,j)}$$
$$+ \left(\left(\nabla \mathbf{L}^{(i,k)} \right)_0 \mathbf{R}_0^{(i,j)} \right)^{tr} \frac{d\mathbf{u}_0}{dt} + \mathbf{L}_0^{(i,k)} \mathbf{r}_0^{(i,j)} \left(\left(\nabla \lambda^{(i)} \right)_0 \mathbf{u}_{0,x} + \lambda_{0,x}^{(i)} \right)$$
$$+ \left(\nabla \left(\mathbf{L}^{(i,k)} \mathbf{b} \right) \right)_0 \mathbf{R}_0^{(i,j)}, \quad k, p = 1, 2, \ldots, m_i.$$

which enable us to determine the amplitudes $\alpha_k^{(i)}$.

4.3.2 Reflected and transmitted amplitudes

In order to study the amplitudes of the reflected and transmitted C^1 discontinuities, when an incident C^1 wave comes in contact with a strong discontinuity, we shall need to consider generalized conservative systems which are a direct consequence of the original system (4.3.1), and have the forms

$$\mathbf{G}_{,t}\,(x,t,\mathbf{u}) + \mathbf{F}_{,x}\,(x,t,\mathbf{u}) = \mathbf{H}(x,t,\mathbf{u}), \tag{4.3.9}$$

$$\mathbf{G}^*_{,t}(x,t,\mathbf{u}^*) + \mathbf{F}^*_{,x}(x,t,\mathbf{u}^*) = \mathbf{H}^*(x,t,\mathbf{u}^*), \tag{4.3.10}$$

respectively, to the left and to the right of the discontinuity curve, $dx/dt = s$, which propagates with speed s. In (4.3.9) and (4.3.10) $\mathbf{u}$ and $\mathbf{u}^*$ are the solution vectors to the left and to the right of the discontinuity line.

Let $P(x_p, t_p)$ be the point at which fastest C^1 discontinuity $\mathbf{\Pi}_1$ of (4.3.9), moving along the characteristic $\phi_1(x,t) = 0$ and originating from the point (x_0, t_0), intersects the discontinuity line.

At P, the C^1 discontinuities $\mathbf{\Pi}^{(R)}_{p-q+i}, i = 1, 2, \ldots, q$ (respectively; $\mathbf{\Pi}^{*(T)}_i, i = 1, 2, \ldots, q^*$) are reflected (respectively; transmitted) along q (respectively; q^*) characteristics belonging to the system (4.3.9) (respectively; (4.3.10)) which as time increases, enter into the region to the left (respectively; right) of the discontinuity line. Thus the speed s of the discontinuity line separates q^* transmitted wave speeds ahead from q reflected wave speeds behind. As a direct consequence of this we have $\lambda^{(p)} < \lambda^{(p-1)} < \ldots < \lambda^{(p-q+1)} < s < \lambda^{*(q)} < \ldots < \lambda^{*(1)}$. Let $\mathbf{u}^{(R)}$ (respectively; $\mathbf{u}^{*(T)}$) be the solution in the reflected (respectively; transmitted) wave region, and $\mathbf{u}_0$ (respectively; $\mathbf{u}^*_0$) be the known solution vector ahead of the leading incident wave (respectively; leading transmitted wave) $\phi_1(x,t) = 0$ (respectively; $\phi^*_1(x,t) = 0$). Once the initial values $\mathbf{\Pi}^{(R)}_i(P)$ and $\mathbf{\Pi}^{*(T)}_i(P)$ of the reflected and transmitted vectors are known, the discontinuity vectors $\mathbf{\Pi}^{(R)}_i$ and $\mathbf{\Pi}^{*(T)}_i$ for $t > t_p$ can be determined from the system (4.3.2) – (4.3.4) after using the initial conditions $X^{(R)}(P) = 0$ and $X^{*(T)}(P) = 0$.

It should be noted that at the point P

$$\mathbf{\Pi}_1(P) + \sum_{i=p-q+1}^{p} \mathbf{\Pi}^{(R)}_i(P) = \mathbf{u}^{(R)}_{,\phi_{p-q+1}}(P) - \mathbf{u}_{0,\phi_1}(P),$$

$$\sum_{i=1}^{q^*} \mathbf{\Pi}^{*(T)}_i(P) = \mathbf{u}^{*(T)}_{,\phi^*_{q^*}}(P) - \mathbf{u}^*_{0,\phi^*_1}(P). \tag{4.3.11}$$

If we introduce the following parameterizations to ϕ_1, ϕ_{p-q+i} and ϕ^*_i,

$$\phi_1(x,t_0) = x - x_0, \quad \phi_{p-q+i}(x,t_p) = x - x_p, \quad \phi^*_i(x,t_p) = x - x_p, \tag{4.3.12}$$

equations (4.3.11) can be written as

$$\mathbf{\Pi}_1(P) + \sum_{i=p-q+1}^{p} \mathbf{\Pi}^{(R)}_i(P) = \mathbf{u}^{(R)}_{,x}(P) - [x_{,\phi_1}]_0\,\mathbf{u}_{0,x}(P),$$

$$\sum_{i=1}^{q^*} \mathbf{\Pi}^{*(T)}_i(P) = \mathbf{u}^{*(T)}_{,x}(P) - \mathbf{u}^*_{0,x}(P). \tag{4.3.13}$$

Across the discontinuity line the following generalized jump conditions (generalized Rankine Hugoniot conditions) hold

$$s\left(\mathbf{G} - \mathbf{G}^*\right) - \left(\mathbf{F} - \mathbf{F}^*\right) = 0.$$

On applying the displacement derivative $\overset{\bullet}{(\ )}= (\ )_{,t} + s(\ )_{,x}$ to these jump conditions, we obtain

$$s\left(\overset{\bullet}{\mathbf{G}} - \overset{\bullet}{\mathbf{G}}^*\right) + \overset{\bullet}{s}\left(\mathbf{G} - \mathbf{G}^*\right) - \left(\overset{\bullet}{\mathbf{F}} - \overset{\bullet}{\mathbf{F}}^*\right) = 0. \tag{4.3.14}$$

Owing to the interaction of the incident wave with a shock like discontinuity at $P(x_p, t_p)$, the shock undergoes an acceleration. The relations (4.3.14) hold at $t = t_{p+}$ (after the interaction) and at $t = t_{p-}$ (before the interaction), and we obtain

$$\left\{\overset{\bullet}{s}\left(\mathbf{G}_0 - \mathbf{G}_0^*\right) + s\left(\overset{\bullet}{\mathbf{G}}_0 + \left(\nabla_0\mathbf{G}_0\right)\overset{\bullet}{\mathbf{u}}_0\right) - s\left(\overset{\bullet}{\mathbf{G}}{}^*_{a_0} + \left(\nabla_0^*\mathbf{G}_0^*\right)\overset{\bullet}{\mathbf{u}}{}^*_0\right)\right.$$
$$\left. - \left(\overset{\bullet}{\mathbf{F}}_0 + \left(\nabla_0\mathbf{F}_0\right)\overset{\bullet}{\mathbf{u}}_0\right) + \left(\overset{\bullet}{\mathbf{F}}{}^*_0 + \left(\nabla_0^*\mathbf{F}_0^*\right)\overset{\bullet}{\mathbf{u}}{}^*_0\right)\right\}_{t=t_{p-}} = 0, \tag{4.3.15}$$

$$\left\{\overset{\bullet}{s}\left(\mathbf{G}^{(R)} - \mathbf{G}^{*(T)}\right) + s\left(\overset{\bullet}{\mathbf{G}}{}^{(R)} + \left(\nabla^{(R)}\mathbf{G}^{(R)}\right)\overset{\bullet}{\mathbf{u}}{}^{(R)}\right)\right.$$
$$- s\left(\overset{\bullet}{\mathbf{G}}{}^{*(T)} + \left(\nabla^{*(T)}\mathbf{G}^{*(T)}\right)\overset{\bullet}{\mathbf{u}}{}^{*(T)}\right) - \left(\overset{\bullet}{\mathbf{F}}{}^{(R)} + \left(\nabla^{(R)}\mathbf{F}^{(R)}\right)\overset{\bullet}{\mathbf{u}}{}^{(R)}\right)$$
$$\left. + \left(\overset{\bullet}{\mathbf{F}}{}^{*(T)} + \left(\nabla^{*(T)}\mathbf{F}^{*(T)}\right)\overset{\bullet}{\mathbf{u}}{}^{*(T)}\right)\right\}_{t=t_{p+}} = 0, \tag{4.3.16}$$

where $\nabla_0, \nabla_0^*, \nabla^{(R)}$ and $\nabla^{*(T)}$ are, respectively, the gradient operators with respect to the components of $\mathbf{u}_0, \mathbf{u}_0^*, \mathbf{u}^{(R)}$ and $\mathbf{u}^{*(T)}$. The dot derivatives of $\mathbf{F}_0, \mathbf{G}_0, \mathbf{F}_0^*, \mathbf{G}_0^*, \mathbf{F}^{(R)}, \mathbf{G}^{(R)}, \mathbf{F}^{*(T)}$ and $\mathbf{G}^{*(T)}$ are evaluated keeping the corresponding vector argument functions $\mathbf{u}_0, \mathbf{u}_0^*, \mathbf{u}^{*(R)}$ and $\mathbf{u}^{*(T)}$ constant. On subtracting (4.3.15) from (4.3.16) and keeping in mind that

$$s(t_{p-}) = s(t_{p+}), \quad \mathbf{G}_0(t_{p-}) = \mathbf{G}^{(R)}(t_{p+}), \quad \mathbf{G}_0^*(t_{p-}) = \mathbf{G}^{*(T)}(t_{p+}),$$

$$\mathbf{F}_0(t_{p-}) = \mathbf{F}^{(R)}(t_{p+}), \quad \mathbf{F}_0^*(t_{p-}) = \mathbf{F}^{*(T)}(t_{p+}), \quad \overset{\bullet}{\mathbf{G}}_0(t_{p-}) = \overset{\bullet}{\mathbf{G}}{}^{(R)}(t_{p+}),$$

$$\overset{\bullet}{\mathbf{G}}{}^*_0(t_{p-}) = \overset{\bullet}{\mathbf{G}}{}^{*(T)}(t_{p+}), \quad \overset{\bullet}{\mathbf{F}}_0(t_{p-}) = \overset{\bullet}{\mathbf{F}}{}^{(R)}(t_{p+}), \quad \overset{\bullet}{\mathbf{F}}{}^*_0(t_{p-}) = \overset{\bullet}{\mathbf{F}}{}^{*(T)}(t_{p+}),$$

$$\nabla_0\mathbf{F}_0(t_{p-}) = \nabla^{(R)}\mathbf{F}^{(R)}(t_{p+}), \quad \nabla_0^*\mathbf{F}_0^*(t_{p-}) = \nabla^{*(T)}\mathbf{F}^{*(T)}(t_{p+}),$$

$$\nabla_0\mathbf{G}_0(t_{p-}) = \nabla^{(R)}\mathbf{G}^{(R)}(t_{p+}), \quad \nabla_0^*\mathbf{G}_0^*(t_{p-}) = \nabla^{*(T)}\mathbf{G}^{*(T)}(t_{p+}),$$

$$\overset{\bullet}{\mathbf{u}}{}^{(R)} - \overset{\bullet}{\mathbf{u}}_0 = (s\mathbf{I} - \mathbf{A}_0)(\mathbf{u}^{(R)}{}_{,x} - \mathbf{u}_{0,x}), \quad \overset{\bullet}{\mathbf{u}}{}^{*(T)} - \overset{\bullet}{\mathbf{u}}{}^*_0 = (s\mathbf{I} - \mathbf{A}_0^*)(\mathbf{u}^{*(T)}{}_{,x} - \mathbf{u}^*_{0,x}),$$

where $\mathbf{I}$ is the unit matrix, we obtain the following system of equations valid at $t = t_p$.

$$[\![\dot{s}]\!] (\mathbf{G} - \mathbf{G}^*)_0 + (\nabla \mathbf{G})_0 (s\mathbf{I} - \mathbf{A}_0)^2 \left(\mathbf{u}_{,x}^{(R)} - \mathbf{u}_{0,x} \right)$$
$$- (\nabla^* \mathbf{G}^*)_0 (s\mathbf{I} - \mathbf{A}_0^*)^2 \left(\mathbf{u}_{,x}^{*(T)} - \mathbf{u}_{0,x}^* \right) = 0, \tag{4.3.17}$$

where $[\![\dot{s}]\!] = \dot{s} (t_{p+}) - \dot{s} (t_{p-})$ is the jump in the shock acceleration at $t = t_p$.

Let $\boldsymbol{\Lambda}_i^{(R)}$ be the jump in $\mathbf{u}_{,x}^{(R)}$ across the ith characteristic. Then equation (4.3.5), in view of the initial condition $(4.3.12)_2$ which implies $x_{,\phi_i}^{(R)} = 1$ and $X_i^{(R)}(P) = 0$, yields the result $\boldsymbol{\Pi}_i(P) = \boldsymbol{\Lambda}_i^{(R)}$; following a similar argument, it follows from (4.3.5) and (4.3.12) that $\boldsymbol{\Pi}_i^{*(T)}(P) = \boldsymbol{\Lambda}_i^{*(T)}(P)$, and $\boldsymbol{\Pi}_1(P) = \boldsymbol{\Lambda}_1(P) + \mathbf{u}_{0,x} (1 - (x_{,\phi_1})_0)$. Consequently, equations (4.3.13) become

$$\boldsymbol{\Lambda}_1(P) + \sum_{i=p-q+1}^{p} \boldsymbol{\Lambda}_i(P) = \mathbf{u}_{,x}^{(R)}(P) - \mathbf{u}_{0,x}(P),$$
$$\sum_{i=1}^{q^*} \boldsymbol{\Lambda}_i^{*(T)}(P) = \mathbf{u}_{,x}^{*(T)}(P) - \mathbf{u}_{0,x}^*(P). \tag{4.3.18}$$

System (4.3.17), in view of equations (4.3.18), becomes

$$[\![\dot{s}]\!] (\mathbf{G} - \mathbf{G}^*)_0 + (\nabla \mathbf{G})_0 (s\mathbf{I} - \mathbf{A}_0)^2 \sum_{i=p-q+1}^{p} \boldsymbol{\Lambda}_i^{(R)}(P)$$
$$- (\nabla^* \mathbf{G}^*)_0 (s\mathbf{I} - A_0^*)^2 \sum_{i=1}^{q^*} \boldsymbol{\Lambda}_i^{*(T)}(P) = - (\nabla \mathbf{G})_0 (s\mathbf{I} - \mathbf{A}_0)^2 \boldsymbol{\Lambda}_1. \tag{4.3.19}$$

In view of the relation (4.3.6), we may write,

$$\boldsymbol{\Lambda}_1(P) = \sum_{k=1}^{m_1} \alpha_k^{(1)}(t_p) \mathbf{R}_0^{(1,k)}, \quad \boldsymbol{\Lambda}_i^{(R)}(P) = \sum_{k=1}^{m_i} \alpha_k^{(i)}(t_p) \mathbf{R}_0^{(i,k)},$$
$$\boldsymbol{\Lambda}_i^{*(T)}(P) = \sum_{k=1}^{m_i^*} \beta_k^{(i)}(t_p) \mathbf{R}_0^{*(i,k)}, \tag{4.3.20}$$

which are valid on the discontinuity line at P; here $\alpha_k^{(i)}$ and $\beta_k^{(i)}$ are constant coefficients, and $\mathbf{R}^{*(i,k)}$ are the right eigenvectors of $\mathbf{A}^*$ corresponding to the eigenvalue $\lambda^{*(i)}$ with multiplicity m_i^*. Because of (4.3.20), equations (4.3.19) become

$$[\![\dot{s}]\!] (\mathbf{G} - \mathbf{G}^*)_0 + (\nabla \mathbf{G})_0 \sum_{i=p-q+1}^{p} \left(\sum_{k=1}^{m_i} \alpha_k^{(i)} \left(s - \lambda^{(i)} \right)^2 \mathbf{R}_0^{(i,k)} \right)$$
$$- (\nabla^* \mathbf{G}^*)_0 \sum_{j=1}^{q^*} \left(\sum_{k=1}^{m_j^*} \beta_k^{(j)} \left(s - \lambda^{*(j)} \right)^2 \mathbf{R}_0^{*(j,k)} \right)$$
$$= - (\nabla \mathbf{G})_0 \sum_{k=1}^{m_1} \alpha_k^{(1)} \left(s - \lambda^{(1)} \right)^2 \mathbf{R}_0^{(1,k)}, \tag{4.3.21}$$

where all quantities are evaluated at $t = t_p$. Matrix equation (4.3.21) represents a system of n inhomogeneous algebraic equations for the unknowns $[\![\dot{s}]\!], \alpha_k^{(i)}$ and $\beta_k^{(j)}$. In the case in which the discontinuity is an evolutionary shock [101], there exists an integer ℓ in the interval $1 \leq \ell \leq p$ such that the following inequalities hold

$$\lambda^{(p)} < \lambda^{(p-1)} < \ldots < \lambda^{(\ell+1)} < s < \lambda^{(\ell)} < \ldots < \lambda^{(1)},$$
$$\lambda^{*(p)} < \lambda^{*(p-1)} < \ldots < \lambda^{*(\ell)} < s < \lambda^{*(\ell-1)} < \ldots < \lambda^{*(1)}. \qquad (4.3.22)$$

In this situation, the system (4.3.21), with $q = \ell + 1$ and $q^* = \ell - 1$, can have at most n unknowns, and admits a unique solution provided the rank of the coefficient matrix and that of the augmented matrix are equal to the number of unknowns. For a strictly hyperbolic system, equation (4.3.21) with $q = \ell + 1, q^* = \ell - 1$ and $p = n$ becomes

$$[\![\dot{s}]\!] \, (\mathbf{G} - \mathbf{G}^*)_0 + (\nabla \mathbf{G})_0 \sum_{i=n-\ell}^{p} \alpha^{(i)} \left(s - \lambda^{(i)} \right)^2 \mathbf{R}_0^{(i)} \qquad (4.3.23)$$

$$- (\nabla^* \mathbf{G}^*)_0 \sum_{j=1}^{\ell-1} \beta^{(j)} \left(s - \lambda^{*(j)} \right)^2 \mathbf{R}_0^{*(j)} = -(\nabla \mathbf{G})_0 \alpha^{(1)} \left(s - \lambda^{(1)} \right)^2 \mathbf{R}_0^{(1)},$$

which for $\mathbf{G}(\mathbf{u}, x, t) \equiv \mathbf{u}$, reduces to the result obtained by Brun [26] and Boillat and Ruggeri [22]. Once the coefficients $\alpha_k^{(i)}$ and $\beta_k^{(j)}$ have been determined, equations (4.3.20) yield the initial strength of the reflected and transmitted waves.

4.4   Weak Discontinuities in Radiative Gasdynamics

In this section, we illustrate the applicability of singular surface theory to investigate the evolutionary behavior of weak discontinuities headed by wave fronts of arbitrary shape in a thermally radiating inviscid gas flow, governed by the system (1.2.30). The governing system of PDEs clearly shows the existence of radiation induced waves, which are followed by modified gasdynamic waves. The transport equations, representing the rate of change of discontinuities in the normal derivatives of the flow variables, show that the radiation induced waves are ultimately damped. Taking into account the unsteady behavior of the flow ahead of a modified gasdynamic wave, it is shown that it is natural to consider the transport of discontinuities along bicharacteristic curves in the characteristic manifold of the governing system of PDEs. An explicit criterion for the growth and decay of modified gasdynamic waves along bicharacteristics curves is given, and the spatial reference is made of diverging and converging waves. It is found that all compressive waves, except in one special case of converging waves in which both the initial principal curvatures are positive and equal, grow without bound only if the magnitude of the initial discontinuity associated with the wave exceeds a critical value. In this case, it is shown that a compressive wave, no matter how weak initially, always grows into a shock before the formation of the focus; emphasis is given on a description of the gasdynamic phenomenon involved rather than on numerical results. In this context, we refer to the reader the work carried out on weak discontinuity waves by Thomas [196], Helliwell [70], Elcrat [52], and Sharma and Shyam [167].

Let us consider a moving singularity surface Σ, across which the flow parameters are essentially continuous but discontinuities in their derivatives are permitted; it thus follows that the quantities $p, \rho, u_i, q_i, E_R, S$ and T are continuous across Σ, and they have their subscript o values at the wave head. Taking jumps across Σ, in equations (1.2.30), we find, on using (4.1.7)$_1$, and (4.1.7)$_2$ that

$$(G - u_{n_o})\zeta = \rho_o\lambda_i n_i; \quad \rho_o(G - u_{n_o})\lambda_i = \xi n_i + (\theta/3)n_i,$$
$$(4/3)E_{R_o}\lambda_i n_i = \rho_o T_o(G - u_{n_o})\chi - u_{n_o}\theta, \tag{4.4.1}$$
$$G\theta = \epsilon_i n_i; \quad G\epsilon_i = (c^2/3)\theta n_i,$$

where $\lambda_i = [u_{i,j}]n_j$, $\xi = [p_{,i}]n_i$, $\zeta = [\rho_{,i}]n_i$, $\chi = [S_{,i}]n_i$, $\theta = [E_{R,i}]n_i$, $\epsilon_i = [q_{i,j}]n_j$ and $u_{n_o} = u_{i_o}n_i$ are the quantities defined on Σ. Also, the equations of the state $p = p(\rho, S)$ and $T = T(\rho, S)$, yield

$$\xi = p_{,\rho_o}\zeta + p_{,S_o}\chi; \quad \eta = T_{,\rho_o}\zeta + T_{,S_o}\chi, \tag{4.4.2}$$

where $\eta = [T_{,i}]n_i$, $p_{,\rho_o} \equiv (\partial p/\partial \rho)_o$, $p_{,S_o} \equiv (\partial p/\partial S)_o$, $T_{,\rho_o} \equiv (\partial T/\partial \rho)_o$ and $T_{,S_o} \equiv (\partial T/\partial S)_o$.

Equations (4.4.1) and (4.4.2) constitute a set of ten homogeneous equations in ten unknowns $\lambda_i, \epsilon_i, \xi, \zeta, \chi$ and θ. Hence, the necessary condition for the existence of nontrivial solutions is that the determinant of the coefficient matrix must vanish; this yields

$$G = u_{n_o}, \quad G = \pm c/\sqrt{3}, \quad G = u_{n_o} \pm a_o,$$

where $a_o^2 = p_{,\rho_o} + (4E_{R_o}/3T_o\rho_o^2)p_{,S_o}$ is the square of modified sound speed. The case $G = u_{n_o}$, in which the surface moves with the fluid, is discarded as having no physical interest. For an advancing wave surface, we shall take G to be positive. We thus find that there are two types of waves present in the gas; one that propagates with the speed $G = c/\sqrt{3}$ has attributes which are basically due to radiation, and is referred to as a radiation induced wave. The other, which propagates with the speed $G = u_{n_o} + a_o$ is essentially a forward moving modified gasdynamic wave.

4.4.1 Radiation induced waves

It is convenient to take the flow upstream from the radiation induced wave Σ to be uniform and at rest. In this case, equations (4.4.1), and (4.4.2), using

$G = c/\sqrt{3}$, yield

$$
\begin{aligned}
&\theta = \epsilon\sqrt{3}/c; \quad \xi = (\epsilon a_o^2\sqrt{3}/c^3)\{1 - (3a_o^2/c^2)\}^{-1}; \\
&\lambda = (\epsilon/\rho_o c^2)\{1 - (3a_o^2/c^2)\}^{-1}; \\
&\zeta = (\epsilon\sqrt{3}/c^3)\{1 - (3a_o^2/c^2)\}^{-1}; \\
&\chi = (\epsilon\sqrt{3}/c^3)\{a_o^2 - p_{,\rho_o}\}\{(1 - (3a_o^2/c^2))p_{,S_o}\}^{-1}; \\
&\eta = (\epsilon\sqrt{3}/c^3)\{(T_{,S}/p_{,S})_o(a_o^2 - a_o^{*2})\}\{1 - (3a_o^2/c^2)\}^{-1},
\end{aligned} \tag{4.4.3}
$$

where $\epsilon = \epsilon_i n_i$, $\lambda = \lambda_i n_i$ and $a_o^{*2} = p_{,\rho_o} - (p_{,S}T_{,\rho}/T_{,S})_o$.

When $(1.2.30)_5$ is differentiated partially with respect to t and use is made of $(1.2.30)_4$, we get

$$
\left(\frac{3}{c^2}\right)\frac{\partial^2 q_j}{\partial t^2} + \left(\frac{6k_p}{c}\right)\frac{\partial q_j}{\partial t} - q_{j,ii} + 16\sigma k_p T^3 T_{,j} + 3k_p^2 q_j = 0.
$$

If we multiply this equation by n_j and take jumps across $\sum$, we find on using equations $(4.1.7)_2$, $(4.1.7)_3$, $(4.1.7)_4$, $(4.1.2)$, $(4.4.3)_4$ and $(4.4.3)_6$ that

$$
\frac{\delta\zeta}{\delta t} = -(\Lambda_1 - \Lambda_2)\zeta, \tag{4.4.4}
$$

where $\Lambda_1 = \{k_p - (\Omega/\sqrt{3})\}c$ and $\Lambda_2 = (8\sigma k_p T_o^3/c^3)(T_{,S}/p_{,S})_o(a_o^2 - a_o^{*2})\{1 - (3a_o^2/c^2)\}^{-1}$. Equation $(4.4.4)$ governs the propagation of ζ associated with a radiation induced wave; its behavior can, of course, be readily established. The mean curvature Ω at any point of the wave surface $\sum$ has the representation (Thomas [197])

$$
\Omega = (\Omega_o - K_o Gt)/(1 - 2\Omega_o Gt + K_o G^2 t^2), \tag{4.4.5}
$$

where $\Omega_o = (\kappa_1 + \kappa_2)/2$ and $K_o = \kappa_1\kappa_2$ are respectively the mean and Gaussian curvatures of $\sum$ at $t = 0$ with κ_1 and κ_2 being the principal curvatures, and $G = c/\sqrt{3}$.

When k_1 and k_2 are nonpositive, the wave is divergent. On the other hand if one or both the principal curvatures are positive, then it corresponds to the case of a convergent wave. Integration of $(4.4.4)$, after using $(4.4.5)$ yields

$$
\zeta = \zeta_o I \exp(-k_p ct), \tag{4.4.6}
$$

where $I = \{(1 - (\kappa_1 ct/\sqrt{3}))(1 - (\kappa_2 ct/\sqrt{3}))\}^{-1/2}$, and ζ_o is the value of ζ at $t = 0$. The term Λ_2 in $(4.4.4)$ has been neglected in comparison with Λ_1 in presenting the result $(4.4.6)$.

For diverging waves, it is apparent from $(4.4.6)$ that $\zeta \to 0$ as $t \to \infty$; this means that the radiation induced waves are ultimately damped out and the formation of a front, carrying discontinuities in flow quantities, is not possible from a continuous flow. From the expressions $(4.4.3)$, it follows that the quantities $\lambda, \xi, \zeta, \chi$ and η are small compared with ϵ and θ as it should be in radiation induced waves.

For converging waves, it follows from $(4.4.6)$ that there exists a finite time $t*$ given by the smallest positive root of $(\sqrt{3} - \kappa_1 ct)(\sqrt{3} - \kappa_2 ct) = 0$ such that $|\zeta| \to \infty$ as $t \to t*$; this corresponds to the formation of a focus.

4.4.2 Modified gasdynamic waves

In general, if the flow ahead of a modified gasdynamic wave Σ, which propagates with the speed $G = u_{n_o} + a_o$, is nonuniform, equations (4.4.1) and (4.4.2) yield

$$a_o\zeta = \rho_o\lambda, \quad \rho_o a_o\lambda = \xi, \quad \chi = (4E_{R_o}/3T_o^2\rho_o^2)\zeta,$$
$$\eta = (T_{,S}/p_{,S})_o(a_o^2 - a_o^{*2})\zeta, \quad \theta = \epsilon = 0. \tag{4.4.7}$$

If we differentiate equations (1.2.30) partially with respect to x_k, take jumps across Σ, use equations $(4.1.1)_3$, $(4.1.1)_4$, (4.1.2) and (4.4.7), and then eliminate from the resulting equations the quantities involving jumps in second order derivatives, we obtain

$$\rho_o\left(\tfrac{\delta\lambda}{\delta t} + u_o^\alpha\lambda_{,\alpha}\right) + (p_{,\rho_o}/a_o)\left(\tfrac{\delta\zeta}{\delta t} + u_o^\alpha\zeta_{,\alpha}\right)$$
$$+ (p_{,S_o}/a_o)\left(\tfrac{\delta\chi}{\delta t} + u_o^\alpha\chi_{,\alpha}\right) + P\zeta + Q\zeta^2 = 0, \tag{4.4.8}$$

where P and Q are known functions depending on the state-ahead variables and their derivatives along with the orientation and the mean curvature of the wave front (see Sharma and Shyam [167]).

The terms involving surface derivatives in (4.4.8) cause some difficulty in its interpretation; but if we transform (4.4.8) to a differential equation along bicharacteristics curves, this difficulty disappears. Thus, using (4.1.5) in (4.4.8) and substituting λ and χ in terms of ζ from (4.4.7), we get

$$\frac{d\zeta}{dt} + \left\{\frac{1}{2}\frac{d}{dt}\log\Lambda + \frac{P}{2a_o}\right\}\zeta + \frac{Q}{2a_o}\zeta^2 = 0, \tag{4.4.9}$$

where $\Lambda = \{a_o^3/(\rho_o p_{,S_o})\}(p_{,S}/p_{,\rho})^{(p_{,\rho_o}/a_o^2)}$.

Equation (4.4.9) is the transport equation governing the propagation of the discontinuity ζ, and it has the same form as the equation for the growth of discontinuities in the unsteady flow of a perfect gas (see Elcrat [52]). One can easily verify that in the limit of vanishing radiation energy and flux, it reduces to the equation derived by Elcrat [52].

Equation (4.4.9) can be integrated to yield

$$\zeta = \zeta_i(\Lambda/\Lambda_i)^{-1/2}\exp(-f(t))/\{1 + \Lambda_i^{1/2}\zeta_i I(t)\}, \tag{4.4.10}$$

where $f(t) = \int_0^t(P(\tau)/2a_o(\tau))d\tau$, $I(t) = \int_0^t(Q(\tau)/2a_o\Lambda^{1/2})\exp(-f(\tau))d\tau$, and the subscript i indicates an initial value at $t = 0$. As (4.4.10) turns out to be a special case of (4.2.8), the general results presented in subsection 4.2.2 can be applied to yield conditions under which the wave ultimately damps out, or takes a stable wave form, or, forms a shock or a focus. Thus, if both $f(t)$ or $I(t)$ are continuous for $0 \le t < \bar{t}$ and have finite limits $f(\bar{t}), I(\bar{t})$ as $t \to \bar{t}$ and if $\text{sign}\zeta_i = \text{sign}I(t)$, then the right-hand side of (4.4.10) will not only remain continuous throughout $0 \le t < \bar{t}$ but will also approach a finite limit as $t \to \bar{t}$.

Also if $\text{sign}\zeta_i = -\text{sign}I(t)$, (4.3.10) will remain finite throughout $[0, \bar{t}]$ provided $|\zeta_i| < \zeta_c^*$, where ζ_c^* is a positive quantity defined as $\zeta_c^* = (\Lambda_i^{1/2}|I(\infty)|)^{-1}$. But if $\text{sign}\zeta_i = -\text{sign}I(t)$, and $|\zeta_i| > \zeta_c^*$, then it follows from (4.4.10) that there exists a finite time $t^* < \bar{t}$, given by $I(t^*) = -1/\zeta_i\Lambda_i^{1/2}$, such that $\zeta \to \infty$ as $t \to t^*$. This signifies the appearance of a shock wave at an instant t^*. Further, if $|\zeta_i| = \zeta_c^*$ and $\text{sign } \zeta_i = -\text{sign}I(t)$, then we find that ζ is continuous for t in $[0, \bar{t})$ but approaches infinity as $t \to \bar{t}$.

4.4.3   Waves entering in a uniform region

We have already noticed in subsection 4.3.1 that for radiation induced waves the quantities $\xi, \lambda, \zeta, \chi$ and η are much smaller as compared to ϵ and θ, and hence the main effects of the radiative transfer on the growth and decay behavior of the gasdynamic wave $\sum$ can be seen for the special case in which $u_i = 0$ and p, ρ, T are constant ahead of the wave $\sum$. For such a wave, propagating through a uniform flow of an ideal gas, equation (4.4.8) becomes

$$\frac{\delta\zeta}{\delta t} + (\omega - a_o\Omega)\zeta + \psi\zeta^2 = 0, \tag{4.4.11}$$

where ω, a_o, and ψ are positive constants given by

$$\omega = (8\sigma k_p T_o^3/\rho_0 a_o^2 R)(\gamma - 1)(a_o^2 - a_o^{*2})\left\{1 + (a_o^2/(\gamma - 1))(c^2 - 3a_0^2)^{-1}\right\},$$
$$a_0^2 - (\gamma p_o + 4(\gamma - 1)p_{R_o})/\rho_o, \quad \psi = (\gamma + 1)a_o/(2\rho_o),$$

with R as the gas constant, γ the adiabatic index, a_o^* the isothermal speed of sound defined as $a_o^* = (p_o/\rho_o)^{1/2}$, and p_{R_o} the radiation pressure, which is one third of the radiation energy E_R. The term $(a_o^2/(\gamma - 1))/(c^2 - 3a_o^2)$ in the expression for ω will be neglected eventually in order to be consistent with the nonrelativistic form of equations (1.2.30); here, it is retained temporarily since it throws some light on the physical nature of the problem.

Equation (4.4.11), after substituting from (4.4.5), can be integrated to yield

$$\zeta = \zeta_o \exp(-\omega t)I_1(t)/(1 + \psi I_2(t)\zeta_o), \tag{4.4.12}$$

where

$$I_1 = \{(1 - \kappa_1 a_o t)(1 - \kappa_2 a_o t)\}^{-1/2}, \quad I_2(t) = \int_o^t I_1(\tau)\exp(-\omega\tau)d\tau, \tag{4.4.13}$$

and ζ_0 is the value of ζ at $t = 0$.

Diverging waves: For diverging waves both κ_1 and κ_2 are nonpositive, and it is apparent from (4.4.13) that I_1 and I_2 both converge to finite limits as $t \to \infty$. Hence, if $\zeta_o > 0$ (i.e., an expansion wave front), then $\zeta \to 0$ as $t \to \infty$. This means that all expansion waves decay and damp out ultimately. But, if $\zeta_o < 0$ (i.e., a compression wave), then there exists a positive critical value,

ζ_c, given by $\zeta_c = 1/\{\psi \int_0^\infty I_1(\tau) \exp(-\omega\tau)d\tau\}$ such that waves with initial discontinuity $|\zeta_o| < \zeta_c$ damp to zero (i.e., $\zeta \to 0$ as $t \to \infty$) and the waves with initial discontinuity $|\zeta_o| > \zeta_c$ grow without bound in a finite time t_c (i.e., $|\zeta| \to \infty$ as $t \to t_c$) given by $I_2(t_c) = 1/\psi|\zeta_o|$. Further, if $\zeta_o < 0$ and $|\zeta_o| = \zeta_c$ then it immediately follows that $\zeta \to (\omega/\psi)$ as $t \to \infty$. Thus, a diverging compression wave for $|\zeta_o| = \zeta_c$ can neither terminate into a shock wave nor can it ever completely damp out. Since $\partial\zeta_c/\partial\omega > 0$, the critical value of the initial discontinuity increases with ω, and thus the effect of thermal radiation is to delay the onset of a shock wave. Also $\partial\zeta/\partial|\kappa_\alpha| = 0, \alpha = 1, 2$, which imply that the initial principal curvatures have a stabilizing effect on the tendency of a wave surface to grow into a shock in the sense that an increase in the value of the initial curvatures causes an increase in the critical value ζ_c. For waves with plane ($\kappa_1 = 0 = \kappa_2$), cylindrical ($\kappa_1 = -1/R_o, \kappa_2 = 0$) and spherical ($\kappa_1 = \kappa_2 = -1/R_o$) geometry, where R_o is the radius of the wave front at $t = 0$, we find that the critical value of the initial discontinuity and the time taken for the shock formation are given by the following relations:

Plane wave: $\qquad \zeta_c = \omega/\psi, \ t_c = \omega^{-1} \log\{1 - (\omega/\psi|\zeta_o|)\}^{-1}$,

Cylindrical wave: $\zeta_c = \dfrac{1}{\psi}\left(\dfrac{\pi R_o}{\omega a_o}\right)^{-1/2} \dfrac{\exp(-\omega R_o/a_o)}{erfc(\omega R_o/a_o)^{1/2}}$,
$$\mathrm{erfc}(\omega t_c + (\omega R_o/a_o))^{1/2} = (1 - (\zeta_c/|\zeta_o|))erfc(\omega R_o/a_o)^{1/2},$$

Spherical wave: $\qquad \zeta_c = \dfrac{a_o \exp(-\omega R_o/a_o)}{\psi R_o E_i(\omega R_o/a_o)}$,
$$E_i(\omega t_c + (\omega R_o/a_o)) = (1 - (\zeta_c/|\zeta_o|))E_i(\omega R_o/a_o),$$

where $\mathrm{erfc}(x) = (2/\sqrt{\pi})\int_x^\infty \exp(-t^2)dt$ and $E_i(x) = \int_x^\infty t^{-1}\exp(-t)dt$ are, respectively, the complementary error function and the exponential integral.

Converging waves:

(i) When only one of the initial principal curvatures is positive or both are positive and $k_1 \neq k_2$, in that case, there exists a finite time $t*$ given by the smallest positive root of $(1 - \kappa_1 a_o t)(1 - \kappa_2 a_o t) = 0$, such that $I_1(t) \to \infty$ as $t \to t*$, whereas $I_2(t)$ tends to a finite value as $t \to t*$. The fact that $I_2(t*)$ is bounded follows from the argument that the singularity at $t*$ of the integrand in $I_2(t)$ is of the form $z^{-1/2}g(z)$ as $z \to 0$, where $g(z)$ is bounded. This can be easily seen by a suitable transformation $z = t - t*$. Hence, it follows from (4.4.12) that for $\zeta_o > 0, |\zeta| \to \infty$ as $t \to t*$, i.e., a converging expansive wave forms a focus. But, if $\zeta_o < 0$, then it follows from (4.4.12) that there exists a positive critical value $\hat{\zeta}_c$ given by $\hat{\zeta}_c = 1/\psi I_2(t*)$, such that if $|\zeta_o| < \hat{\zeta}_c$, then $|\zeta| \to \infty$ as $t \to t*$, which corresponds to the formation of a focus but not the shock. On the other hand, if $|\zeta_o| > \hat{\zeta}_c$, then we find that $|\zeta| \to \infty$ as $t \to \hat{t}_c$, where $\hat{t}_c$ is finite time given by $I_2(\hat{t}_c) = I/\psi|\zeta_o|$. It is evident that $\hat{t}_c < t*$. Thus, when $|\zeta_o| > \hat{\zeta}_c$, we find that the shock wave is formed before the focus can. The case $|\zeta_o| = \hat{\zeta}_c$ corresponds to the simultaneous formation of a shock and a focus.

In this situation, it is interesting to note that not all converging compressive waves will grow into shock waves.

(ii) When both the initial principal curvatures are positive and $\kappa_1 = \kappa_2$ in that case, the singularity at $t*$ of the integrand in I_2 is of the form $z^{-1}r(z)$ as $z \to 0$, where $r(z)$ is bounded away from zero. Hence, $I_2 \to \infty$ as $t \to t*$. Thus, it follows from (4.4.12) that if $\zeta_o > 0$, then $|\zeta| \to \infty$ as $t \to t*$, i.e., a focus is formed within a finite time $t*$. But if $\zeta_o < 0$, then there exists a finite time $\tilde{t} < t*$ given by $I_2(\tilde{t}) = I/\psi|\zeta_o|$ for which the denominator of (4.4.12) vanishes whereas the numerator remains finite. This means that in this particular situation a converging compressive wave, no matter how weak initially, always grows into a shock before the formation of the focus (see Sharma and Menon [175]).

4.5   One-Dimensional Weak Discontinuity Waves

A comprehensive review relating to the propagation of one-dimensional waves in a nonequilibrium medium has been given by Becker [12], Chu [36], and Clarke and McCheseny [37]. In what follows we shall use different methods of approach to study this problem in one-dimension; the waves may be thought of as being produced by a moving piston. The basic equations governing the motion are same as in (1.2.21); for a one-dimensional motion with plane ($m = 0$), cylindrical ($m = 1$) or spherical ($m = 2$) symmetry, these equations may be written in the form

$$p_{,t} + up_{,x} + \rho a_f^2 (u_{,x} + (mu/x)) = -\omega a_f^2 (\rho_{,q} + (\alpha/T)\rho_{,S}),$$
$$u_{,t} + uu_{,x} + (1/\rho)p_{,x} = 0, \quad S_{,t} + uS_{,x} = \alpha\omega/T, \quad q_{,t} + uq_{,x} = \omega. \tag{4.5.1}$$

where the spatial coordinate x is either axial (in flows with planar geometry) or radial (in cylindrically or spherically symmetric flows), $\rho_{,q} = \partial\rho/\partial q$, $\rho_{,S} = \partial\rho/\partial S$, and the remaining symbols have the same meaning as in (1.2.21). We assume that the medium ahead of the wave front is in a state of complete thermodynamic equilibrium and at rest, i.e., the pressure p_o, the density ρ_o, entropy S_o and the progress variable q_o ahead of the wave front are constant, and the velocity u_o is zero.

4.5.1   Characteristic approach

Equations (4.5.1) constitute a hyperbolic system with four families of characteristics. Two of these characteristics, $dx/dt = u \pm a_f$, represent waves propagating in the $\pm x$ directions with the frozen sound speed a_f, and the remaining two, $dx/dt = u$, form a set of double characteristics representing the particle path or trajectory.

In studying wave phenomena governed by hyperbolic equations, it is usually more natural and convenient to use the characteristics of the governing system as the reference coordinate system. We thus introduce two characteristic variables ϕ and ψ such that

$$\phi_{,t} + (u + a_f)\phi_{,x} = 0, \quad \psi_{,t} + u\psi_{,x} = 0.$$

The leading characteristic front can be represented by $\phi = 0$, and if a material particle crosses the front at time t, its path will be represented by $\psi = t$. Keeping in view the properties of ϕ and ψ, it is obvious that the functions $x(\phi, \psi)$ and $t(\phi, \psi)$ satisfy the following differential equations.

$$x_{,\phi} = ut_{,\phi}, \quad x_{,\psi} = (u + a_f)t_{,\psi}. \tag{4.5.2}$$

The transformation from the (x, t) plane to the (ϕ, ψ) plane will be one-to-one if, and only if, the Jacobian

$$J = x_{,\phi}t_{,\psi} - x_{,\psi}t_{,\phi}, \tag{4.5.3}$$

does not vanish or become undefined anywhere. Hence, if F denotes any of the flow variables, p, ρ, u, S, or q, then

$$F_{,x} = (F_{,\phi}t_{,\psi} - F_{,\psi}t_{,\phi})/J, \quad F_{,t} + uF_{,x} = -a_f F_{,\phi}t_{,\psi}/J. \tag{4.5.4}$$

Since the overlapping of fluid particles is prohibited from physical considerations, it implies that $t_{,\psi} \neq 0$. Consequently $J = 0$ if, and only if, $t_{,\phi} = 0$ which corresponds to the situation when two adjoining characteristics merge into a shock wave.

In terms of the characteristic coordinates, the system of equations (4.5.1) can be transformed into the following equivalent system.

$$p_{,\psi} + \rho a_f u_{,\psi} + a_f^2 w((\rho_{,q} + (\alpha/T)\rho_{,S})t_{,\psi} + (m\rho u a_f^2/x)t_{,\psi} = 0,$$
$$(p_{,\phi} - \rho a_f u_{,\phi})t_{,\psi} - p_{,\psi}t_{,\phi} = 0, \tag{4.5.5}$$
$$S_{,\phi} = (\alpha w/T)t_{,\phi}, \quad q_{,\phi} = t_{,\phi}w,$$

where use has been made of (4.5.3) and (4.5.4).

Consider the case in which the wave front is an outgoing characteristic; then the flow variables are continuous across it, and the boundary conditions which hold on it can be written as

$$p = p_o, \quad \rho = \rho_o, \quad S = S_o, \quad q = q_o, \quad u = 0, \quad t = \psi \quad \text{at} \quad \phi = 0. \tag{4.5.6}$$

The last condition is, of course, a consequence of the particular method of labeling the particle paths; indeed, conditions (4.5.6) demand that

$$p_{,\psi} = u_{,\psi} = \rho_{,\psi} = S_{,\psi} = q_{,\psi} = 0, \quad t_{,\psi} = 1 \quad \text{at} \quad \phi = 0. \tag{4.5.7}$$

Also, equations (4.5.2), (4.5.5)$_2$, (4.5.5)$_3$, and (4.5.5)$_4$, when evaluated at the wave front $\phi = 0$, yield

$$p_{,\phi} = \rho_o a_{f_o} u_{,\phi}, \quad S_{,\phi} = 0 = q_{,\phi}, \quad x_{,\phi} = 0, \quad x_{,\psi} = a_{f_o}. \tag{4.5.8}$$

Now to compute $s = u_{,x}$ at the wave front $\phi = 0$, we invoke (4.5.4)$_1$, (4.5.3) and (4.5.7)$_2$ to obtain

$$s = -u_{,\phi}/(t_{,\phi} a_{f_o}). \tag{4.5.9}$$

Differentiating (4.5.2)$_2$ and (4.5.5)$_1$ with respect to ϕ and (4.5.2)$_1$ and (4.5.5)$_2$ with respect to ψ and using the foregoing results, we find that at the wave front $\phi = 0$

$$u_{,\phi\psi} = -\left\{\mu_o + \frac{m a_{f_o}}{2x}\right\} u_{,\phi}, \tag{4.5.10}$$

$$t_{,\phi\psi} = -\frac{1}{2a_{f_o}}(1 + \Gamma_o)u_{,\phi}, \tag{4.5.11}$$

where $\mu_o = (a_f^2 \rho_{,q} \omega_{,p})_o/2$ and $\Gamma_o = 1 + \rho_o((a_f^2)_{,p})_o$; in the derivation of (4.5.10), we have used the fact that the state ahead of the wave front is in complete thermodynamic equilibrium, i.e., $\alpha_o = 0 = \omega_o$ (see [37]).

Differentiating (4.5.9) with respect to ψ and using (4.5.10) and (4.5.11), we find that s, given by (4.5.9), satisfies the following Bernoulli-type differential equation:

$$\frac{ds}{dt} + \left(\mu_0 + \frac{m a_{f_o}}{2x}\right)s + \left(\frac{1 + \Gamma_o}{2}\right)s^2 = 0, \tag{4.5.12}$$

where $x = x(0) + a_{f_o}t$ with $x(0)$ as the value of x at $t = 0$. In obtaining (4.5.12), we have used the fact that $t = \psi$ at $\psi = 0$.

Equation (4.5.12) is of the form (4.2.8) with $\mu = \mu_o + (m a_{f_o}/2x)$ and $\beta = (1 + \Gamma_o)/2$. Note that in (4.5.12), Γ_o and μ_o are positive constants. Indeed, for an ideal gas, it may be readily verified that $\Gamma = \gamma$, the adiabatic index; similarly in an equilibrium state in which $\omega_o = 0 = \alpha_o$, it readily transpires that

$$\mu_o = \frac{1}{2}a_{f_o}^2 (\rho_{,q} \omega_{,p})_o = \frac{1}{2\tau_o}\left(\frac{a_{f_o}^2}{a_{eo}^2} - 1\right), \tag{4.5.13}$$

where $\tau_o = -(\omega_{,q})_o^{-1}$ is a relaxation time of the medium and a_{eo}^2 is the equilibrium speed of sound given by $a_{eo}^2 = ((\partial p/\partial \rho)_{S,q=q_e})_o$ (see [37]). Since $a_{f_o} > a_{eo}$, it follows that in a complete equilibrium state, $\mu_o > 0$.

The behavior of the solution of (4.5.12) can now be obtained by applying our general results presented in subsection 4.2.2. Thus, when sgn $s(0) = -$ sgn $\beta = -1$ (i.e., an expansion wave front), then $\lim\limits_{t \to \infty} s(t) = 0$ (i.e., the wave damps out). However, when sgn $s(0) = $ sgn $\beta = +1$ (i.e., a compression wave front), then there exists a positive critical amplitude s_c given by

$$s_c = \left[\frac{1 + \Gamma_o}{2}\int_0^\infty (1 + \kappa_o a_{f_o}\tau)^{-m/2} \exp[-\mu_o\tau]d\tau\right]^{-1},$$

where $\kappa_0 = 1/x(0)$ is the initial wave front curvature, such that, for $s(0) < s_c$, $s \to 0$ as $t \to \infty$, for $s(0) > s_c$, $s \to \infty$ as $t \to t_c$ (i.e., the wave culminates into a shock wave in a finite time t_c), where t_c is given by the solution of

$$\int_0^{t_c} (1 + \kappa_0 a_{f_o} t)^{-m/2} \exp[-\mu_o t] dt = 2/(s(0)(1 + \Gamma_o)),$$

and for $s(0) = s_c$, $s \to 2\mu_o/(1 + \Gamma_o)$ as $t \to \infty$ (i.e., the wave ultimately takes a stable wave form).

4.5.2 Semi-characteristic approach

The method introduced by Jeffrey [80] for determining the evolution law of weak discontinuities consists in introducing a change of variables from x and t to semi-characteristic coordinates. The system of linear equations arising from this method appears formally to be quite different from the Bernoulli equation, which is found when the equation governing the amplitude of the weak discontinuity is derived directly. This problem has been considered by Boillat and Ruggeri [21], who showed them to be equivalent. Here, we illustrate his method by confining ourselves to the one-dimensional system (4.5.1), which using vector-matrix notation can be written as

$$\mathbf{u}_{,t} + \mathbf{A}(\mathbf{u})\mathbf{u}_{,x} = \mathbf{b}, \tag{4.5.14}$$

where $\mathbf{u}$ is a column vector with components p, u, S and q, and the square matrix $\mathbf{A}$ and the column vector $\mathbf{b}$ can be read off by inspection of (4.5.1). Let $\lambda^{(j)}, j = 1, 2, 3, 4$, be the eigenvalues of $\mathbf{A}$, and $\mathbf{L}^{(j)}$ the corresponding left eigenvectors; then

$$\begin{aligned}
\lambda^{(1)} &= u + a_f, & \mathbf{L}^{(1)} &= [\ 1 \quad \rho a_f \quad 0 \quad 0\ ]^{tr}, \\
\lambda^{(2)} &= u - a_f, & \mathbf{L}^{(2)} &= [\ 1 \quad -\rho a_f \quad 0 \quad 0\ ]^{tr}, \\
\lambda^{(3)} &= u, & \mathbf{L}^{(3)} &= [\ 0 \quad 0 \quad 1 \quad 0\ ]^{tr}, \\
\lambda^{(4)} &= u, & \mathbf{L}^{(4)} &= [\ 0 \quad 0 \quad 0 \quad 1\ ]^{tr}.
\end{aligned}$$

Equations $dx/dt = \lambda^{(1,2)}$ represent waves propagating in the $\pm x$ directions, and the equations $dx/dt = \lambda^{(3,4)}$ represent particle path or trajectory. Let $\phi(x, t) = 0$ be the wave front or the forward facing characteristic corresponding to the eigenvalue $\lambda^{(1)}$; the unperturbed field ahead of the wave front, which we denote by using the subscript o, is assumed to be at rest and in a state of complete thermodynamical equilibrium. Let us introduce the new variables $\phi = \phi(x, t), t' = t$ such that the Jacobian $\partial(\phi, t')/\partial(x, t) \equiv x_{,\phi} = 1/\phi_{,x}$ and its inverse do not vanish. The leading characteristic $x = x_o + a_{f_o} t$, issuing out from the point $(x_o, 0)$, is taken as a member of the family given by $\phi_{,t} + \lambda^{(1)}\phi_{,x} = 0$, $\phi(x, 0) = x - x_o$ so that it is expressed by the equation $\phi(x, t) = 0$.

Transforming from (x, t) to (ϕ, t') coordinates, pre-multiplying by $\mathbf{L}^{(j)}$ and using the relation $\mathbf{L}^{(j)}\mathbf{A} = \lambda^{(j)}\mathbf{L}^{(j)}$, equation (4.5.14) yields

$$\mathbf{L}^{(j)}\{x_{,\phi}\mathbf{u}_{,t'} + (\lambda^{(j)} - \lambda^{(1)})\mathbf{u}_{,\phi}\} + \mathbf{L}^{(j)}\mathbf{b}x_{,\phi} = 0, \tag{4.5.15}$$

where $\mathbf{u}_{,t'} \equiv \mathbf{u}_{,t} + \lambda_o^{(1)} \mathbf{u}_{,x}$ is the derivative along the wave front, which we shall denote subsequently by the symbol $d\mathbf{u}/dt$.

Across the wave front $\phi = 0$, $\mathbf{u}$ and $\mathbf{u}_{,t'}$ are continuous whilst $\mathbf{u}_{,\phi}$ and $x_{,\phi}$ are discontinuous and have jumps $\mathbf{\Pi}$ and X, respectively; it may be noticed that at the rear of $\phi = 0$, we have

$$s = u_{,x} = u_{,\phi}/x_{,\phi}. \tag{4.5.16}$$

Since in the region ahead of the wave front $(\mathbf{u}_{,\phi})_o = 0$, $(\mathbf{u}_{,t'}) = 0$, and $\mathbf{b}_o = 0$, equation (4.5.15) with $\lambda^{(j)} \neq \lambda^{(1)}$ yields on differencing across $\phi = 0$ that $\mathbf{L}_o^{(j)} \mathbf{\Pi} = 0$, $(j \neq 1)$, which in view of the left eigenvectors $\mathbf{L}^{(2)}, \mathbf{L}^{(3)}$ and $\mathbf{L}^{(4)}$, yields the following relations

$$\pi_1 = \rho_o a_{f_o} \pi_2, \quad \pi_3 = 0 = \pi_4, \tag{4.5.17}$$

where π_1, π_2, π_3 and π_4 are components of the column vector $\mathbf{\Pi}$. We now set $j = 1$ in equation (4.5.15), differentiate the resulting equation with respect to ϕ, and then form the jumps in the usual way across $\phi = 0$ to obtain

$$\mathbf{L}_o^{(1)} \mathbf{\Pi}_{,t'} + \left\{ \nabla(\mathbf{L}^{(1)} \mathbf{b}) \right\}_o \mathbf{\Pi} = 0, \tag{4.5.18}$$

where ∇ stands for the gradient operator with respect to the components of the vector $\mathbf{u}$. Equation (4.5.18), in view of the eigenvector $\mathbf{L}^{(1)}$, the column vector $\mathbf{b}$, and the relation (4.5.17) yields.

$$(\pi_2)_{,t'} + \left\{ \mu_o + \frac{m a_{f_o}}{2x} \right\} \pi_2 = 0, \tag{4.5.19}$$

where μ_o is same as in (4.5.13). Also, along the wave front $\phi = 0$ we have $x_{,t'} = \lambda^{(1)}$, which on differentiation with respect to ϕ, and on forming jumps across $\phi = 0$, yields

$$\frac{dX}{dt} = (\nabla \lambda^{(1)})_o \mathbf{\Pi} = \frac{(1 + \Gamma_o)}{2} \pi_2, \tag{4.5.20}$$

where Γ_o is same as in (4.5.11); in obtaining (4.5.20), we have made use of (4.5.17).

If we differentiate (4.5.16) with respect to t' and use (4.5.18) and (4.5.20), we find that s satisfies (4.5.12); thus, the law of propagation of weak discontinuities obtained using Jeffrey's method is in agreement with the Bernoulli's law found using the characteristic approach.

4.5.3   Singular surface approach

Among various approaches, the use of singular surface theory quickly leads to the results of general significance. This is an alternative approach, without marked difference in analytical extent, which leads to the same basic result concerning the evolutionary behavior of weak discontinuity waves. Here, we demonstrate the applicability of this method to study the development of jump discontinuities associated with the one-dimensional system (4.5.1).

Let $x = x(t)$, or for brevity, $\sum(t)$, denote the weak discontinuity wave across which the flow variables are essentially continuous but discontinuities in their derivatives are permitted. The description of the wave front $\sum(t)$ is such that its speed of propagation $G = dx/dt$ is always positive. The unperturbed field ahead of the wave is assumed to be uniform and at rest, and in a state of complete thermodynamic equilibrium. We infer that

$$p = p_o, u = u_o, S = S_o, q = q_o, a_f = a_{f_o}, \alpha = \alpha_o = 0, \omega = \omega_o = 0, \quad (4.5.21)$$

where a subscript o indicates a value in the medium just ahead of the wave front.

In a one-dimensional case, the geometric and kinematical conditions (4.1.1) of first and second order, stated in Section 4.1, reduce to

$$[f_{,x}] = d, \quad [f_{,t}] = -Gd, \quad [f_{,xx}] = \bar{d}, \quad [f_{,xt}] = -G\bar{d} + \frac{\delta d}{\delta t}, \quad (4.5.22)$$

where the quantity f may represent any of the variables p, u, S, q, α and ω. The square brackets stand for the value of the quantity enclosed immediately behind minus its value just ahead of the wave front $\sum(t)$. The quantities d and $\bar{d}$ are defined on the wave front $\sum(t)$, and the δ-time derivative is defined as $\delta f/\delta t = f_{,t} + Gf_{,x}$. Thus, the δ-time derivative of any quantity, which is defined on $\sum(t)$ is identical with the time derivative, d/dt, of that quantity following the wave front.

Taking jumps, across $\sum(t)$, in (4.5.1) and making use of (4.5.21), we get

$$G\xi = \rho_o a_{f_o}^2 s, \quad \xi = \rho_o Gs, \quad \zeta = 0 = \eta, \quad (4.5.23)$$

where $\xi = [p_{,x}]$, $s = [u_{,x}], \zeta = [S_{,x}]$ and $\eta = [q_{,x}]$ are quantities defined on $\sum(t)$.

The case $G = 0$, which corresponds to a material surface is discarded as uninteresting; we therefore assume that $G \neq 0$. Then it follows from (4.5.23) that $G = \pm a_{f_o}$; for an advancing wave, we shall take

$$G = a_{f_o}. \quad (4.5.24)$$

This is the speed with which the wave $\sum(t)$ propagates into the medium ahead. Equation $(4.5.23)_2$, in view of (4.5.24) yields

$$\xi = \rho_o a_{f_o} s. \quad (4.5.25)$$

If we differentiate $(4.5.1)_1$ and $(4.5.1)_2$ with respect to x, take jumps across $\sum(t)$ and then make use of (4.5.22) – (4.5.25), we get

$$\frac{\delta s}{\delta t} = \frac{1}{\rho_o}(\bar{\xi} - \rho_o a_{f_o}\bar{s}) - 2s(\mu_o + \frac{ma_{f_o}}{2x}) - (1 + \Gamma_o)s^2, \quad (4.5.26)$$

$$\frac{\delta s}{\delta t} = -\frac{1}{\rho_o}(\bar{\xi} - \rho_o a_{f_o}\bar{s}), \quad (4.5.27)$$

where $\overline{\xi} = [p_{,xx}], \overline{s} = [u_{,xx}]$, and μ_o and Γ_o are the same as in (4.5.10) and (4.5.11). Eliminating the term $(\overline{\xi} - \rho_o a_{f_o}\overline{s})$ from (4.5.26) and (4.5.27), and recalling the definition of δ-time derivative, we find that s satisfies the same Bernoulli's law (4.5.12).

Remark 4.5.1: We thus observe that the various approaches used for studying the propagation law of weak discontinuities are in agreement with the familiar Bernoulli's law. However, the singular surface method is more general in the sense that it is applicable to higher dimensions and quickly leads to the basic results concerning wave propagation.

4.6   Weak Nonlinear Waves in an Ideal Plasma

The propagation of one dimensional waves in a gaseous medium may be thought of as being produced by a moving piston. When the piston recedes or advances into the gas, not all parts of the gas are affected instantaneously. A wave proceeds from the piston into the gas, and the particles which have been reached by the wave front are disturbed from their initial state of rest. If this wave represents a continuous motion caused by the receding piston, the wave front propagates into the gas with local sound speed. But, if the piston moves into the gas, the characteristics, in general, coalesce to form a shock wave during propagation. An impulsive motion of the piston may produce an instantaneous unsteady shock which may grow or decay with time, depending on the condition of the undisturbed gas and the behavior of the piston. Here, we envisage the one dimensional planar motion of a plasma, which is assumed to be an ideal gas with infinite electrical conductivity and to be permeated by a magnetic field orthogonal to the trajectories of gas particles; the motion is conceived of as being produced by a piston moving with a small velocity as compared to the magnetoacoustic speed. We follow Chu [36] to study the main features of weak nonlinear waves, namely the expansion waves and shock waves. The trajectories of these waves and particle paths are determined, and the effects of magnetic field strength and the adiabatic heat exponent on the wave propagation are discussed.

The basic equations governing the motion are same as in (1.2.24) with $m = 0 = n$, and can be written as

$$\begin{aligned}
&\rho_{,t} + u\rho_{,x} + \rho u_{,x} = 0, \\
&u_{,t} + uu_{,x} + \rho^{-1}(p+h)_{,x} = 0, \\
&p_{,t} + up_{,x} + \rho a^2 u_{,x} = 0, \\
&h_{,t} + uh_{,x} + 2hu_{,x} = 0,
\end{aligned} \qquad (4.6.1)$$

where $h = \mu H^2/2$; all other symbols have the same meaning as in (1.2.24). We assume that the medium ahead of the wave front is in a uniform state at rest, and use the subscript o to denote this state.

Equations (4.6.1) constitute a hyperbolic system with four families of characteristics. Two of these characteristics, $dx/dt = u \pm c$, represent waves propagating with the magnetoacoustic speed $c = (a^2 + b^2)^{1/2}$ with $b = (2h/\rho)^{1/2}$ as the Alfven speed, and the remaining two, $dx/dt = u$, form a set of double characteristics representing the particle path or trajectory. We introduce the characteristic variables ϕ and ξ such that

$$\phi_{,t} + (u + c)\phi_{,x} = 0, \quad \psi_{,t} + u\psi_{,x} = 0.$$

Keeping in view the properties of ϕ and ψ, it immediately follows that

$$x_{,\phi} = ut_{,\phi}, \quad x_{,\psi} = (u + c)t_{,\psi}. \tag{4.6.2}$$

The transformation from x-t plane to the ϕ-ξ plane will be one to one if, and only if, the Jacobian

$$J = x_{,\phi}t_{,\psi} - x_{,\psi}t_{,\phi}, \tag{4.6.3}$$

does not vanish or become undefined anywhere. Hence, if F denotes any of the flow variables, p, ρ, u or h, then

$$F_{,x} = (F_{,\phi}t_{,\psi} - F_{,\psi}t_{,\phi})/J, \quad F_{,t} + uF_{,x} = -cF_{,\phi}t_{,\psi}/J. \tag{4.6.4}$$

Since the overlapping of fluid particles is prohibited from physical considerations, it implies that $t_{,\psi} \neq 0$. Consequently $J = 0$ if, and only if, $t_\phi = 0$ which corresponds to the situation when two adjoining characteristics merge into a shock wave.

In terms of the characteristic variables, the system (4.6.1) can be transformed into the following equivalent system

$$
\begin{aligned}
(\rho u_{,\phi} - c\rho_{,\phi})t_{,\psi} - \rho u_{,\psi}t_{,\phi} &= 0, \\
((p + h)_{,\phi} - \rho cu_{,\phi})t_{,\psi} - (p + h)_{,\psi}t_{,\phi} &= 0, \\
(cp_{,\phi} - \rho a^2 u_{,\phi})t_{,\psi} + \rho a^2 u_{,\psi}t_{,\phi} &= 0, \\
(ch_{,\phi} - 2hu_{,\phi})t_{,\psi} + 2hu_{,\psi}t_{,\phi} &= 0.
\end{aligned}
\tag{4.6.5}
$$

In equations (4.6.1), we nondimensionalize time t by an appropriate time t^* that characterizes the change in piston motion. It is easy to check that the form of equations (4.6.1) remains unchanged if the velocities u and c are nondimensionalized by c_o, the distance x by $c_o t^*$, the density ρ by ρ_o, the pressure p and magnetic pressure h by $\rho_o c_o^2$, and the sound speed a by c_o. Thus, without introducing new notations, the variables x, t, ρ, u, p, h etc., appearing in (4.6.1) will henceforth be regarded as dimensionless. The characteristic variables ϕ and ψ will also be dimensionless to begin with; moreover, in the undisturbed medium, we have $c = c_o = 1$ and $\rho = \rho_o = 1$.

Indeed, a wave is weak if the magnitudes of the dimensionless velocity, pressure, etc. are small compared to 1; in particular, the piston speed is small compared to c_o. Along the piston path, $x = F(t)$, we have $u = F'(t)$ where $F' = dF/dt$. We assume that F' is small and take it as $F'(t) = \epsilon f'(t)$ where

$\epsilon(<< 1)$ characterizes the amplitude of the disturbance and $f'(t) \sim O(1)$; here, ϵ may be regarded as the maximum piston velocity divided by c_o. The boundary conditions at the piston are

$$x = \epsilon f(\phi), \quad t = \phi, \quad u = \epsilon f'(\phi) \quad \text{at } \psi = 0. \tag{4.6.6}$$

If the wave front is an outgoing characteristic, the boundary conditions on it can be written as

$$\rho = \rho_o = 1, \quad c = c_o = 1, \quad u = 0, \quad a = a_o \; p = p_o, \quad h = h_o, \quad t = \psi \quad \text{at } \phi = 0. \tag{4.6.7}$$

As ϵ is a small parameter, we seek the solution of the piston problem in the form

$$z = z^{(o)}(\phi, \psi) + \epsilon z^{(1)}(\phi, \psi) + O(\epsilon^2), \tag{4.6.8}$$

where z may denote any of the dependent variables $\rho, u, p, h, x, t, a, c$ and $z^{(o)}$ represents the value of z in the undisturbed uniform region. Writing the dependent variables in (4.6.2), (4.6.6) and (4.6.7) in the form (4.6.8), and collecting terms of the zeroth order in ϵ, we get

$$\rho^{(o)} = 1, \quad u^{(o)} = 0, \quad p^{(o)} = p_o, \quad h^{(o)} = h_o, \quad a^{(o)} = a_o,$$

$$c^{(o)} = c_o = 1, \quad x^{(o)}_{,\phi} = 0, \quad x^{(o)}_{,\psi} = t^{(o)}_{,\psi}, \tag{4.6.9}$$

which, together with the boundary conditions

$$\begin{aligned} t^{(o)} &= \phi, \quad x^{(o)} = 0 \quad \text{at } \psi = 0, \\ t^{(o)} &= \psi \quad \text{at } \phi = 0, \end{aligned} \tag{4.6.10}$$

yield the zeroth order solution as

$$x^{(o)} = \psi, \quad t^{(o)} = \phi + \psi. \tag{4.6.11}$$

To this order, the characteristics in the physical plane are

$$x = \psi, \quad t = x + \phi. \tag{4.6.12}$$

In order to find the solution valid to the first order, we use (4.6.8) in (4.6.2) and (4.6.5), and collect terms of the order of ϵ; we thus obtain

$$x^{(1)}_{,\phi} = u^{(1)}, \quad x^{(1)}_{,\psi} = t^{(1)}_{,\psi} + u^{(1)} + c^{(1)},$$

$$\rho^{(1)}_{,\phi} - u^{(1)}_{,\phi} + u^{(1)}_{,\psi} = 0, \quad u^{(1)}_{,\phi} - (p^{(1)} + h^{(1)})_{,\phi} + (p^{(1)} + h^{(1)})_{,\psi} = 0, \tag{4.6.13}$$

$$p^{(1)}_{,\phi} - a_o^2 u^{(1)}_{,\phi} + u^{(1)}_{,\psi} = 0, \quad h^{(1)}_{,\phi} - 2h_o u^{(1)}_{,\phi} + 2h_o u^{(1)}_{,\psi} = 0.$$

It may be noticed that

$$a_o^2 + b_o^2 = 1, \quad c^{(1)} = a_o a^{(1)} + b_o b^{(1)}, \quad b_o^2 = 2h_o,$$

$$b_o b^{(1)} = h^{(1)} - h_o \rho^{(1)}, a^{(1)} = (a_o/2)((p^{(1)}/p_o) - \rho^{(1)}). \tag{4.6.14}$$

The first order boundary conditions from (4.6.6) are

$$x^{(1)} = f(\phi), \quad t^{(1)} = 0, \quad u^{(1)} = f'(\phi) \quad \text{at } \psi = 0. \tag{4.6.15}$$

Similarly, in view of (4.6.2) and (4.6.7), the first order boundary conditions at the characteristic front are

$$t^{(1)} = 0, \quad u^{(1)} = 0 \; p^{(1)} = 0, \quad \rho^{(1)} = 0, \quad h^{(1)} = 0 \quad \text{at } \phi = 0. \tag{4.6.16}$$

Solution of the system (4.6.13), satisfying (4.6.15) and (4.6.16) may be constructed by the method of Laplace transforms. If we define the Laplace transform of $z(\phi, \psi)$ with respect to ϕ by

$$\hat{z}(\xi, \psi) = \int_0^\infty z(\phi, \psi) e^{-\xi\phi} d\phi,$$

then we find that

$$\xi \hat{x}^{(1)} = \hat{u}^{(1)}, \quad \frac{d\hat{x}^{(1)}}{d\psi} = \frac{d}{d\psi} \hat{t}^{(1)} + \hat{u}^{(1)} + \hat{c}^{(1)}, \tag{4.6.17}$$

$$\xi(\hat{\rho}^{(1)} - \hat{u}^{(1)}) = -\frac{d}{d\psi} \hat{u}^{(1)}, \quad \xi(\hat{u}^{(1)} - \hat{p}^{(1)} - \hat{h}^{(1)}) = -\frac{d}{d\psi}(\hat{p}^{(1)} + \hat{h}^{(1)}), \tag{4.6.18}$$

$$\xi(\hat{p}^{(1)} - a_o^2 \hat{u}^{(1)}) = -\frac{d}{d\psi} \hat{u}^{(1)}, \quad \xi(\hat{h}^{(1)} - 2h_o \hat{u}^{(1)}) = -2h_o \frac{d\hat{u}^{(1)}}{d\psi}. \tag{4.6.19}$$

The transformed boundary conditions (4.6.15), in view of (4.6.17)$_1$ yield

$$\hat{x}^{(1)} = \hat{f}(\xi), \quad \hat{t}^{(1)} = 0, \quad \hat{u}^{(1)} = \xi \hat{f}(\xi) \quad \text{at } \psi = 0. \tag{4.6.20}$$

From equations (4.6.14), (4.6.18)$_1$ and (4.6.19), we obtain

$$\hat{h}^{(1)} = b_o^2 \hat{\rho}^{(1)}, \quad \hat{b}^{(1)} = (b_o/2) \hat{\rho}^{(1)}, \quad \hat{p}^{(1)} = \hat{\rho}^{(1)} - (1 - a_o^2) \hat{u}^{(1)}, \tag{4.6.21}$$

which together with (4.6.18) yield, on eliminating $\hat{\rho}^{(1)}$ and $d\hat{\rho}^{(1)}/d\psi$, the following second order linear homogeneous differential equation

$$\frac{d^2 \hat{u}^{(1)}}{d\psi^2} - \lambda \frac{d\hat{u}^{(1)}}{d\psi} = 0, \tag{4.6.22}$$

where $\lambda = \xi(1 + (1 + b_o^2)^{-1})$. The solution of (4.6.22), which satisfies (4.6.20) and remains bounded as $\psi \to \infty$, is

$$\hat{u}^{(1)} = \xi \hat{f}(\xi),$$

and hence (4.6.17)$_1$, (4.6.18)$_1$ and (4.6.21) imply that

$$\hat{x}^{(1)} = \hat{f}(\xi), \quad \hat{\rho}^{(1)} = \xi \hat{f}(\xi), \quad \hat{p}^{(1)} = a_o^2 \xi \hat{f}(\xi),$$

$$\hat{h}^{(1)} = b_o^2 \xi \hat{f}(\xi), \quad \hat{b}^{(1)} = (b_o/2) \xi \hat{f}(\xi). \tag{4.6.23}$$

Finally, equation $(4.6.17)_2$, in view of $(4.6.14)_2$ and $(4.6.20)_2$, implies that $\hat{t}^{(1)} = -\Lambda_o \xi \hat{f}(\xi) \psi$, where $\Lambda_o = \{(1+\gamma) + (2-\gamma)b_o^2\}/2$. Thus, we can easily obtain the complete solution for p, ρ, u, h, x and t; the solutions for u, x and t are

$$u(\phi, \psi) = \frac{\epsilon}{2\pi i} \int_{\alpha - i\infty}^{\alpha + i\infty} \xi \hat{f}(\xi) e^{\xi \phi} d\xi + O(\epsilon^2),$$

$$x = \psi + \frac{\epsilon}{2\pi i} \int_{\alpha - i\infty}^{\alpha + i\infty} \hat{f}(\xi) e^{\xi \phi} d\xi + O(\epsilon^2), \qquad (4.6.24)$$

$$t = \phi + \psi - \frac{\epsilon \Lambda_o \psi}{2\pi i} \int_{\alpha - i\infty}^{\alpha + i\infty} \xi \hat{f}(\xi) e^{\xi \phi} d\xi + O(\epsilon^2),$$

where α is a positive number, satisfying the conditions that all the singularities of these integrals are to the left of the line $\xi = \alpha$ in the complex plane. The trajectories of the outgoing waves and particle paths in the x-t plane are described by equations $(4.6.24)_2$ and $(4.6.24)_3$, which clearly exhibit the bending of the characteristics. If we retain only the zeroth order terms, equation $(4.6.24)_1$ becomes

$$u(x, t) = \frac{\epsilon}{2\pi i} \int_{\alpha - i\infty}^{\alpha + i\infty} \xi \hat{f}(\xi) e^{\xi(t-x)} d\xi + O(\epsilon^2),$$

which is obviously the solution of the linearized equations $(4.6.1)$.

4.6.1 Centered rarefaction waves

Here, we shall study the rarefaction waves caused by a receding piston motion; indeed, a piston receding from a gas at rest with speed which never decreases causes a rarefaction wave of particles moving toward the piston. Of particular interest is the case in which the acceleration of the piston from rest to a constant velocity takes place in an infinitely small time interval, i.e., instantaneously; then the family of characteristics forming the simple wave degenerates into a centered rarefaction wave. The problem can be considered as the special case of a receding piston $x = \epsilon f(t)$ with $f(t) = -\frac{1}{2}kt^2$ for $t < 1/k$, and $f(t) = g(t)$ for $t > 1/k$, where $g(1/k) = -1/(2k)$, and k is the dimensionless piston acceleration; in the limit $k \to \infty$, the piston velocity will change instantaneously from 0 to $-\epsilon$. We introduce a new characteristic level β, defined as $\beta = k\phi$ for $\phi < 1/k$, so that as ϕ varies from 0 to $1/k$, β changes from 0 to 1, and seek a solution to the problem in which an instantaneous change in the piston velocity generates a concentrated expansion wave that spreads out into an expansion fan in the course of propagation.

In terms of β, the governing equations for the zeroth order variables become

$$x_{,\beta}^{(o)} = 0, \quad x_{,\psi}^{(o)} = t_{,\psi}^{(o)}, \qquad (4.6.25)$$

along with the boundary conditions

$$t^{(o)} = \beta/k, \quad x^{(o)} = 0 \quad \text{at } \psi = 0 \quad \text{for } \beta < 1,$$
$$t^{(o)} = \psi \text{ at } \beta = 0. \tag{4.6.26}$$

Equations (4.6.25) and (4.6.26) admit the following solution

$$t^{(o)} = \psi + (\beta/k), \quad x^{(o)} = \psi. \tag{4.6.27}$$

The governing equations (4.6.13) for the first order variables, and the boundary conditions (4.6.15) and (4.6.16), when written in terms of β, become

$$x^{(1)}_{,\beta} = u^{(1)}/k, \quad x^{(1)}_{,\psi} = t^{(1)}_{,\psi} + u^{(1)} + c^{(1)}, \rho^{(1)}_{,\beta} - u^{(1)}_{,\beta} + k^{-1}u^{(1)}_{,\psi} = 0,$$
$$u^{(1)}_{,\beta} - (p^{(1)} + h^{(1)})_{,\beta} + k^{-1}(p^{(1)} + h^{(1)})_{,\psi} = 0,$$
$$p^{(1)}_{,\beta} - a_o^2 u^{(1)}_{,\beta} + k^{-1}u^{(1)}_{,\psi} = 0, \quad h^{(1)}_{,\beta} - 2h_o u^{(1)}_{,\beta} + (2h_o/k)u^{(1)}_{,\psi} = 0, \tag{4.6.28}$$

$$x^{(1)} = -\beta^2/2k, \quad t^{(1)} = 0, \quad u^{(1)} = -\beta \quad \text{at } \psi = 0 \text{ for } \beta < 1, \tag{4.6.29}$$

$$t^{(1)} = 0, \quad u^{(1)} = 0, \quad p^{(1)} = 0, \quad \rho^{(1)} = 0, \quad h^{(1)} = 0 \quad \text{at } \beta = 0. \tag{4.6.30}$$

The solution of the system (4.6.28) – (4.6.30) in the limit $k \to \infty$ is determined as

$$x^{(1)} = 0, \quad u^{(1)} = -\beta, \quad \rho^{(1)} = -\beta, \quad p^{(1)} = -a_o^2\beta, \quad h^{(1)} = -b_o^2\beta, \quad t^{(1)} = \Lambda\beta\psi, \tag{4.6.31}$$

where

$$\Lambda = \{(\gamma + 1) + (2 - \gamma)b_o^2\}/2. \tag{4.6.32}$$

Thus, the complete solution valid up to the first order of ϵ is

$$\rho = 1 - \epsilon\beta + O(\epsilon^2), \quad u = -\epsilon\beta + O(\epsilon^2), \quad p = p_o - \epsilon a_o^2\beta + O(\epsilon^2),$$

$$a = a_o(1 - \tfrac{\gamma-1}{2}\epsilon\beta) + O(\epsilon^2), \quad h = h_o - \epsilon b_o^2\beta + O(\epsilon^2), \tag{4.6.33}$$

$$x = \psi + O(\epsilon^2), \quad t = \psi(1 + \epsilon\beta\Lambda) + O(\epsilon^2).$$

Equation of the characteristics in the expansion fan, and the flow field in the physical plane can be obtained from (4.6.33) as

$$x = t(1 - \epsilon\beta\Lambda) + O(\epsilon^2), \quad u = \tfrac{1}{\Lambda}\left(\tfrac{x}{t} - 1\right) + O(\epsilon^2),$$

$$\rho = 1 + \tfrac{1}{\Lambda}\left(\tfrac{x}{t} - 1\right) + O(\epsilon^2), \quad a = a_o\left(1 + \tfrac{\gamma-1}{2\Lambda}\left(\tfrac{x}{t} - 1\right)\right) + O(\epsilon^2), \tag{4.6.34}$$

$$p = p_o + \tfrac{a_o^2}{\Lambda}\left(\tfrac{x}{t} - 1\right) + O(\epsilon^2), \quad h = h_o + \tfrac{b_o^2}{\Lambda}\left(\tfrac{x}{t} - 1\right) + O(\epsilon^2).$$

In the absence of magnetic field, $b_o \to 0$ and $\Lambda \to (\gamma + 1)/2$, equations $(4.6.34)_2$ and $(4.6.34)_4$ become

$$u = \frac{2}{\gamma + 1}\left(\frac{x}{t} - 1\right) + O(\epsilon^2), \quad a = a_o\left(\frac{\gamma - 1}{\gamma + 1}\frac{x}{t} + \frac{2}{\gamma - 1}\right) + O(\epsilon^2),$$

which, in terms of the dimensionless variables defined earlier, is precisely the solution for expansion waves obtained in [210].

4.6.2   Compression waves and shock front

Let us now consider the case when a piston, initially at rest, for $t \leq 0$, is pushed forward with a uniform velocity $\epsilon c_o(> 0)$ for $t > 0$. The fan is reversed and forms a multivalued region, which can be envisaged as a fold in the $x - t$ plane; indeed, it corresponds to an immediate breaking, and must be replaced by a shock wave propagating into the undisturbed medium. The differential equations governing the flow remain the same as in (4.6.1); in particular the first order variables $\rho^{(1)}, u^{(1)}, p^{(1)}, h^{(1)}, x^{(1)}$ and $t^{(1)}$ are governed by (4.6.13). The boundary conditions at the piston are given by

$$x^{(1)} = f(\phi), \quad t^{(1)} = 0, \quad u^{(1)} = 1 \text{ at } \psi = 0, \tag{4.6.35}$$

instead of (4.6.14), and the boundary conditions at the shock front, $\phi = \Phi(\psi)$ are the R-H conditions given by (see [96] and [27])

$$[\rho(u - s)] = 0, \quad [\rho u(u - s) + p + h] = 0,$$
$$\left[(u - s)(\tfrac{1}{2}\rho u^2 + \tfrac{1}{\gamma-1}p + h) + u(p + h)\right] = 0, \quad [h(u - s)^2] = 0, \tag{4.6.36}$$

where s is the shock speed. Since the shock path in the physical plane is represented by $x = x(\Phi, \psi), t = t(\Phi, \psi)$, the shock velocity s and the function $\Phi(\psi)$ are related to each other by

$$s = (x_{,\phi}\Phi'(\psi) + x_{,\psi})/(t_{,\phi}\Phi'(\psi) + t_{,\psi}), \tag{4.6.37}$$

where $\Phi'(\psi) \equiv d\Phi/d\psi$. It may be recalled that the variables ρ, u, p, h, x, t, etc., are all dimensionless. It is easy to see that the equations (4.6.36) remain unchanged if the shock velocity s is nondimensionalized by c_o. Thus, if $\delta \equiv [\rho]/\rho_o$, is the density strength of the shock, then it follows from $(4.6.36)_1$, $(4.6.36)_2$ and $(4.6.36)_4$ that

$$\rho = 1 + \delta = s/(s - u), \quad h = h_o(1 + \delta)^2, \quad p = p_o - h_o\delta(2 + \delta) + s^2\delta(1 + \delta)^{-1}. \tag{4.6.38}$$

Using (4.6.38) in the energy equation $(4.6.36)_3$, we find that

$$s^2 = \frac{2(1 + \delta)}{2 - (\gamma - 1)\delta}\left\{1 + \left(\frac{2 - \gamma}{2}\right)b_o^2\delta\right\}. \tag{4.6.39}$$

In the limit of vanishing magnetic field, i.e., $b_o \to 0$, (4.6.39) reduces to the corresponding gasdynamic result (see Whitham [210]). For a weak shock wave $\delta = 0(\epsilon)$, and consequently, to the first order of approximations in ϵ, we have

$$s = 1 + (\Lambda\delta/2) + O(\epsilon^2), \qquad (4.6.40)$$

where Λ is same as in (4.6.32). Thus, the shock conditions (4.6.38) valid up to the first order of approximation in ϵ are

$$\rho = 1 + \delta + O(\epsilon^2), \quad u = \delta + O(\epsilon^2),$$
$$h = (b_o^2/2)(1 + 2\delta) + O(\epsilon^2), \quad p = p_0 + \delta a_o^2 + O(\epsilon^2). \qquad (4.6.41)$$

Along with (4.6.41), the requirement that $t = \psi$, and (4.6.37) must be satisfied at the shock $\phi = \Phi(\psi)$, which is yet to be determined.

Now if $\epsilon \to 0$, the shock front tends to the characteristic $\phi = 0$; indeed, the equation of the shock path in the characteristic plane may be written as

$$\phi = \epsilon\Phi^{(1)}(\psi) + O(\epsilon^2), \qquad (4.6.42)$$

and a quantity $z(\phi, \psi)$ at the shock front, assuming that it has a Taylor expansion in ϕ, can be evaluated at $\phi = 0$, i.e.,

$$z|_{\phi=\Phi(\psi)} = z^{(o)}(0,\psi) + \epsilon\{z^{(1)}(0,\psi) + z_{,\phi}^{(0)}(0,\psi)\Phi^{(1)}(\psi)\} + O(\epsilon^2).$$

In view of (4.6.11), (4.6.37), (4.6.40), (4.6.41) and (4.6.42), it follows that the boundary conditions for the first order variables at the shock are

$$\rho^{(1)} = u^{(1)} = 2(s-1)/\Lambda, \quad p^{(1)} = 2a_o^2(s-1)/\Lambda, \quad h^{(1)} = 2b_o^2(s-1)/\Lambda,$$
$$t^{(1)} = 0, \quad x_{,\psi}^{(1)} = t_{,\psi}^{(1)} + (\Lambda/2)\rho^{(1)} + d\Phi^{(1)}/d\psi \text{ at } \phi = 0. \qquad (4.6.43)$$

Equations (4.6.13) together with the boundary conditions (4.6.35) and (4.6.43) can be solved by using Laplace transforms; indeed, the solution of (4.6.22), which satisfies the transformed boundary condition (4.6.35), i.e., $\hat{u}^{(1)} = 1/\xi$, and remains bounded as $\psi \to \infty$, is

$$\hat{u}^{(1)} = 1/\xi, \qquad (4.6.44)$$

and the complete solution, similar to (4.6.24), can easily be obtained. The shock locus can be obtained from (4.6.43)$_5$, which together with (4.6.13)$_2$, yields

$$d\Phi^{(1)}/d\psi = (2 - \Lambda)(u^{(1)}/2) + c^{(1)} \text{ at } \phi = 0. \qquad (4.6.45)$$

In view of the equations (4.6.14), equations (4.6.43) yield $c^{(1)} = (\Lambda - 1)u^{(1)}$ and, thus, (4.6.45) becomes

$$d\Phi^{(1)}/d\psi = (\Lambda/2)u^{(1)} \text{ at } \phi = 0. \qquad (4.6.46)$$

The value of $u^{(1)}$ at $\phi = 0$ can be obtained from (4.6.44) by making use of the Tauberian relation, i.e., $\lim_{\phi \to 0} u^{(1)} = \lim_{\xi \to \infty} \xi \hat{u}^{(1)}(\xi, \psi) = 1$, and thus (4.6.46)

yields $\Phi^{(1)} = (\Lambda/2)\psi$; the shock path in the characteristic plane is, therefore,

$$\phi = \epsilon(\Lambda/2)\psi. \tag{4.6.47}$$

The shock path in the physical plane can also be obtained by noting that $x = \psi + \epsilon x^{(1)} + O(\epsilon^2)$ and $t = \phi + \psi + \epsilon t^{(1)} + O(\epsilon^2)$, which imply that

$$\phi = t - x - \epsilon(t^{(1)} - x^{(1)}) + O(\epsilon^2), \tag{4.6.48}$$

and similarly $(4.6.13)_2$, in view of the foregoing results, implies that

$$t^{(1)} - x^{(1)} = -\Lambda\psi. \tag{4.6.49}$$

Thus, using (4.6.47) and (4.6.49) in (4.6.48), we obtain the shock path, to the first order in ϵ, in the $x - t$ plane as

$$x = t(1 + \epsilon\frac{\Lambda}{2}) + O(\epsilon^2), \tag{4.6.50}$$

which in the absence of magnetic field, i.e., $b_o \to 0$ reduces exactly to the one obtained in [210]. Since the quantity Λ depends on the dimensionless Alfven speed b_o and the specific heat ratio γ, it may be remarked that for $\gamma < 2$, an increase in b_o (i.e., the magnetic field strength) causes an increase in the shock speed to the first order, relative to what it would have been in the absence of the magnetic field; however for $\gamma > 2$, the behavior is just the opposite. It is interesting to note that for $\gamma = 2$, the shock speed to the first order is not at all influenced by the magnetic field strength. It may be pointed out that $\gamma = 2$ corresponds to an ideal plasma with transverse magnetic field, whereas $\gamma > 2$ is suitable for barotropic fluids in relativistic astrophysics and cosmology.

4.7   Relatively Undistorted Waves

Varley and Cumberbatch [206] introduced the method of relatively undistorted waves to take into account nonlinear phenomena which are governed by nonlinear equations. Based on this method, which places no restriction on the wave amplitude, Dunwoody [51] has discussed high frequency plane waves in ideal gases with internal dissipation. This method, which has been discussed in detail by Seymour and Mortell [165], proposes an expansion scheme that accounts for amplitude dispersion and shock formation.

The advantage of this method, which makes no assumption on the magnitude of a disturbance, lies in the fact that the solution can be obtained by solving ordinary differential equations; a study of this problem has been made by Radha and Sharma [146]. In order to illustrate this method, we consider disturbances in a one-dimensional unsteady flow of a relaxing gas in a duct of cross sectional area $A(x)$, where x is the distance along the duct. The gas molecules have only one lagging internal mode (i.e., vibrational relaxation) and the various transport effects are negligible. The governing equations, using summation convention, can be written in the form (see Clarke [37])

$$u^i_{,t} + A_{ij} u^j_{,x} + B_i = 0, \quad i,j = 1,2,3,4 \tag{4.7.1}$$

where u^i and $\mathcal{B}_i$ are the components of column vectors $\mathbf{u}$ and $\mathbf{B}$ defined as $\mathbf{u} = (\rho, u, \sigma, p)^{tr}$, $\mathbf{B} = (\rho u \Omega, 0, -Q, \rho a^2 u \Omega + (\gamma - 1)\rho Q)^{tr}$ with $\Omega = A'/A$. A_{ij} are the components of 4×4 matrix $\mathcal{A}$ with nonzero components $A_{11} = A_{22} = A_{33} = A_{44} = u$, $A_{12} = \rho$, $A_{24} = 1/\rho$ and $A_{42} = \rho a^2$. Here u is the particle velocity along the x-axis, t the time, ρ the density, p the pressure, σ the vibrational energy, γ the frozen specific heat ratio, $a = (\gamma p/\rho)^{1/2}$ the frozen speed of sound in the gas, and $A' \equiv dA/dx$. The quantity Q, which is a known function of p, ρ and σ, denotes the rate of change of vibrational energy. The situation $Q = 0$, corresponds to a physical process involving no relaxation; indeed, it includes both the cases in which the vibrational mode is either inactive or follows the translational mode according as the flow is either frozen ($\sigma = $ constant) or in equilibrium ($\sigma = \sigma^*$), where σ^* is the equilibrium value of σ evaluated at local p and ρ. Following Scott and Johannesen [163], the entity Q is given by

$$Q = (\sigma^* - \sigma)/\tau, \quad \sigma^* = \sigma_0 + c(\rho \rho_0)^{-1}(p\rho_0 - \rho p_0), \tag{4.7.2}$$

where the suffix 0 refers to the initial rest condition, and the quantities τ and c, which are respectively the relaxation time and the ratio of vibrational specific heat to the specific gas constant, are assumed to be constant. The solution vector $\mathbf{u}$ is said to define a relatively undistorted wave, if there exists a family of propagating wavelets $\phi(x,t) = $ constant, such that the magnitude of the rate of change of $\mathbf{u}$ moving with the wavelet is small compared with the magnitude of rate of change of $\mathbf{u}$ at fixed t. Let us consider the transformation, $x = x$, $t = T(x, \phi)$, from (x,t) to (x, ϕ) coordinate system, and let $\mathbf{u}(x,t) = \mathbf{U}(x, \phi)$. Then a relatively undistorted wave is defined by the relation $\|\mathbf{U}_{,x}\| \ll \|\mathbf{u}_{,x}\|$, where $\|\cdot\|$ denotes the Euclidean norm of a vector. Further, since $\mathbf{U}_{,x} = \mathbf{u}_{,x} + T_{,x}\mathbf{u}_{,t}$, in a relatively undistorted wave $\mathbf{u}_{,x} \simeq -T_{,x}\mathbf{u}_{,t}$ holds, and consequently

$$\|\mathbf{U}_{,x}\| \ll \|\mathbf{u}_{,t}\|. \tag{4.7.3}$$

A special example of a relatively undistorted wave is an acceleration front, where $[\mathbf{U}_{,x}] = 0$, while in general $[\mathbf{u}_{,t}] \neq 0$, and hence (4.7.3) is trivially satisfied.

In terms of independent variables (x, ϕ), equation (4.7.1) becomes

$$(A_{ij}(T_{,x}) - \delta_{ij})\, u^j_{,t} = A_{ij} U^j_{,x} + B_i, \tag{4.7.4}$$

where δ_{ij} is the Kronecker function. Equations (4.7.3) and (4.7.4) are compatible if $\|\mathcal{B}\| = O(\|\mathcal{A}U_{,x}\|)$, while $(T_{,x})^{-1}$ is an eigenvalue of $\mathcal{A}$. Otherwise, (4.7.4) would completely determine the $u^i_{,t}$ as linear forms in the $U^i_{,x}$, and (4.7.3) could not hold. Accordingly, in such waves, where $\det\left(\mathcal{A}_{ij} - \delta_{ij}(T_{,x})^{-1}\right) = 0$, U^i must satisfy the compatibility condition

$$L_i\left(\mathcal{A}_{ij}U^j_{,x} + \mathcal{B}_i\right) = 0, \tag{4.7.5}$$

for every left eigenvector $\mathbf{L}$ of $\mathcal{A}$ corresponding to the eigenvalue $(T_{,x})^{-1}$. Thus, the essential idea underlying this method is based on a scheme of successive approximations to the system (4.7.4) which, to a first approximation, is replaced by

$$\left(\mathcal{A}_{ij} - \delta_{ij}(T_{,x})^{-1}\right)U^j_{,\phi} = 0. \tag{4.7.6}$$

Then, if $\mathbf{R} = (R_i)$ is the right eigenvector of $\mathcal{A}$ corresponding to the eigenvalue $(T)^{-1}_{,x}$, equation (4.7.6) implies that to a first approximation

$$U^i_{,\phi} = k(\phi, x)R_i, \tag{4.7.7}$$

for some scalar $k(\phi, x)$. It is important to appreciate that equations (4.7.7) are approximate and, in general, cannot be integrated to obtain relations in U^i which are uniformly valid for all time. The terms which have been neglected in arriving at (4.7.7) will, in general, ultimately produce first order contributions to U^i.

The matrix $\mathcal{A}$ has eigenvalues $u \pm a$ and u (with multiplicity two); here we are concerned with the solution in the region $x > x_0$, where a motion consisting of only one component wave, associated with the eigenvalue $(T_{,x})^{-1} = u + a$, is perturbed at the boundary $x = x_0$ by an applied pressure $p(x_0, t) = \Pi(t)$. It may be noticed that for nonlinear systems, there is, in general, no superposition principle so that when more than one wave mode is excited, the propagation of the individual component waves cannot be calculated independently. Consequently, the problem involving nonlinear interaction of component waves needs a different approach. The left and right eigenvectors of $\mathcal{A}$ corresponding to the eigenvalue

$$(T_{,x})^{-1} = u + a \tag{4.7.8}$$

are

$$\mathbf{L} = (0, \rho a, 0, 1), \quad \mathbf{R} = (\rho, a, 0, \rho a^2)^{tr}. \tag{4.7.9}$$

4.7.1   Finite amplitude disturbances

Let us consider the situation when the disturbance, which is headed by the front $\phi(x, t) = 0$, is moving into a region, where prior to its arrival the gas is in a uniform state at rest with $u = 0$, $p = p_0$, $\rho = \rho_0$ and $\sigma = \sigma_0$. It is possible to choose the label, ϕ, of each wavelet so that $\phi = t$ on $x = x_0$; consequently the boundary conditions for p and T become

$$p = \Pi(\phi), \quad T = \phi \quad \text{at} \quad x = x_0. \tag{4.7.10}$$

To a first approximation, conditions in the wave region, associated with $(T_{,x})^{-1} = u + a$, are determined by the differential relations (4.7.7), which can be formally integrated subject to the uniform reference values $u = 0$, $p = p_0$, $\rho = \rho_0$ and $\sigma = \sigma_0$ on the leading front $\phi = 0$. Thus we have

$$\rho = \rho_0(p/p_0)^{1/\gamma}, \quad u = 2a_0(\gamma - 1)^{-1}\left\{(p/p_0)^{(\gamma-1)/2\gamma} - 1\right\},$$
$$a = a_0(p/p_0)^{(\gamma-1)/2\gamma}, \quad \sigma = \sigma_0, \tag{4.7.11}$$

which hold at any x on any wavelet $\phi = $ constant. Equation (4.7.5), in view of (4.7.9)$_1$ and (4.7.11), provides the following transport equations for the variation of p and T at each wavelet $\phi = $ constant.

$$p_{,x} + \frac{(\gamma - 1)\rho_0}{4a_0F(p)}\left(\frac{p}{p_0}\right)^{\frac{1}{\gamma}}\left\{\frac{2a_0^3\Omega}{\gamma - 1}\left(\frac{p}{p_0}\right)^{\frac{\gamma-1}{\gamma}}\left(\left(\frac{p}{p_0}\right)^{\frac{(\gamma-1)}{2\gamma}} - 1\right) + (\gamma - 1)Q\right\} = 0,$$
$$\tag{4.7.12}$$

$$T_{,x} = (\gamma - 1)/2a_0F(p), \tag{4.7.13}$$

where $F(p) = \{((\gamma + 1)/2)(p/p_0)^{(\gamma-1)/2\gamma} - 1\}$. Equation (4.7.12), on using (4.7.2) and (4.7.11), may be integrated using the boundary condition (4.7.10)$_1$ to give $p = P(x, \Pi(\phi))$; once p is known, equation (4.7.13) may be integrated, subject to (4.7.10)$_2$, to determine $t = T(x, \phi)$. Subsequently $p(x, t)$ may be obtained, and hence $\rho(x, t), u(x, t)$ and $a(x, t)$. As a matter of fact, the integration of (4.7.13) leads to the determination of the location of ϕ wavelets in the form $T = \phi + \int_{x_0}^{x} \Lambda(s, \Pi(\phi))ds$, where $\Lambda = (\gamma - 1)/(2a_0F(p(s, \Pi(\phi))))$. From this result, it follows immediately that a shock forms on the wavelet ϕ_s at the point x_s, where $1 + \Pi' \int_{x_0}^{x_s} \Lambda_{,\Pi}(X, \Pi(\phi_s))dX = 0$.

The above results indicate that both the amplitude dispersion and shock formation along any wavelet depend on the amplitude, $\Pi(\phi)$, carried by that wavelet.

4.7.2 Small amplitude waves

In the small amplitude limit, equations (4.7.12) and (4.7.13) can be linearized about the uniform reference state $p = p_0, \rho = \rho_0, \sigma = \sigma_0, u = 0$, to yield

$$p_{1,x} + (\alpha + \Omega/2)p_1 = 0, \quad T_{,x} = \left\{1 - (\gamma + 1)(2\rho_0a_0^2)^{-1}p_1\right\}a_0^{-1}, \tag{4.7.14}$$

where p_1 is the small perturbation of the equilibrium value p_0 and

$$\alpha = (\gamma - 1)^2c/2\gamma a_0\tau,$$

which serves as the amplitude attenuation rate on account of relaxation; it may be noted that α is identical with the absorption rate defined by Johannesen and Scott [87] and Scott and Johannesen [163].

The ideal gas case corresponds to $\alpha = 0$. The boundary conditions for p_1 and T which follow from (4.7.10) can be rewritten as

$$p_1 = \hat{\Pi}(\phi), \quad T = \phi \quad \text{at} \quad x = x_0, \tag{4.7.15}$$

where $\left|\hat{\Pi}(\phi)\right| = |\Pi(\phi) - p_0| \ll 1$. Equations (4.7.14), together with (4.7.15), yield on integration

$$p_1 = \hat{\Pi}(\phi)\psi(x), \tag{4.7.16}$$
$$a_0(T - \phi) = x - x_0 - (\gamma + 1)(\hat{\Pi}(\phi)/2\rho_0 a_0^2)J(x_0, x, \alpha), \tag{4.7.17}$$

where $J(x_0, x, \alpha) = \displaystyle\int_{x_0}^{x} (A(s)/A_0)^{-1/2} \exp(-\alpha(s - x_0))\, ds$, and

$$\psi(x) = (A/A_0)^{-1/2} \exp(-\alpha(x - x_0)).$$

Equation (4.7.16) implies that for each wavelet $\phi = $ constant, which decays exponentially, the attenuation factor is independent of the amplitude $\hat{\Pi}(\phi)$ carried by the wavelet. However, conditions at any x on a wavelet $\phi_1 = $ constant are determined by the signal carried by ϕ_1, and are independent of the precursor wavelets $0 \le \phi < \phi_1$.

It may be recalled that for plane ($m = 0$) and radially symmetric ($m = 1, 2$) flow configurations, $(A/A_0) = (x/x_0)^m$; and therefore the integral J converges to a finite limit J_0 as $x \to \infty$, i.e.,

$$\lim_{x \to \infty} J(x_0, x, \alpha) \equiv J_0 = \begin{cases} 1/\alpha, & \text{(plane)} \\ (\pi x_0/\alpha)^{1/2} \exp(\alpha x_0) erfc(\sqrt{\alpha x_0}), & \text{(cylindrical)} \\ x_0 E(\alpha x_0) \exp(\alpha x_0), & \text{(spherical)} \end{cases} \tag{4.7.18}$$

where $erfc(x) = 2\pi^{-1/2} \int_x^{\infty} \exp(-t^2)\, dt$ and $E(x) = \int_x^{\infty} t^{-1} \exp(-t)\, dt$ are, respectively, the complementary error function and the exponential integral. Thus J can be expressed as

$$J = J_0(1 - K(x)), \tag{4.7.19}$$

where

$$K(x) = \begin{cases} \exp(-\alpha(x - x_0)), & \text{(plane)} \\ erfc(\sqrt{\alpha x})/erfc(\sqrt{\alpha x_0}), & \text{(cylindrical)} \\ E(\alpha x)/E(\alpha x_0), & \text{(spherical)}. \end{cases}$$

Equation (4.7.17) indicates that a shock first forms at (x_s, ϕ), where the minimum value of x (i.e., x_s) is given by the solution of

$$H(x) \equiv 1 - (\hat{\Pi}'(\phi)/b)(1 - K(x)) = 0, \tag{4.7.20}$$

where $b = 2\rho_0 a_0^3/(\gamma + 1)J_0 > 0$ and $\hat{\Pi}'(\phi) \equiv d\hat{\Pi}/d\phi$; since the expression $(1 - K(x))$ monotonically increases from 0 to 1 as x increases from x_0 to ∞, it follows that a shock can only occur if $\hat{\Pi}' > b > 0$, and subsequently $H(x)$

first vanishes at that wavelet ϕ where $\hat{\Pi}'(\phi)$ is greatest. Expression for the shock formation distance on the leading wavefront $\phi = 0$, where (4.7.20) is exact, was pointed out by Johannesen and Scott [87]. It may be noted that the relatively undistorted approximation is valid only if

$$\left| \hat{\Pi}' / \hat{\Pi} \right| \gg \left| (\alpha + \Omega/2) \left\{ 1 + (\gamma + 1)(2\rho_0 a_0^2)^{-1} \hat{\Pi} \psi(x) \right\} a_0 H \right|, \qquad (4.7.21)$$

which, in fact, corresponds to the slow modulation approximation (see Varley and Cumberbatch [204]). As discussed above, a shock wave may be initiated in the flow region, and once it is formed, it will propagate by separating the portions of the continuous region. When the shock is weak, its location can be found from the equal area rule (see Whitham [210])

$$2 \int_{\phi_1}^{\phi_2} \hat{\Pi}(t) \, dt = (\phi_2 - \phi_1) \left\{ \hat{\Pi}(\phi_1) + \hat{\Pi}(\phi_2) \right\}, \qquad (4.7.22)$$

where ϕ_1 and ϕ_2 are the wavelets ahead of and behind the shock. For a weak shock propagating into an undisturbed region, where $\hat{\Pi}(\phi_1) = 0$ for $\phi_1 \leq 0$, equation (4.7.22), on using (4.7.17) and (4.7.19), becomes

$$\int_0^{\phi_2} \hat{\Pi}(t) \, dt = (\gamma + 1)\hat{\Pi}^2(\phi_2)(1 - K(x))J_0/4\rho_0 a_0^3. \qquad (4.7.23)$$

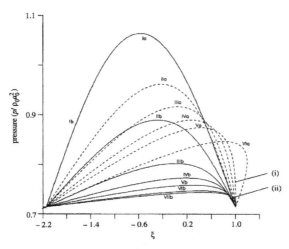

Figure 4.7.1: The variation of the dimensionless pressure against the dimensionless variable ξ (defined as $(x - a_o t)/x_o$, using the initial profile Ia (defined in (4.7.25)). The distortion of the profile is delineated at various distances before and after shock formation on the leading wavelet, $\hat{\phi} = 0$ in cylindrically ($m = 1$) and spherically ($m = 2$) symmetric flow configurations of a nonrelaxing gas ($\hat{\alpha} = 0$) for $\hat{\delta} = 0.35$ and $\gamma = 1.4$. Shock forms at (i) $\hat{x}_s = 4.79$ (cylindrical) and (ii) $\hat{x}_s = 10.81$ (spherical). For $m = 1$, $\hat{x} = 1$ (Ia), $\hat{x} = 2$ (IIa), $\hat{x} = 3$ (IIIa), $\hat{x} = 4$ (IVa), $\hat{x} = 4.79$ (Va), $\hat{x} = 7$ (VIa); for $m = 2$, $\hat{x} = 1$ (Ib), $\hat{x} = 2$ (IIb), $\hat{x} = 4$ (IIIb), $\hat{x} = 6$ (IVb), $\hat{x} = 8$ (Vb), $\hat{x} = 10.81$ (VIb), $\hat{x}_s = 12$ (VIIb).

This equation, in the limit $\alpha \to 0$, yields a result which agrees fully with the result obtained by Whitham [210] for nonrelaxing gases. Assuming that the integral on the left-hand side of (4.7.23) is bounded, it follows that for sufficiently large x, $\hat{\Pi}(\phi_2)$ stays in proportion to $J^{-1/2}$. Consequently, in view of (4.7.13) and (4.7.16), the pressure jump, $[p]$, across the shock, which is defined as the measure of shock amplitude, decays like

$$[p] \propto \begin{cases} \exp(-\hat{\alpha}\hat{x}), & \text{(plane)} \\ \hat{x}^{-1/2} \exp(-\hat{\alpha}\hat{x}), & \text{(cylindrical)} \\ \hat{x}^{-1} \exp(-\hat{\alpha}\hat{x}), & \text{(spherical)} \end{cases}$$

where $\hat{\alpha} = \alpha x_0$ and $\hat{x} = x/x_0$. For a nonrelaxing gas ($\alpha = 0$), we find that as $x \to \infty$, the shock decays like

$$[p] \sim x_0 \begin{cases} \hat{x}^{-1/2}, & \text{(plane)}; \\ \hat{x}^{-3/4}, & \text{(cylindrical)}; \\ \hat{x}^{-1}(\ln \hat{x})^{-1/2}, & \text{(spherical)}. \end{cases} \qquad (4.7.24)$$

These asymptotic results for nonrelaxing gases are in full accord with the earlier results (see Whitham [210]).

In order to trace the early history of shock decay after its formation on the leading wave front $\phi = 0$, we consider a special case in which the disturbance at the boundary $x = x_0$ is a pulse defined as

$$\hat{\Pi}(\phi) = \begin{cases} 0, & \phi < 0 \\ \delta \sin(a_0\phi/x_0), & 0 < \phi < \pi x_0/a_0 \\ 0, & \phi > \pi x_0/a_0 \end{cases} \quad ; \quad \delta > 0. \qquad (4.7.25)$$

It may be recalled that $H(x)$, in equation (4.7.20), first vanishes on that wavelet for which $\hat{\Pi}'(\phi)$ has a maximum value greater than b, i.e., $\delta a_0/x_0 > b$; consequently the shock first forms on the wavelet $\phi = 0$ at a distance $x = x_s$, nearest to x_0, given by the solution of

$$E(x) \equiv ((\gamma + 1)/2)(1 - K(x))\hat{\delta}\hat{J}_0 = 1, \qquad (4.7.26)$$

where $\hat{\delta} = \delta/\rho_0 a_0^2$ and $\hat{J}_0 = J_0/x_0$ are the dimensionless constants. The distortion of the pulse, defined in (4.7.25), is shown in Figs 1 and 2. The usual steepening of the compressive phase and flattening of the expansive phase of the wavelet are quite evident from the distorted profiles. The depression and flattening of the peaks with increasing $\hat{x}$, which become even more pronounced on account of relaxation or the wavefront geometry, indicate that the disturbance is undergoing a general attenuation. Equation (4.7.20) shows that for $\hat{\alpha} = 0$, the pressure profile develops a vertical slope, thereby indicating the appearance of a shock at $\hat{x} = 4.79$ and 10.81 in cylindrically and spherically symmetric flows respectively, while for $\hat{\alpha} = 0.05$, shocks in respective flow configurations develop at $\hat{x} = 5.31$ and 20.92. Thus, the presence of relaxation or an increase in the wavefront curvature both serve to delay the onset of a shock. Equation (4.7.23), in view of (4.7.25), then simplifies to give ϕ_2 on the shock by the following relation,

$$\sin \hat{\phi}_2 = 2(E(x) - 1)^{1/2}/E(x), \tag{4.7.27}$$

where $\hat{\phi}_2 = \phi_2 a_0/x_0$ is the dimensionless variable. Equation (4.7.27), together with (4.7.16) and (4.7.25), implies

$$[\hat{p}] = 2\hat{\delta}\hat{x}^{-m/2}\exp(-\hat{\alpha}(\hat{x} - 1))(E(x) - 1)^{1/2}/E(x), \tag{4.7.28}$$

where $\hat{p} = p/\rho_0 a_0^2$ is the dimensionless pressure. Equation (4.7.28) shows that the shock after its formation on $\phi = 0$, at $x = x_s > x_0$, grows to a maximum strength at $x = x_1 > x_s$, where x_1 is given by the solution of $E(x) = 2$, and then decays ultimately in proportion to $x^{-m/2}\exp(-\alpha x)$ as concluded above.

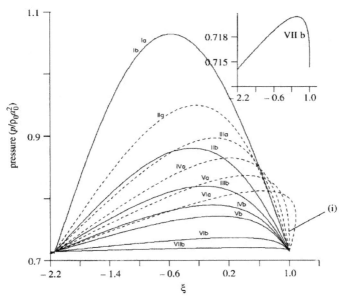

Figure 4.7.2: Development of the dimensionless pressure against ξ, using the initial profile Ia (defined in (4.7.25)). The distortion of the profile is delineated at various distances before and after shock formation on the leading wavelet for $\hat{\delta} = 0.35$, $\hat{\alpha} = 0.05$ and $\gamma = 1.4$. The figure in the inset shows the development on the wavehead $\phi = 0$, for $m = 2$ when the profile Vb develops a vertical slope signifying the appearance of a shock at $\hat{x}_s = 20.92$ in a relaxing gas. Shock forms at (i) $\hat{x}_s = 5.31$ (cylindrical). For $m = 1$, $\hat{x} = 1$ and (ii) $\hat{x}_s(Ia)$, $\hat{x} = 2$ (IIa), $\hat{x} = 3$ (IIIa), $\hat{x} = 4$(IVa), $\hat{x} = 5.31$ (Va), $\hat{x} = 7$ (VIa); for $m = 2$, $\hat{x} = 1$ (Ib), $\hat{x} = 2$ (IIb), $\hat{x} = 3$ (IIIb), $\hat{x} = 4$ (IVb), $\hat{x} = 5$ (Vb), $\hat{x} = 10$ (VIb), $\hat{x}_s = 20.92$ (VIIb).

Let us now consider a special case in which the small disturbance at the boundary, $x = x_0$, has a periodic wave form given by

$$\hat{\Pi}(\phi) = \bar{\delta}\sin(\hat{\phi}), \tag{4.7.29}$$

where $\overline{\delta} < 0$ and $\hat{\phi} = a_0\phi/x_0$, and consider the development over one cycle $0 \le \hat{\phi} \le 2\pi$. In this case, the shock first forms on the wavelet $\hat{\phi} = \pi$ at a distance $x = x_s$, nearest to x_0, given by the solution of (4.7.20); this can occur, of course, if $|\overline{\delta}|a_0/x_0 > b$. Equations (4.7.17) and (4.7.22) are satisfied on the shock, if $\hat{\phi}_1 + \hat{\phi}_2 = 2\pi$ and $\hat{\phi}_1 - \hat{\phi}_2 = 2\theta$, where θ is given by the solution of

$$\theta/\sin\theta = (\gamma+1)(1 - K(x))\hat{J}_0|\tilde{\delta}|/2, \qquad (4.7.30)$$

with $\tilde{\delta} = \overline{\delta}/\rho_0a_0^2$. The discontinuity in $\hat{p}$ at the shock is, therefore, given by

$$[\hat{p}] = 2|\tilde{\delta}|(x/x_0)^{-m/2}\exp(-\alpha(x - x_0))\sin\theta, \qquad (4.7.31)$$

which shows that the shock starts with zero strength corresponding to $\theta \to 0$ at $x = x_s$, given by the solution of (4.7.30). The shock amplitude decays to zero as $\theta \to \theta_m$, where θ_m is given by the solution of $\theta_m/\sin(\theta_m) = (\gamma+1)|\tilde{\delta}|\hat{J}_0/2$. The shock strength grows for θ lying in the interval $0 < \theta < \theta_*$, whereas it decays over the interval $\theta_* < \theta < \theta_m$, thus exhibiting a maximum corresponding to $\theta = \theta_*$ at a distance $x = x_*$, where both θ_* and x_* can be determined using (4.7.30) and the relation

$$4(\hat{\alpha} + m/2\hat{x}_*)(\sin\theta_* - \theta_*\cos\theta_*) = (\gamma+1)|\tilde{\delta}|\hat{x}_*^{-m/2}\sin(2\theta_*)\exp(-\hat{\alpha}(\hat{x}_* - 1)).$$

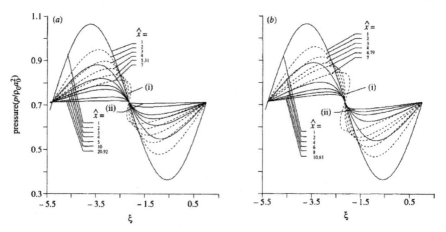

Figure 4.7.3: (a) The variation of the dimensionless pressure against ξ using the initial profile defined in (4.7.29). The distortion of the profile is exhibited at various distances before and after shock formation on the wavelet, $\hat{\phi} = \pi$, in cylindrically and spherically symmetric flow in a relaxing gas for $\hat{\delta} = -0.35, \hat{\alpha} = 0.05$ and $\gamma = 1.4$. Shock forms at (i) $\hat{x}_s = 5.31$ (cylindrical) and (ii) $\hat{x}_s = 20.92$ (spherical). (b) development of the pressure profile against ξ using the same initial profile (as in (a)), at the wavehead $\hat{\phi} = \pi$, for $\hat{\delta} = -0.35, \hat{\alpha} = 0$ and $\gamma = 1.4$. Shock forms at (i) $\hat{x}_s = 4.79$ (cylindrical) and (ii) $\hat{x}_s = 10.81$ (spherical).

The shock decays ultimately with $\theta = \theta_m$, $x \to \infty$, according to the law

$$[\hat{p}] \sim \lambda\hat{x}^{-m/2}\exp(-\hat{\alpha}\hat{x}), \qquad (4.7.32)$$

where $\lambda = 4\theta_m/(\gamma + 1)\hat{J}_0$. However, in the absence of relaxation, it follows from (4.7.30) and (4.7.31) that as $\theta \to \pi$, $x \to \infty$, the shock decays like

$$[\hat{p}] \sim 4\pi(\gamma + 1)^{-1} \begin{cases} \hat{x}^{-1}, & \text{(plane)}; \\ (2\hat{x})^{-1} & \text{(cylindrical)}; \\ (\hat{x}\ln\hat{x})^{-1}, & \text{(spherical)}. \end{cases} \tag{4.7.33}$$

The development of the pressure profile with initial disturbance given by (4.7.29), and the subsequent shock formation at the wavehead $\hat{\phi} = \pi$ are exhibited in Figs. 3a (with relaxation) and 3b (without relaxation), showing in effect that the influence of relaxation is to delay the onset of a shock wave. The evolutionary behavior of the pressure profile before and after the shock formation, exhibited in Fig. 4.7.3a, follows a slightly different pattern from that depicted in Fig. 4.7.3b, in the sense that the profile, which eventually folds into itself, develops concavity with the peak slightly advanced. The evolutionary behavior of shocks evolving from profile (4.7.29) are depicted in Fig. 4.7.4; indeed, the shock after its formation at $x = x_s$ grows to a maximum strength at $x = x_*$, and then decays according to the law (4.7.32) or (4.7.33) depending on whether the gas is relaxing or nonrelaxing respectively. However, in the absence of relaxation, a shock resulting from (4.7.29) decays faster than that evolving from the pulse (4.7.25).

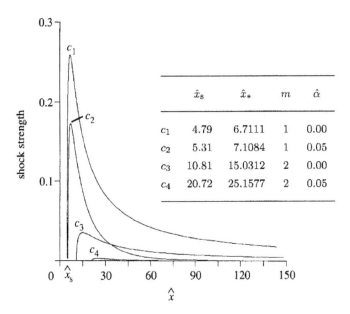

	$\hat{x}_s$	$\hat{x}_*$	m	$\hat{\alpha}$
c_1	4.79	6.7111	1	0.00
c_2	5.31	7.1084	1	0.05
c_3	10.81	15.0312	2	0.00
c_4	20.72	25.1577	2	0.05

Figure 4.7.4: Growth and decay of a shock wave which appears first at $\hat{x} = \hat{x}_s$ on the wavelet $\hat{\phi} = \pi$; comparison is made with the behavior from the same initial and boundary data for a nonrelaxing ($\hat{\alpha} = 0$) gas. The effects of relaxation and the wavefront curvature on the shock formation distance ($\hat{x}_s$), and the distance $\hat{x}_*$ at which the shock strength attains maximum strength are exhibited; $\hat{\delta} = -0.35$.

4.7.3   Waves with amplitude not-so-small

We now seek to explore the predictions of subsection 4.6.1 by considering the amplitude limit to be not so small; in fact, here we extend the analysis of the preceding subsection to the next order by including the nonlinear quadratic terms in the perturbed flow quantities p_1, ρ_1 etc, which were otherwise ignored in the subsection 4.6.2 while writing equations (4.7.14). In this instance, equations (4.7.12) and (4.7.13) reduce to

$$p_{1,x} + (\alpha + \Omega/2)p_1 - q(x)p_1^2 = 0, \tag{4.7.34}$$

$$T_{,x} = \{1 - (\gamma + 1)(2\rho_0 a_0^2)^{-1}p_1 + 3(\gamma + 1)^2(8\rho_0^2 a_0^4)^{-1}p_1^2\}/a_0, \tag{4.7.35}$$

where $q(x) = \{\alpha\gamma + (3 - \gamma)\Omega/4\}/2\rho_0 a_0^2$. Equations (4.7.34) and (4.7.35), together with the boundary conditions for p_1 and T at $x = x_0$, yield on integration

$$p_1 = \hat{\Pi}(\phi)\psi(x)\left\{1 - \hat{\Pi}(\phi)M(x)\right\}^{-1}, \tag{4.7.36}$$

$$a_0(T - \phi) = (x - x_0) + \frac{\gamma + 1}{2\rho_0 a_0^2}\int_{x_0}^{x}\left(\frac{3(\gamma + 1)}{4\rho_0 a_0^2}p_1^2 - p_1\right)dx, \tag{4.7.37}$$

where $M(x) = \int_{x_0}^{x}\psi(x')q(x')\,dx'$ and ϕ is the parameter distinguishing the wavelets from each other. Equations (4.7.36) and (4.7.37) indicate that in contrast to the results of subsection 4.6.2, both the rate at which the amplitude varies on any wavelet and the time taken to form a shock are influenced by the amplitude of the signal carried by the wavelet. Indeed, for $\hat{\Pi}M < 0$ (respectively, > 0) the amplitude decays more rapidly (respectively, slowly) than that predicted in subsection 4.6.2; the computed results are shown in Figs 5(a,b). The results computed for small amplitude disturbances in subsection 4.6.2 are also incorporated into Figures 5(a) and 5(b) for the sake of comparison and completeness. The numerical results indicate that the shock arrival time on a particular wavelet increases as compared to the case discussed in subsection 4.6.2. Behind the shock the wavelets are determined by (4.7.37). According to Pfriem's rule, the shock velocity $dT/dx\,|_s$ of a weak shock is given by the average of the characteristic speeds ahead of and behind the shock. Thus, to the present approximation

$$dT/dx\,|_s = \{1 - (\gamma + 1)(4\rho_0 a_0^2)^{-1}(p_1 - 3(\gamma + 1)p_1^2/4\rho_0 a_0^2)\}/a_0, \tag{4.7.38}$$

where $p_1 = p_1(x, \phi_s)$ denotes the value at position x on the shock. Evaluating (4.7.37) along the shock, we obtain the shock trajectory $T = T(x)$ and, thus, an alternative expression for the shock speed

$$\frac{dT}{dx}\bigg|_s = \frac{d\phi_s}{dx}\left\{1 - \frac{\gamma + 1}{2\rho_0 a_0^3}\int_{x_0}^{x}p_{1,\phi}\,dx + \frac{3(\gamma + 1)^2}{4\rho_0^2 a_0^5}\int_{x_0}^{x}p_1 p_{1,\phi}\,dx\right\}$$

$$+ \frac{1}{a_0}\left\{1 - \frac{\gamma + 1}{2\rho_0 a_0^2}p_1 + \frac{3(\gamma + 1)^2}{8\rho_0^2 a_0^4}p_1^2\right\}. \tag{4.7.39}$$

Eliminating $dT/dx\,|_s$ between (4.7.38) and (4.7.39), we obtain a differential equation for the unknown ϕ_s, which can be integrated numerically subject to the condition, $x = x_s$ at $\phi=0$. Having determined ϕ_s at different locations x, the shock strength at these respective locations, and hence the evolutionary behavior of shock decay, can be determined from (4.7.36).

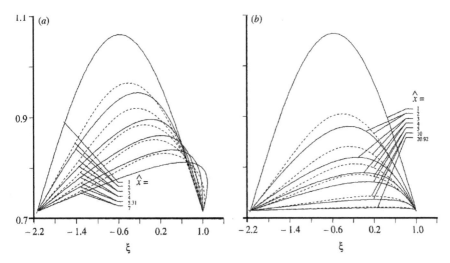

Figure 4.7.5: (a) solution for pressure in the pulse region of a cylindrically symmetric flow of a relaxing gas at various distances when the wave amplitude is not-so-small (see the dashed lines). Comparison is shown with the corresponding situation when the wave amplitude is small (see the solid lines); the initial profile is defined in (4.7.25). (b) Solution for pressure in the pulse region of a spherically symmetric flow of a relaxing gas at various distances when the disturbance amplitude is not-so-small (see the dashed lines). Comparison is shown with the corresponding situations when the wave amplitude is small (see the solid lines); the initial profile is defined in (4.7.25). For both (a) and (b), $\hat{\alpha} = 0.05, \hat{\delta} = 0.35, \gamma = 1.4$.

Chapter 5

Asymptotic Waves for Quasilinear Systems

In this chapter, we shall assume the existence of appropriate asymptotic expansions, and derive asymptotic equations for hyperbolic PDEs. These equations display the qualitative effects of dissipation or dispersion balancing nonlinearity, and are easier to study analytically. In this way, the study of complicated models reduces to that of models of asymptotic approximations expressed by a hierarchy of equations, which facilitate numerical calculations; this is often the only way that progress can be made to analyze complicated systems. Essential ideas underlying these methods may be found in earlier publications; see for example Boillat [19], Taniuti, Asano and their coworkers ([7], [194]), Seymour and Varley [166], Germain [61], Roseau [155], Jeffrey and Kawahara [82], Fusco, Engelbrecht and their coworkers ([58], [59]), Cramer and Sen [43], Kluwick and Cox [94], and Cox and Kluwick [42]. An account of some of the rigorous results which deal with convergence of such expansions may be found in [47].

5.1 Weakly Nonlinear Geometrical Optics

Here we shall discuss the behavior of certain oscillatory solutions of quasi-linear hyperbolic systems based on the theory of weakly nonlinear geometrical optics (WNGO), which is an asymptotic method and whose objective is to understand the laws governing the propagation and interaction of small amplitude high frequency waves in hyperbolic PDEs; this method describes asymptotic expansions for solutions satisfying initial data oscillating with high frequency and small amplitude. The propagation of the small amplitude waves, considered over long time-intervals, is referred to as weakly nonlinear. Most waves are of rather small amplitude; indeed, small amplitude high frequency short waves are frequently encountered. Linearized theory satisfactorily describes such waves only for a finite time; after a sufficiently long time-interval, the cumulative nonlinear effects lead to a significant change in the wave field. the basic principle involved in the methodology used for treating weakly nonlinear waves lies in a systematic use of the method of multiple scales. Each dissipative mechanism present in the flow, such as rate dependence of the medium, or geometrical dissipation due to nonplanar wave fronts, or inhomogeneity of the medium, defines a local characteristic length (or time) scale, that arises in a natural way. Indeed, WNGO is based on the assumption that the wave length of the wave is much smaller than any other characteristic length scale in the problem. When this assumption is satisfied, i.e., the time (or length) scale defined by the dissipative mechanism is large compared with the time (or length) scale associated with the boundary data, the wave is referred to as a short wave, or a high frequency wave. For instance, consider the motion excited by the boundary data $u(0, t) = af(t/\tau)$ on $x = 0$, where a represents the size of the boundary data, while τ is the applied period or pulse length. When we normalize the boundary data, by a suitable nondimensional quantity and t by the time scale τ_r, which is defined by the dissipative mechanism present in the flow, the geometrical acoustics limit then corresponds to the high frequency condition $\epsilon = \tau/\tau_r \ll 1$; in fact, there is a region in the neighborhood of the front where the nonlinear convection associated with the high frequency characteristics is important. For both significant nonlinear distortion and dissipation, in this high frequency limit, we must have the nondimensional amplitude $|a| = O(\epsilon)$. In terms of these normalized variables, the boundary condition becomes $u(0, t) = \epsilon g(t/\epsilon)$, which describes oscillations of small amplitude ϵ and of frequency ϵ^{-1}. For data of size ϵ, the method of WNGO yields approximations, which are valid on time intervals typically of size ϵ^{-1}. Following the pioneering work of Landau [99], Lighthill [108], Keller [91] and Whitham [210], a vast amount of literature has emerged on the development, both formal and rigorous, in the theory of WNGO. However, the development of these methods through systematic self-consistent perturbation schemes in one and several space dimensions is due to Chouqet-Bruhat [34], Varley and Cumberbatch ([204], [206]), Parker [140], Mortell and Varley [128], Seymour and Mortell [165], Hunter and Keller [75], Fusco [60], Mazda and Rosales [115], and Joly, Metivier and Rauch [88]. For results based on numerical computations of model equations of weakly nonlinear ray theory, the reader is referred to Prasad [145]. Using the systematic procedure, alluded to and applied by the above mentioned authors, we study here certain aspects of WNGO, which are largely based on our papers ([146], [171], [173], [190], and [192]).

5.1.1   High frequency processes

Consider a quasilinear system of hyperbolic PDEs in a single space variable

$$\mathbf{u}_{,t} + \mathbf{A}(\mathbf{u})\mathbf{u}_{,x} + \mathbf{b}(\mathbf{u}) = 0, \, -\infty < x < \infty, t > 0 \qquad (5.1.1)$$

where $\mathbf{u}$ and $\mathbf{b}$ are n-component vectors and $\mathbf{A}$ is a $n \times n$ matrix. Let $\mathbf{u}_o$ be a known constant solution of (5.1.1), such that $\mathbf{b}(\mathbf{u}_o) = 0$. We consider small amplitude variations in $\mathbf{u}$ from the equilibrium state $\mathbf{u} = \mathbf{u}_o$, which are of the size ϵ, described earlier; and look for a small amplitude high frequency wave solution of (5.1.1), representing a single wave front, and valid for times of the order of $O(1/\epsilon)$. Then the WNGO ansatz for the situation described above is the formal expansion for the solution of (5.1.1)

$$\mathbf{u} = \mathbf{u}_o + \epsilon \mathbf{u}_1(x, t, \theta) + \epsilon^2 \mathbf{u}_2(x, t, \theta) + \dots \tag{5.1.2}$$

where θ is a fast variable defined as $\theta = \phi(x,t)/\epsilon$ with ϕ as the phase function to be determined. The wave number k and wave frequency ω are defined by $k = \phi_{,x}$ and $\omega = -\phi_{,t}$. By considering the Taylor expansion of $\mathbf{A}$ and $\mathbf{b}$ in the neighborhood of $\mathbf{u}_o$, and taking into account (5.1.2), equation (5.1.1) implies that

$$O(\epsilon^o) : (\mathbf{A}_o - \lambda \mathbf{I})\mathbf{u}_{1,\theta} = 0$$
$$O(\epsilon^1) : (\mathbf{A}_o - \lambda \mathbf{I})\mathbf{u}_{2,\theta} = -(\nabla \mathbf{A})_o \mathbf{u}_1 \mathbf{u}_{1,\theta} - \{\mathbf{u}_{1,t} + \mathbf{A}_o \mathbf{u}_{1,x} + (\nabla \mathbf{b})_o \mathbf{u}_1\}\phi_{,x}^{-1},$$
$$\tag{5.1.3}$$

where the subscript o refers to the evaluation at $\mathbf{u} = \mathbf{u}_o$, $\mathbf{I}$ is the $n \times n$ unit matrix, $\lambda = -\phi_{,t}/\phi_{,x}$, and ∇ is the gradient operator with respect to the components of $\mathbf{u}$. The equation $(5.1.3)_1$ implies that for a particular choice of the eigenvalue λ_o (assuming that it is simple),

$$\mathbf{u}_1 = \pi(x, t, \theta)\mathbf{R}_o, \tag{5.1.4}$$

where π is a scalar oscillatory function to be determined, and $\mathbf{R}_o$ is the right eigenvector of $\mathbf{A}_o$ corresponding to the eigenvalue λ_o. The phase $\phi(x,t)$ is determined by

$$\phi_{,t} + \lambda_o \phi_{,x} = 0. \tag{5.1.5}$$

It may be noticed that if $\phi(x,0) = x$, then (5.1.5) implies that $\phi(x,t) = x - \lambda_o t$.

Let $\mathbf{L}_o$ be the left eigenvector of $\mathbf{A}_o$ corresponding to its eigenvalue λ_o, satisfying the normalization condition $\mathbf{L}_o \cdot \mathbf{R}_o = 1$. Then if $(5.1.3)_2$ is multiplied by $\mathbf{L}_o$ on the left, we obtain along the characteristic curves associated with (5.1.5) the following transport equation for π ;

$$\pi_{,\tau} + \nu(\pi^2/2)_{,\theta} = -\mu\pi, \tag{5.1.6}$$

where $\partial/\partial\tau = \partial/\partial t + \lambda_o \partial/\partial x$, and

$$\mu = \mathbf{L}_o \cdot \nabla \mathbf{b}|_o \cdot \mathbf{R}_o, \quad \nu = \mathbf{L}_o \cdot \nabla \mathbf{A}|_o \cdot \mathbf{R}_o \mathbf{R}_o. \tag{5.1.7}$$

If the initial condition for π is specified, i.e., $\pi|_{\tau=0} = \pi_o(x_o, \theta_o)$ with $x_o = x|_{\tau=o}$ and $\theta_o = \epsilon^{-1}\phi|_{\tau=0}$, then the minimum time taken for the smooth solution to breakdown can be computed explicitly. In fact, the initial conditions lead to a shock only when $\nu\partial\pi_o/\partial\theta_o < 0$ and $-\nu\partial\pi_o/\partial\theta_o > \mu$; in passing, we remark that the presence of source term $\mathbf{b}$ in (5.1.1) makes the solution exist for a longer time relative to what it would have been in the absence of a source term. Further, if (5.1.1) has an associated conservative form, then (5.1.6), which is the proper conservation form, can be used to study the propagation of weak shocks (see [75]).

5.1.2 Nonlinear geometrical acoustics solution in a relaxing gas

The foregoing asymptotic analysis can be used to study the small amplitude high frequency wave solution to the one dimensional unsteady flow of a relaxing gas described by the system (4.6.1), i.e., $u^i{}_{,t} + \mathcal{A}_{ij} u^j{}_{,x} + \mathcal{B}_i = 0$, $i, j = 1, 2, 3, 4$, where the symbols have the same meaning, defined earlier. Here we are concerned with the motion consisting of only one component wave associated with the eigenvalue $\lambda = u + a$. The medium ahead of the wave is taken to be uniform and at rest. The left and right eigenvectors of $\mathcal{A}$ corresponding to this eigenvalue are

$$\mathbf{L} = (0, 1/(2a), 0, 1/(2\rho a^2)), \quad \mathbf{R} = (\rho, a, 0, \rho a^2), \tag{5.1.8}$$

and the phase function $\phi(x, t)$ is given by $\phi(x, t) = x - a_o t$, where the subscript o refers to the uniform state $\mathbf{u}_o = (\rho_o, 0, \sigma_o, p_o)$ ahead of the wave. The transport equation for the wave amplitude π is given by (5.1.6) with coefficients μ and ν determined as

$$\mu = (\alpha + (a_o \Omega)/2) \text{ and } \nu = (\gamma + 1)a_o/2, \tag{5.1.9}$$

where α is the same as in $(4.6.14)_1$. The characteristic field associated with π in (5.1.6) is defined by the equations

$$dx/dt = a_o, \quad d\theta/dt = ((\gamma + 1)a_o \pi)/2. \tag{5.1.10}$$

In view of (5.1.10), equation (5.1.6) can be written as $d\pi/dt = -a_o(\alpha + (\Omega/2))\pi$ along any characteristic curve belonging to this field, and yields on integration

$$\pi = \pi^o(x_o, \theta_o)(A/A_o)^{-1/2} \exp\{-\alpha a_o t\}, \tag{5.1.11}$$

along the rays $x - a_o t = x_o$ (constant). We look for an asymptotic solution of the hyperbolic system (4.6.1)

$$\mathbf{u} = \mathbf{u}_o + \epsilon \mathbf{u}_1(x, t, \theta) + O(\epsilon^2), \tag{5.1.12}$$

satisfying the small amplitude oscillatory initial condition

$$u(x, 0) = \epsilon g(x, x/\epsilon) + O(\epsilon^2), \tag{5.1.13}$$

where g is smooth with a compact support; indeed, expansion (5.1.12) with $\mathbf{u}_1$ given by (5.1.4), where the wave amplitude π is given by (5.1.11), is uniformly valid to the leading order until shock waves have formed in the solution (see Majda [116]). Using (5.1.11) in $(5.1.10)_2$ we obtain the family of characteristic curves parametrized by the fast variable θ_o as

$$\theta - \frac{(\gamma + 1)}{2} a_o \pi^o(x_o, \theta_o) \int_0^t \left(\frac{A(x_o + a_o \hat{t})}{A(x_o)} \right)^{-1/2} \exp\left(-\alpha a_o \hat{t}\right) d\hat{t} = \theta_o. \tag{5.1.14}$$

It may be noticed that equations (5.1.11) and (5.1.14) are similar to the equations (4.6.16) and (4.6.17), and therefore the discussion with regard to the shock formation and its subsequent propagation follows on parallel lines. It may be recalled that for plane $(m = 0)$ and radially symmetric $(m = 1, 2)$ flow configurations, $(A/A_o) = (x/x_o)^m$, and therefore the integral in (5.1.14) converges to a finite limit as $t \to \infty$. Thus, an approximate solution (5.1.12), satisfying (5.1.13), is given by

$$u = u_0 + \epsilon \mathbf{R}_o \pi^o(x_o, \theta_o)(A(x_o + a_o t)/A(x_o))^{-1/2} \exp(-a_o \alpha t),$$

where $\mathbf{R}$ is given by $(5.1.8)_2$; the fast variable θ_o, given by (5.1.14), is chosen such that $\theta_o = x/\epsilon$ at $t = 0$, and the initial value of π is determined from (5.1.13) as $\pi^o(x, \xi) = g(x, x/\epsilon)/a_o$. This completes the solution of (4.6.1) and (5.1.13); any multivalued overlap in this solution has to be resolved by introducing shocks into the solution.

5.2   Far Field Behavior

When the characteristic time τ associated with the boundary data is large compared with the time scale τ_r defined by the dissipative mechanism present in the medium (i.e., $\delta = \tau_r/\tau \ll 1$), the situation corresponds to the low frequency propagation condition. This means that the time and distances considered are large in comparison to the relaxation time or relaxation length. Since at large distances, away from the source, any nonlinear convection is associated with the low frequency characteristics, and as the principal signal in this region is centered on the equilibrium or low frequency characteristics, it is possible to introduce a reduced system of field equations which provides an approximate description of the wave process. Based on Whitham's ideas, it was shown by Fusco [60] that for a quasilinear first order system of PDEs with several space variables involving a source term, it is possible to introduce a reduced system which, in an asymptotic way, brings out the dissipative effects produced by the source term against the typical nonlinear steepening of the waves. In fact, the wave motion is asymptotically described by a transport equation of Burger's type that holds along the characteristic rays of the reduced system. We illustrate this procedure for the one dimensional nonequilibrium gas flow described by the system (4.6.1).

As the principal signal in the far field region is centered on the equilibrium characteristic, the system (4.6.1) is approximated by the following reduced system.

$$\rho_{,t} + u\rho_{,x} + \rho u_{,x} + \Omega \rho u = 0, \quad u_{,t} + u u_{,x} + \rho^{-1} p_{,x} = 0,$$
$$p_{,t} + u p_{,x} + \rho a_*^2(u_{,x} + \Omega u) = 0, \quad Q(p, p, \sigma) = 0, \tag{5.2.1}$$

where $a_* = (\partial p / \partial \rho)|^{1/2}_{S, \sigma = \sigma_*}$ is the equilibrium speed of sound with S as specific entropy of the gas.

In order to study the influence of nonequilibrium relaxation in (4.6.1) on the wave motion, associated with (5.2.1), we consider the following stretching of the independent variables $\tilde{x} = \delta^2 x$, $\tilde{t} = \delta^2 t$. When expressed in terms of $\tilde{x}$ and $\tilde{t}$, and then suppressing the tilde sign, the system (4.6.1) yields the following set of equations

$$\rho_{,t} + u\rho_{,x} + \rho u_{,x} + \Omega \rho u = 0, \quad u_{,t} + uu_{,x} + \rho^{-1} p_{,x} = 0,$$
$$p_{,t} + up_{,x} + \gamma p(u_{,x} + \Omega u) = -(\gamma - 1)\rho Q \delta^{-2}, \quad (\sigma_{,t} + u\sigma_{,x}) = Q\delta^{-2}.$$
$$(5.2.2)$$

In the limit $\delta \to 0$, the above system yields the reduced system (5.2.1), and, therefore, the transformed system (5.2.2) may be regarded as a perturbed problem of an equilibrium state characterized by (5.2.1); see Fusco [60].

We now look for an asymptotic solution of (5.2.2) exhibiting the character of a progressive wave, i.e.,

$$f(x, t) = f_0 + \delta f_1(x, t, \theta) + \delta^2 f_2(x, t, \theta) + \cdots, \qquad (5.2.3)$$

where f may denote any of the dependent variables ρ, u, σ and p; f_0 refers to the known constant state, $\theta = \phi(x, t)/\delta$ is a fast variable, and $\phi(x, t)$ is the phase function to be determined.

By introducing (5.2.3) and the Taylor's expansion of Q about the uniform state $(\rho_0, 0, \sigma_0, p_0)$ into the transformed equations (5.2.2)$_1$ and (5.2.2)$_2$ and canceling the coefficients of δ^0, δ^1, we obtain the following system of first order partial differential equations for the first and second order variables

$$O(\delta^0) : \rho_{1,\theta}\phi_{,t} + \rho_0 u_{1,\theta}\phi_{,x} = 0, \quad \rho_0 u_{1,\theta}\phi_{,t} + p_{1,\theta}\phi_{,x} = 0,$$
$$O(\delta^1) : \rho_{2,\theta}\phi_{,t} + \rho_0 u_{2,\theta}\phi_{,x} = -\rho_{1,t} - \rho_0 u_{1,x} - \frac{m\rho_0 u_1}{x} + (u_1\rho_{1,\theta} + \rho_1 u_{1,\theta})\phi_{,x},$$
$$O(\delta^1) : \rho_0 u_{2,\theta}\phi_{,t} + p_{2,\theta}\phi_{,x} = -\rho_1 u_{1,\theta}\phi_{,t} - \rho_0 u_1 u_{1,\theta}\phi_{,x} - p_{1,x} - \rho_0 u_{1,t}.$$
$$(5.2.4)$$

Similarly, equations (5.2.2)$_3$ and (5.2.2)$_4$, on equating the coefficients of δ^0 and δ^{-1}, yield the following equations

$$O(\delta^0) : p_{1,\theta}\phi_{,t} + \rho_0 a_o^2 u_{1,\theta}\phi_{,x} = -(\gamma - 1)\rho_0 Q_2, \quad \sigma_{1,\theta}\phi_{,t} = Q_2,$$
$$O(\delta^{-1}) : \sigma_1 = c(p_1 - (a_o^2/\gamma)\rho_1)/\rho_0, \qquad (5.2.5)$$

where $Q_2 = \left\{ \dfrac{c}{\tau \rho_o^2} \left(\dfrac{p_o \rho_1^2}{\rho_o} - p_1 \rho_1 \right) + \dfrac{cp_2}{\tau \rho_o} - \dfrac{cp_o \rho_2}{\tau \rho_o^2} - \dfrac{\sigma_2}{\tau} \right\}.$

Eliminating Q_2 between (5.2.5)$_1$ and (5.2.5)$_2$, and using (5.2.5)$_3$ in the resulting equation, we get

$$\gamma\{1 + (\gamma - 1)c\}p_{1,\theta}\phi_{,t} + \rho_0 a_o^2 \gamma u_{1,\theta}\phi_{,x} - (\gamma - 1)ca_o^2 \rho_{1,\theta}\phi_{,t} = 0. \qquad (5.2.6)$$

Equation (5.2.6), together with (5.2.4)$_1$ and (5.2.4)$_2$, constitutes a system of three equations for the unknowns $\rho_{1,\theta}, u_{1,\theta}$ and $p_{1,\theta}$. The necessary and sufficient condition for this system to have a nontrivial solution is that the determinant of the coefficient matrix must vanish, i.e., $-\phi_t/\phi_x = 0, \pm\Gamma$, where $\Gamma = a_0 \left\{(c\gamma + \gamma - c)/(c\gamma^2 - c\gamma + \gamma)\right\}^{1/2}$. For $-\phi_t/\phi_x = +\Gamma$, the phase function ϕ is determined as $\phi(x,t) = t - ((x - x_o)/\Gamma)$ when $\phi(x_o, t) = t$; as the vector of unknowns is collinear to the right null vector of the coefficient matrix, we get

$$\rho^{(1)} = \Gamma^{-2}p^{(1)}, \quad u^{(1)} = (\rho_0\Gamma)^{-1}p^{(1)}, \quad \sigma^{(1)} = (\gamma-1)c\left\{\rho_0(\gamma + (\gamma - 1)c)\right\}^{-1}p^{(1)}.$$
$$(5.2.7)$$

It may be noticed that Γ is a characteristic speed related to the reduced system (5.2.1). The system of equations for the second order variables, on multiplying by the left null vector of the coefficient matrix, corresponding to the value $-\phi_{,t}/\phi_{,x} = \Gamma$, and taking into account the relations (5.2.7), yields the following transport equation for $p^{(1)}$ in the moving set of coordinates X and ξ,

$$\partial_X p^{(1)} - \lambda\, p^{(1)}\partial_\xi p^{(1)} + mp^{(1)}/2X = \omega\partial_{\xi\xi}^2 p^{(1)}, \qquad (5.2.8)$$

where $\partial_X = \partial_x + \Gamma^{-1}\partial_t$ and the nonlinear and dissipation coefficients λ and ω are given by

$$\lambda = (\gamma/\rho_0\Gamma\Lambda_1)\left\{\gamma + 1 + 2c(\gamma - 1)\right\}, \quad \omega = 2c\tau\gamma\Gamma a_0^2\left\{(\gamma - 1)/\Lambda_1\right\}^2,$$

with $\Lambda_1 = 2a_0^2(c\gamma + \gamma - c)$.

Equation (5.2.8) is known as the generalized Burgers equation which allows us to study in detail various effects that appear in the propagation of plane ($m = 0$), cylindrical ($m = 1$) and spherical ($m = 2$) waves in a dissipative medium with a quadratic nonlinearity.

In contrast to the high frequency nonlinear solution, discussed earlier, a significant feature of the low frequency solution in the far field is its continuous structure, i.e., any convective steepening is always diffused by the dissipative nature of the relaxation. The reader is referred to Crighton and Scott [44] and Manickam et al. [117] for a detailed discussion of the analytical and numerical solutions of such an equation.

Remarks 5.2.1: In a special case where the coefficient ω vanishes in (5.2.8), the source term $\mathcal{B}$ involved in (4.7.1) does not cause dissipation; in this situation, the asymptotic development (5.2.3) needs to be modified in order to see how the higher order terms in (4.7.1) may influence the wave motion associated with the characteristic speed of the reduced system. The transport equation for the wave amplitude in this case is the generalized Korteveg-de Vries (K dV) equations. While considering the diffraction of a weakly nonlinear high frequency wave in a direction transverse to its rays for a first order quasilinear system involving a source term, it can be shown that the wave amplitude is governed by the Zabolotskaya-Khokhlov equation or by a modified Kadomtsev-Petviashvili (KP) equation (see [59] and [60]).

5.3   Energy Dissipated across Shocks

We have noticed that the classical solutions of nonlinear hyperbolic conservation laws, in one or several unknowns, break down in a finite time beyond which the solutions have to be interpreted in the sense of distributions, known as weak solutions. These weak solutions often contain discontinuities supported along lower dimensional manifolds. The strength of the discontinuity is constrained through the Rankine-Hugoniot formula, which relates the jump in the solution to the unit normal to the discontinuity locus. Recently, motivated by their study on gravity currents, Montgomery and Moodie [125] have analyzed the effect of singular forcing terms on the Rankine-Hugoniot conditions in systems involving one space dimension. The singular forcing function considered, is a sum of a surface density and a volume density and the latter does not contribute to the jump relation. Their results to several space dimensions were generalized in [191] using pull back of distributions; indeed, the analysis in [191] sheds some light on the mechanism through which the terms involving the volume density cancel out, thereby not contributing to the jump conditions. The analysis in this section is related to another problem in the system of conservation laws namely, the manipulation of conservation laws through multiplication by polynomial nonlinearities. The resulting systems contain distributional terms supported along singularity loci; these may be regarded as singular forcing terms considered in [125]. Here we derive an expression for these singular terms which have been interpreted as the energy dissipated at the shocks. As an application, the expression for these singular terms is obtained from the transport equation governing the propagation of small amplitude high frequency waves in hyperbolic conservation laws.

5.3.1   Formula for energy dissipated at shocks

We consider an autonomous scalar conservation law without source term for a scalar valued function u on a domain Ω in $\mathbb{R}^{n+1}$ namely,

$$\sum_{j=0}^{n} \partial_j F_j(u) = 0. \tag{5.3.1}$$

Let the system (5.3.1), when multiplied by $P(u)$, a polynomial or a smooth function of one variable, transform into

$$\sum_{j=0}^{n} \partial_j(\tilde{F}_j(u)) = 0, \tag{5.3.2}$$

for certain densities $\tilde{F}_j(u)$ defined on Ω. The systems (5.3.1) and (5.3.2) are equivalent for smooth solutions u of (5.3.1), namely the identity

$$P(u) \sum_{j=0}^{n} \partial_j F_j(u) = \sum_{j=0}^{n} \partial_j \tilde{F}_j(u),$$

holds for all smooth functions u. However, the class of weak solutions of this systems involving discontinuities, are different since the corresponding Rankine-Hugoniot conditions differ. To restore the equivalence of the two systems for a class of distributional solutions which are piecewise smooth with a jump discontinuity along a locus $\Phi(x) = 0$, we modify the equation (5.3.2) by adding a distributional term to (5.3.2) namely,

$$P(u) \sum_{j=0}^{n} \partial_j F_j(u) = \sum_{j=0}^{n} \partial_j (\tilde{F}_j(u)) + E, \tag{5.3.3}$$

where the distribution E is supported on $\Phi(x) = 0$. In the context of nonlinear geometrical optics, this singular term is interpreted as the energy dissipated across the shock locus $\Phi(x) = 0$. We proceed to determine explicitly the distribution E. Let ϕ be any arbitrary test function, then (5.3.3) gives

$$\sum_{j=0}^{n} \int_{\Omega} P(u) \partial_j F_j(u) \phi = \sum_{j=0}^{n} \langle P(u) \partial_j F_j(u), \phi \rangle$$

$$= \langle E, \phi \rangle + \sum_{j=0}^{n} \langle \partial_j \tilde{F}_j(u), \phi \rangle, \tag{5.3.4}$$

in the sense of distributions. We proceed to simplify the term $\sum_{j=0}^{n} \langle \partial_j \tilde{F}_j(u), \phi \rangle$ on the right-hand side of (5.3.4):

$$\sum_{j=0}^{n} \langle \partial_j \tilde{F}_j(u), \phi \rangle = -\int_{\Omega} \sum_{j=0}^{n} \tilde{F}_j(u) \partial_j \phi$$

$$= -\int_{\Phi=0} \sum_{j=0}^{n} [\tilde{F}_j] n_j \phi dS + \int_{\Omega} \sum_{j=0}^{n} \partial_j \tilde{F}_j(u) \phi, \tag{5.3.5}$$

where $[f(u)]$ denotes the jump of $f(u)$ across the locus of discontinuity $\Phi = 0$, $\hat{n} = (n_1, n_2, \ldots, n_n)$ is the unit normal vector to $\Phi = 0$ and dS is the area measure on the surface. Since the function u is of class C^1 up to the boundary on either side of the discontinuity locus, the left-hand side of (5.3.4) transforms as

$$\sum_{j=0}^{n} \int_{\Omega} P(u) \partial_j F_j(u) \phi = \sum_{j=0}^{n} \int_{\Omega} \partial_j \tilde{F}_j(u) \phi, \tag{5.3.6}$$

and so (5.3.4) and (5.3.5) prove that the distribution E is a smooth density supported on the discontinuity locus given by

$$E : \phi \mapsto \int_{\phi=0} \phi \sum_{j=0}^{n} [\tilde{F}_j(u)] n_j dS. \tag{5.3.7}$$

We now prove a general result concerning the distribution E arising out of multiplying (5.3.1) by $P(u)$.

Theorem 5.3.1 *Assume that the conservation law*

$$\partial_t F_0(u) + \partial_x F_1(u) + G(x)u = 0,$$

on multiplication by $P(u)$ becomes

$$\partial_t \tilde{F}_0(u) + \partial_x \tilde{F}_1(u) + G(x)uP(u) + E = 0.$$

The distribution E, supported on the discontinuous locus $t = \Psi(x)$, is given by the density

$$- \left([\tilde{F}_0] \frac{[F_1]}{[F_0]} - [\tilde{F}_1] \right) \Phi^*(\delta_0), \tag{5.3.8}$$

where $\Phi^*(\delta_0)$ denoted the pull back of δ_0 by Φ with $\Phi(x,t) = t - \Psi(x)$.

Proof: In the special case of one space variable, the discontinuity locus is taken to be the graph of the function $t = \Psi(x)$. So, $\hat{n}dS = (-\Psi', 1)dx$ which on substitution in (5.3.7) and using the Rankine-Hugoniot condition (see (5.3.12) below), gives (5.3.8). □

Remark 5.3.1: The distribution E is of order zero and so involves only the pull back of δ_0. Presumably for higher order equations, the distribution E may be of higher order involving the pull back of derivatives of δ_0.

5.3.2   Effect of distributional source terms

Recall the formula for pull back of a distribution by a submersion [74]

$$\partial_j \Phi^* H = (\partial \Phi_j) \Phi^* \delta_0 = \frac{(\partial_j \Phi) dS}{|\nabla_x \Phi|}, \tag{5.3.9}$$

where ∇_x is the spatial gradient operator, and H denotes the Haviside function so that $\Phi^*(H)$ is the characteristic function of the set $\{x : \Phi(x) > 0\} = \Omega_+$, implying thereby

$$\partial_j \chi_{\Omega_+} = \frac{(\partial_j \Phi) dS}{|\nabla_x \Phi|}. \tag{5.3.10}$$

Similarly, we define $\Omega_- \{x : \Phi(x) < 0\}$. Let us now consider a weak solution u of a scalar conservation law (5.3.1) in n independent variables in a region Ω

enclosing a single discontinuity of the solution u. We write $u = f\chi_{\Omega_+} + g\chi_{\Omega_-}$ for some smooth maps f, g. Then

$$\sum_{j=0}^{n} \partial_j F_j(u) = \sum_{j=0}^{n} \partial_j F_j(f)\chi_{\Omega_+} + \sum_{j=0}^{n} \partial_j F_j(g)\chi_{\Omega_-}$$

$$+ \sum_{j=0}^{n} F_j(f)\partial_j\chi_{\Omega_+} + \sum_{j=0}^{n} F_j(g)\partial_j\chi_{\Omega_-}. \quad (5.3.11)$$

The left-hand side as well as the first two terms on the right-hand side vanish as functions in L^1_{loc}, so that on evaluating the distribution equation (5.3.11) against a test function ϕ gives in view of (5.3.10), the formula

$$\left\langle \sum_{j=0}^{n} (F_j(f) - F_j(g))n_j dS, \phi \right\rangle := \int_{\phi=0} (F_j(f) - F_j(g))n_j dS$$

$$= \int_{\phi=0} \phi \sum_{j=0}^{n} [F_j(u)]n_j dS = 0. \quad (5.3.12)$$

Equation (5.3.12) is the classical Rankine-Hugoniot condition that we have demonstrated in the spirit of distribution theory. Note that if linear source terms $G(x)u$ are present in the conservation law, they cancel out from either side of equation (5.3.11) since $(G(x)f + \sum_{j=1}^{n}\partial_j F_j(f))\chi_{\Omega_+}$ and $(G(x)g + \sum_{j=1}^{n}\partial_j F_j(g))\chi_{\Omega_-}$ both vanish, and similarly the formula is unaffected by integro-differential terms [154]. However, the analysis shows how nonhomogeneous terms, which are distributions singular along $\Psi = 0$, are to be handled.

Theorem 5.3.2 *Let* $\beta(\mathbf{x})$ *be a smooth function defined on the manifold* $\Phi = 0$ *and consider a weak solution* u *for the system*

$$\sum_{j=0}^{n} \partial_j F_j(u) = \beta(\mathbf{x})\Phi^*(\delta_0), \quad (5.3.13)$$

with a jump discontinuity along $\Phi = 0$. *Then the jump in* u *satisfies the following modified Rankine-Hugoniot condition*

$$\int_{\phi=0} \phi \sum_{j=0}^{n} [F_j(u)]n_j dS = \langle \beta(\mathbf{x})\Phi^*(\delta_0), \phi \rangle, \quad (5.3.14)$$

where ϕ *is an arbitrary test function.*

A detailed proof of theorem (5.3.2) is available in [191]. As an application of theorem (5.3.2), recall that in classical potential theory, the electric field vector $\mathbf{E}$ and potential V, satisfy the Poisson's equation

$$div\mathbf{E} = \nabla^2_x V = 4\pi\rho. \quad (5.3.15)$$

The density of the electrostatic medium ρ is often singular along manifolds and, in the case of layer potentials is given by $\rho = c\Phi^*(\delta_0)$.

Applying formula (5.3.14) to the Poisson's equation, we recover as a special case of the Rankine-Hugoniot condition, the following classical theorem [85] giving the jump in the normal derivative of the layer potential

$$\frac{\partial V}{\partial \hat{\mathbf{n}}} = [\mathbf{E}].\hat{\mathbf{n}} = 4\pi\rho. \tag{5.3.16}$$

5.3.3   Application to nonlinear geometrical optics

We now proceed to apply the formula (5.3.8) to the transport equations of nonlinear geometrical optics which, at leading order, govern the propagation of high frequency monochromatic waves propagating into a given background state. Let us consider the small amplitude high frequency wave solutions of a quasilinear hyperbolic system of partial differential equations in N unknowns $\mathbf{u}$ and n space variables:

$$\frac{\partial \mathbf{u}}{\partial t} + \sum_{j=1}^{n} \frac{\partial}{\partial x_j}(\mathbf{F}_j(\mathbf{x}, t, \mathbf{u})) = 0, \tag{5.3.17}$$

where the n flux functions $\mathbf{F}_j$ are assumed to depend smoothly on their arguments. Let $\mathbf{u} = \mathbf{u}_0(\mathbf{x}, t)$ be a smooth solution of (5.3.17) which we call the background state and denote by $\mathbf{A}_j^0$ the derivative $D_u\mathbf{F}_j(\mathbf{x}, t, \mathbf{u})$ evaluated at $\mathbf{u}_0$ and $\lambda_q^0(\mathbf{k}, \mathbf{x}, t, \mathbf{u}_0)$ the qth eigenvalue of $\Sigma_{j=1}^n k_j \mathbf{A}_j^0$; the left and right eigenvectors being respectively $\mathbf{L}_q^0(\mathbf{k}, \mathbf{x}, t, \mathbf{u}_0)$ and $\mathbf{R}_q^0(\mathbf{k}, \mathbf{x}, t, \mathbf{u}_0)$. Without any loss of generality (with densities and fluxes redefined, if necessary), we may assume $\mathbf{u}_0 \equiv 0$. We assume that only one eigen-mode, the qth mode, in the initial data is excited. Motivated by the linear theory we seek an asymptotic series solution in which only the qth eigen mode is present at the leading order namely,

$$\mathbf{u}_\epsilon(\phi/\epsilon, \mathbf{x}, t) = \mathbf{u}_0 + \epsilon\pi(\phi/\epsilon, \mathbf{x}, t)\mathbf{R}_q^0 + \epsilon^2\mathbf{u}_2(\phi/\epsilon, \mathbf{x}, t) + \cdots. \tag{5.3.18}$$

The phase function ϕ satisfies the qth branch of the eikonal equation

$$\frac{\partial \phi}{\partial t} = -\lambda_q^0(\nabla_x\phi, \mathbf{x}, t), \tag{5.3.19}$$

with the wave number vector $\mathbf{k} = \nabla_x\phi$. The validity of (5.3.18) over time scales of order $O(\epsilon^{-1})$ requires the use of multiple time scales and we denote by ξ the fast variable $\xi = \phi/\epsilon$. The equation governing the evolution of the principal amplitude $\pi(\xi, t, \mathbf{x})$ is given by (see [153]):

$$\frac{d\pi}{dt} + \pi\chi + \frac{\partial}{\partial \xi}\left(a\pi + \frac{b\pi^2}{2}\right) = 0, \tag{5.3.20}$$

where $\frac{d}{dt}$ denotes the ray derivative, namely, the directional derivative along the characteristics of (5.3.19), χ is the smooth function $\frac{1}{2}\sum_{j=1}^{n}\frac{\partial(L_q^0 A_j^0 R_q^0)}{\partial x_j}$ that is related to the geometry of the wave front, and a and b are known functions of $\mathbf{x}$ and t. On multiplying (5.3.20) by π, we get

$$\frac{d}{dt}\left(\frac{\pi^2}{2}\right) + \frac{\pi^2}{2}\sum_{j=1}^{n}\frac{\partial(L_q^0 A_j^0 R_q^0)}{\partial x_j} + \frac{\partial(a\pi^2/2 + b\pi^3/3)}{\partial\xi} + E = 0. \qquad (5.3.21)$$

Formula (5.3.8) with $F_0 = \pi$, $F_1 = a\pi + \frac{1}{2}b\pi^2$, $\tilde{F}_0 = \frac{1}{2}\pi^2$ and $\tilde{F}_1 = \frac{1}{2}a\pi^2 + \frac{1}{3}b\pi^3$ gives the following distribution density

$$E = -\left(\frac{b}{4}[\pi^2]^2 - \frac{b}{3}[\pi^3][\pi]\right)\frac{\Phi^*(\delta_0)}{[\pi]} = \frac{b}{12}[\pi]^3\Phi^*(\delta_0). \qquad (5.3.22)$$

Remarks 5.3.2:

(i) The motivation for multiplying (5.3.20) by π to get (5.3.21) lies in the observation that, away from the loci of discontinuities, the expression

$$\frac{d}{dt}\left(\frac{\pi^2}{2}\right) + \frac{\pi^2}{2}\sum_{j=1}^{n}\frac{\partial}{\partial x_j}(L_q^0 A_j^0 R_q^0) + \frac{\partial}{\partial\xi}\left(\frac{a\pi^2}{2} + \frac{b\pi^3}{6}\right),$$

is an exact divergence, namely,

$$\frac{d}{dt}\left(\frac{\pi^2}{2}\right) + \sum_{j=1}^{n}\frac{\partial}{\partial x_j}\left(\frac{\pi^2}{2}L_q^0 A_j^0 R_q^0\right) + \frac{\partial}{\partial\xi}\left(\frac{a\pi^2}{2} + \frac{b\pi^3}{6}\right),$$

implying the blow-up of intensity π^2 as the ray tube collapses.

(ii) In [153], the distribution E given by equation (5.3.22) is interpreted as the energy dissipated across the shock. We note here that (5.3.22) may be recast in zero divergence form and implies the blow-up of the intensity as the ray tube collapses.

(iii) The distributional term E in (5.3.21) may be regarded as a singular forcing term which is yet another motivation for the study carried out in [191].

In certain systems exhibiting anomalous thermodynamic behavior, namely, when the so called fundamental derivative is small [94], the nonlinear distortions in the solution of the Cauchy problem for (5.3.17) are noticeable over much longer time scale of $O(\epsilon^{-2})$. This calls for a different scaling in the fast variable, namely $\xi = \phi/\epsilon^2$. In place of (5.3.20) we get the following equation governing the evolution of π (see [94]):

$$\frac{d\pi}{dt} + \left(\frac{\gamma}{2}\pi^2 + \Sigma_1\pi\right)\frac{\partial\pi}{\partial\xi} + \chi\pi = 0, \qquad (5.3.23)$$

where, γ and Σ_1 are known constants. Equation (5.3.23) can be written as

$$\frac{d}{dt}\left(\frac{\pi^2}{2}\right) + \frac{\pi^2}{2}\sum_{j=1}^{n}\frac{\partial(\mathbf{L}_q^0\mathbf{A}_j^0\mathbf{R}_q^0)}{\partial x_j} + \frac{\partial}{\partial\xi}\left(\frac{\gamma\pi^4}{8} + \frac{\Sigma_1\pi^3}{3}\right) + E = 0.$$

Applying Formula (5.3.8), we get expression for E:

$$E = \left(\frac{\gamma[\pi^2][\pi]^2}{24} + \frac{\Sigma_1[\pi]^3}{12}\right)\Phi^*(\delta_0). \tag{5.3.24}$$

If there are several shocks located at discrete locations $\{\Phi_i = 0\}_i$, the formula (5.3.8) must be modified to

$$-\sum_i\left([\tilde{F}_0]_i\frac{[F_1]_i}{[F_0]_i} - [\tilde{F}_1]_i\right)\Phi_i^*(\delta_0). \tag{5.3.25}$$

Equations (5.3.22) and (5.3.24) get modified to

$$E = \sum_i\frac{b}{12}[\pi]_i^3\Phi_i^*(\delta_0), \tag{5.3.26}$$

and

$$E = \sum_i\left(\frac{\gamma[\pi^2]_i[\pi]_i^2}{24} + \frac{\Sigma_1[\pi]_i^3}{12}\right)\Phi_i^*(\delta_0), \tag{5.3.27}$$

respectively. Note that equation (5.3.27) differs from (5.3.26) by a correction term with coefficient γ, thereby generalizing the result of [153]. Here the notation $[\pi]_i$ denotes the jump across the ith shock with locus $\Phi_i(x,t) = 0$.

5.4 Evolution Equation Describing Mixed Nonlinearity

It has been found that in certain systems with singular thermodynamic behavior, where the so-called fundamental derivative is small, the effect of nonlinearity is perceptible over time scales longer by an order of magnitude, necessitating the use of fast variables of a higher order of magnitude, namely of order $O(\epsilon^{-2})$. This has been studied in an abstract setting by Kluwick and Cox [94], and on applying it to the usual equations of gas-dynamics, the authors have found that the transport equations governing the asymptotic behavior contain, in addition to the usual quadratic nonlinearity, cubic correction terms.

Srinivasan and Sharma [192] have generalized the work of [94] to the case of multiphase expansions. The analysis in [192], which parallels the one in [77], is complicated due to the fact that the amplitudes at the leading order and the next order appear together in the transport equations, and to separate them a two stage averaging process has to be employed rather than the single stage process detailed in [77]. Although this complexity is absent in [77], the requisite analytic apparatus is developed in [77].

In as much as the results obtained in [94] are important and interesting, we have rederived the transport equation (equation (5.4.23) below) for the leading amplitude in a spirit closer to ([115], [77]), expressing the result in terms of the Glimm interaction coefficients Γ^k_{ij}.

Besides, we have employed a method similar to the one used by Cramer and Sen [43] in contrast to the approach in [94], where the perturbative parameter is introduced in the differential equations. In particular, we rederive the coefficients of the nonlinear terms of the (cubic) Burgers equation (5.4.23) in terms of the coefficients Γ^k_{ij}. In the case of a symmetric and isotropic systems, the mean curvature of the wave front, which appears as the coefficient of the linear term in the transport equation, is related to the areal derivative along the bicharacteristics. It is known from physical considerations and proved rigorously in the context of linear geometrical optics that the amplitude near a caustic becomes unbounded inversely as the area of cross section of the ray tube; we provide a short proof in Section 5.4.1 imitating the classical proof of Liouville type theorems on integral invariants. It turns out that the square of the amplitude is a multiplier in the sense of Jacobi–Poincaré [144], implying thereby the blow-up of intensities in a neighborhood of caustics [153]. However, the analysis involves the manipulation of conservation laws through multiplication by polynomials in the unknown which is known to change the Rankine-Hugoniot conditions across discontinuities. It is well known that these manipulations are invalid across discontinuities; in order to restore the validity, one has to modify the equations so obtained by adding singular terms which are distributions supported along discontinuity loci. These singular terms have been interpreted as the energy dissipated across shocks in [153]. To simplify the exposition somewhat, we have assumed throughout that the hyperbolic system of equations are in conservation form, which has the effect of making the coefficients Γ^k_{ij} symmetric in i and j. In the final section we present an application to the system governing the propagation of acoustic waves through a nonuniform medium stratified by gravitational source terms.

5.4.1   Derivation of the transport equations

We derive the transport equation (5.4.23) governing the propagation of the high frequency monochromatic waves, that includes both quadratic and cubic nonlinearities inherent in the system of conservation laws. In the first stage we introduce the fast variable corresponding to the frequency of the wave and set up the stage for the perturbative analysis. In the second stage we obtain an ϵ-approximate equation (equation (5.4.16) below), at the second level of perturbation, balancing the error at the third level resulting in the transport equation (equation (5.4.23) below), satisfied by the primary amplitude π of the wave solution.We have in the last paragraph of this section, briefly compared our derivation of (5.4.23) with that in [94].

We consider the hyperbolic system of conservation laws with a source term,

$$\frac{\partial \mathbf{u}}{\partial t} + \sum_{i=1}^{n} \frac{\partial \mathbf{F}_i\,(\mathbf{x}, t, \mathbf{u})}{\partial x_i} + \mathbf{G}\,(\mathbf{x}, t, \mathbf{u}) = 0. \qquad (5.4.1)$$

The small amplitude high frequency solution $\mathbf{u}$ propagates in a background state $\mathbf{u}_0$, a spatially dependent known solution of (5.4.1), i.e.,

$$\sum_{j=1}^{n} \frac{\partial \mathbf{F}_i(\mathbf{x}, t, \mathbf{u}_0)}{\partial x_j} + \mathbf{G}\,(\mathbf{x}, t, \mathbf{u}_0) \quad = \quad 0.$$

The independent variable $\mathbf{x}$ varies over an open set in $\mathbb{R}^n$, and densities $\mathbf{F}_i$ as well as the unknown $\mathbf{u}$ are $\mathbb{R}^m$ valued. The hyperbolicity of system (5.4.1) means that for each nonzero n-vector $(\vartheta_1, \vartheta_2, ..., \vartheta_n)$, the matrix $\sum_{j=1}^{n} \vartheta_j \mathbf{A}_j^0$, where $\mathbf{A}_j^0 = \mathbf{D}_u \mathbf{F}_j(\mathbf{u}_0)$, is diagonalizable with real eigenvalues. We assume a highly oscillatory initial Cauchy data with one excited eigen-mode, say, the qth,

$$\mathbf{u}\big|_{t=0} = \mathbf{u}_0 + \epsilon \pi^0\left(\mathbf{x}, \frac{\phi}{\epsilon^2}\right) \mathbf{R}_q^0, \qquad (5.4.2)$$

where the initial amplitude π^o is a smooth function which is bounded with bounded first derivative; $\mathbf{R}_j^0$ and $\mathbf{L}_j^0$ are, respectively, the right and left eigenvectors corresponding to the eigenvalue $\lambda_j^0(\vartheta_1, \ldots, \vartheta_n)$, $1 \le j \le n$, and the zero superscript denotes their values at $\mathbf{u} = \mathbf{u}_0$. Denoting by $\mathbf{B}_k[\mathbf{v}_1, \mathbf{v}_2]$ and $\mathbf{C}_k[\mathbf{v}_1, \mathbf{v}_2, \mathbf{v}_3]$, respectively, the second and third derivatives $D^2\mathbf{F}_k[\mathbf{v}_1, \mathbf{v}_2]$ and $D^3\mathbf{F}_k[\mathbf{v}_1, \mathbf{v}_2, \mathbf{v}_3]$, with a zero superscript indicating evaluation at $\mathbf{u} = \mathbf{u}_0$, we expand the fluxes in a Taylor series

$$\mathbf{F}_j = \mathbf{F}_j^0 + \mathbf{A}_j^0(\mathbf{x}, t)(\mathbf{u} - \mathbf{u}_0) + \frac{1}{2}\mathbf{B}_j^0(\mathbf{x}, t)[\mathbf{u} - \mathbf{u}_0, \mathbf{u} - \mathbf{u}_0]$$
$$+ \frac{1}{6}\mathbf{C}_j^0(\mathbf{x}, t)[\mathbf{u} - \mathbf{u}_0, \mathbf{u} - \mathbf{u}_0, \mathbf{u} - \mathbf{u}_0] + O(|\mathbf{u} - \mathbf{u}_0|^4). \qquad (5.4.3)$$

Likewise the source term $\mathbf{G}\,(\mathbf{x}, t, \mathbf{u})$ may be developed as

$$\mathbf{G}\,(\mathbf{x}, t, \mathbf{u}) = \mathbf{G}\,(\mathbf{x}, t, \mathbf{u}_0) + D_u\mathbf{G}\,(\mathbf{x}, t, \mathbf{u}_0)\,(\mathbf{u} - \mathbf{u}_0) + O(|\mathbf{u} - \mathbf{u}_0|^2). \quad (5.4.4)$$

Introducing the fast variable of order $O(\epsilon^{-2})$, $\xi = \theta(\mathbf{x}, t)/\epsilon^2$, where the phase function $\theta(\mathbf{x}, t)$ is the solution, corresponding to λ_q^0, of the eikonal equation

$$Det\left(\sum_{j=1}^{n} \frac{\partial \theta}{\partial x_j}\mathbf{A}_j^0 + \frac{\partial \theta}{\partial t}\mathbf{I}\right) = 0, \qquad (5.4.5)$$

the PDE (5.4.1) may be recast in the form

$$\sum_{j=0}^{n}\left[\mathbf{A}_j^0\frac{\partial \mathbf{v}}{\partial x_j} + \frac{\partial \mathbf{A}_j^0}{\partial x_j}\mathbf{v} + \frac{1}{\epsilon^2}\frac{\partial \theta}{\partial x_j}\frac{\partial}{\partial \xi}\left[\mathbf{A}_j^0\mathbf{v} + \frac{1}{2}\mathbf{B}_j^0[\mathbf{v}, \mathbf{v}] + \frac{1}{6}\mathbf{C}_j^0[\mathbf{v}, \mathbf{v}, \mathbf{v}]\right]\right] + \mathbf{E}_0\mathbf{v} = 0.$$
$$(5.4.6)$$

In Eqs. (5.4.1), (5.4.2), and (5.4.6), the expressions such as $\partial \mathbf{F}_j / \partial x_j$ and $\partial \mathbf{A}_j^0 / \partial x_j$ refer to the x_j partial of the composite function $\mathbf{F}_j(\mathbf{x}, t, \mathbf{u}(\mathbf{x}, t))$ (respectively, $\mathbf{A}_j^0(\mathbf{x}, t, \mathbf{u}_0(\mathbf{x}))$). In (5.4.6) we have retained the terms of order at most $O(\epsilon^3)$, using the notations $\mathbf{v} = \mathbf{u} - \mathbf{u}_0$, $\mathbf{F}_0(\mathbf{v}) \equiv \mathbf{v}$, $x_0 = t$, and $\mathbf{E}_0 = D_{\mathbf{u}} \mathbf{G}(\mathbf{x}, t, \mathbf{u}_0)$. We assume that the multiplicities of the eigenvalues remain constant, which implies that the eigenvalues depend smoothly on $(\vartheta_1, ..., \vartheta_n)$, thereby ensuring that the various branches are algebraic functions of $\vartheta_1, ..., \vartheta_n$. Since the excited eigen-mode is the qth, we take the qth branch namely,

$$\frac{\partial \theta}{\partial t} = -\lambda_q^0(\mathbf{x}, t, \nabla_x \theta). \tag{5.4.7}$$

The characteristics of (5.4.7), are the linear bicharacteristics of (5.4.1), which will be referred to as rays. We seek a solution to (5.4.6) as a perturbation series

$$\mathbf{v} = \mathbf{u} - \mathbf{u}_0 = \epsilon \mathbf{u}_1 + \epsilon^2 \mathbf{u}_2 + \epsilon^3 \mathbf{u}_3 + ... = \epsilon \pi(x, t, \xi) \mathbf{R}_q^0 + \epsilon^2 \mathbf{u}_2 + \epsilon^3 \mathbf{u}_3 + \cdots. \tag{5.4.8}$$

Substituting (5.4.8) into (5.4.6), multiplying through by ϵ, we get at levels $O(\epsilon^k)$ for $k = 1, 2, 3$ the following equations:

$$\sum_{j=0}^{n} \mathbf{A}_j^0 \frac{\partial \theta}{\partial x_j} \frac{\partial \pi}{\partial \xi} \mathbf{R}_q^0 = 0, \tag{5.4.9}$$

$$\sum_{j=0}^{n} \mathbf{A}_j^0 \frac{\partial \theta}{\partial x_j} \frac{\partial \mathbf{u}_2}{\partial \xi} = -\sum_{j=0}^{n} \frac{\partial \theta}{\partial x_j} \mathbf{B}_j^0 [\mathbf{R}_q^0, \mathbf{R}_q^0] \pi \pi_{,\xi}, \tag{5.4.10}$$

$$\sum_{j=0}^{n} \frac{\partial \theta}{\partial x_j} \mathbf{A}_j^0 \frac{\partial \mathbf{u}_3}{\partial \xi} = -\sum_{k=0}^{n} \frac{\partial \theta}{\partial x_k} \frac{\partial}{\partial \xi} \left(\mathbf{B}_k^0 [\mathbf{u}_1, \mathbf{u}_2] + \frac{1}{6} \mathbf{C}_k^0 [\mathbf{u}_1, \mathbf{u}_1, \mathbf{u}_1] \right)$$

$$- \sum_{k=0}^{n} \mathbf{A}_k^0 \frac{\partial \mathbf{u}_1}{\partial x_k} - \mathbf{E}_0 \mathbf{u}_1, \tag{5.4.11}$$

where (5.4.9) holds by the choice of $\theta(\mathbf{x}, t)$.

5.4.2 The ϵ-approximate equation and transport equation

The solvability condition for (5.4.10) obtained by pre-multiplying with $\mathbf{L}_q^0$ is the vanishing of

$$\Sigma_0 := \sum_{j=0}^{n} \frac{\partial \theta}{\partial x_j} \mathbf{L}_q^0 \mathbf{B}_j^0 [\mathbf{R}_q^0, \mathbf{R}_q^0] = 0. \tag{5.4.12}$$

In certain applications to media with mixed nonlinearity, condition (5.4.12) generally holds only approximately with an error of $O(\epsilon)$. This is seen from physical considerations of the parameters involved, and seems to have been the main motivation in [94] for implicitly assuming an ϵ-dependence for the matrices $\mathbf{A}_j^0$ allowing a development of $\mathbf{B}_k$ and $\mathbf{C}_k$ as a series in ϵ. Cramer and Sen [43], in a different context, cope with this problem by manipulating the perturbation expansion by introducing ϵ-corrections to the coefficients of ϵ^2 and ϵ^3 as follows. When (5.4.8) is substituted in (5.4.6) the result is an asymptotic expansion

$$\epsilon Z_1 + \epsilon^2 Z_2 + \epsilon^3 Z_3 + \cdots = 0, \tag{5.4.13}$$

where $Z_1 = 0$ holds since the $O(\epsilon)$ term in (5.4.8) is proportional to $\mathbf{R}_q^0$; however $Z_2 = 0$, which is equation (5.4.10), only holds with an error of order $O(\epsilon)$. Rewriting (5.4.13) as

$$\epsilon Z_1 + \epsilon^2 (Z_2 - \epsilon Z_3') + \epsilon^3 (Z_3 + Z_3') + ... = 0, \tag{5.4.14}$$

where $Z_3' = Z_2/\epsilon$ and $Z_3 = Z_3'$. In other words (5.4.10) is replaced by an approximate equation (5.4.16) given below which we solve exactly, thereby compensating the error at the $O(\epsilon^3)$ level. The smallness of the physical parameters involved makes this procedure licit. Let us write

$$-\sum_{k=0}^{n} \frac{\partial \theta}{\partial x_k} \mathbf{B}_k^0 [\mathbf{R}_q^0, \mathbf{R}_q^0] = \sum_{j=1}^{n} \mu_j \mathbf{R}_j^0, \tag{5.4.15}$$

so that $\mu_q = O(\epsilon)$ and

$$\mu_j \mathbf{L}_j^0 \mathbf{R}_j^0 = -\sum_{k=0}^{n} \frac{\partial \theta}{\partial x_k} \mathbf{L}_j^0 \mathbf{B}_k [\mathbf{R}_q^0, \mathbf{R}_q^0].$$

We now solve exactly the ϵ-approximate equation

$$\sum_{j=0}^{n} \frac{\partial \theta}{\partial x_j} \mathbf{A}_j^0 \frac{\partial \mathbf{u}_2}{\partial \xi} = -\pi \pi_{,\xi} \sum_{j \neq q} \mu_j \mathbf{R}_j^0, \tag{5.4.16}$$

namely,

$$\frac{\partial \mathbf{u}_2}{\partial \xi} = -\sum_{j \neq q} \sum_{k=0}^{n} \frac{\partial \theta}{\partial x_k} \mathbf{L}_j^0 \mathbf{B}_k^0 [\mathbf{R}_q^0, \mathbf{R}_q^0] \mathbf{R}_j^0 \frac{\pi \pi_{,\xi}}{(\lambda_j^0 - \lambda_q^0) \mathbf{L}_j^0 \mathbf{R}_j^0}, \tag{5.4.17}$$

which implies

$$\mathbf{u}_2 = -\sum_{j \neq q} \sum_{k=0}^{n} \frac{\partial \theta}{\partial x_k} \mathbf{L}_j^0 \mathbf{B}_k^0 [\mathbf{R}_q^0, \mathbf{R}_q^0] \mathbf{R}_j^0 \frac{\pi^2}{2(\lambda_j^0 - \lambda_q^0) \mathbf{L}_j^0 \mathbf{R}_j^0}. \tag{5.4.18}$$

These expressions are unique up to an additive multiple of $\mathbf{R}_q^0$. With the notations

$$\Delta_j^0 = \mathbf{L}_j^0 \mathbf{R}_j^0, \quad \text{and} \quad \Gamma_{ij}^k = \sum_{l=0}^{n} \frac{\partial \theta}{\partial x_l} \mathbf{L}_k^0 \mathbf{B}_l^0 [\mathbf{R}_i^0, \mathbf{R}_j^0],$$

the formula for $\mathbf{u}_2$ can be stated as

$$\mathbf{u}_2 = -\sum_{j\neq q} \frac{\Gamma_{qq}^j \mathbf{R}_j^0 \pi^2}{2\Delta_j^0(\lambda_j^0 - \lambda_q^0)}, \qquad \frac{\partial \mathbf{u}_2}{\partial \xi} = -\sum_{j\neq q} \frac{\Gamma_{qq}^j \pi \pi_{,\xi} \mathbf{R}_j^0}{\Delta_j^0(\lambda_j^0 - \lambda_q^0)}. \tag{5.4.19}$$

In terms of Γ_{ij}^k, we get $\Sigma_0 = \Gamma_{qq}^q$. We note here that the general system considered in [94] is not in conservation form and hence the multilinear maps $\mathbf{B}_k^0, \mathbf{C}_k^0$ would be unsymmetric in general. Multiplying (5.4.11) by $\mathbf{L}_q^0$ and using (5.4.8) we get the compatibility condition

$$\sum_{k=1}^n \frac{\partial\theta}{\partial x_k}\frac{\partial}{\partial\xi}(\mathbf{L}_q^0\mathbf{B}_k^0[\mathbf{u}_1, \mathbf{u}_2]) + \frac{1}{2}\sum_{k=0}^n \mathbf{L}_q^0 \frac{\partial\theta}{\partial x_k}\mathbf{C}_k^0[\mathbf{R}_q^0, \mathbf{R}_q^0, \mathbf{R}_q^0]\pi^2\pi_{,\xi}$$

$$+ \sum_{k=0}^n \mathbf{L}_q^0\mathbf{A}_k^0\frac{\partial(\pi\mathbf{R}_q^0)}{\partial x_k} + \mathbf{L}_q^0\mathbf{E}_0\mathbf{R}_q^0\pi = 0. \tag{5.4.20}$$

On Substituting in (5.4.20) the values of $\partial\mathbf{u}_2/\partial\xi$ and $\mathbf{u}_2$ from equation (5.4.19) we get

$$-\sum_{j\neq q} \frac{3\Gamma_{qq}^j\Gamma_{qj}^q\pi^2\pi_{,\xi}}{2\Delta_j^0(\lambda_j^0 - \lambda_q^0)} + \frac{1}{2}\sum_{k=0}^n \mathbf{L}_q^0 \frac{\partial\theta}{\partial x_k}\mathbf{C}_k^0[\mathbf{R}_q^0, \mathbf{R}_q^0, \mathbf{R}_q^0]\pi^2\pi_{,\xi}$$

$$+ \left(\frac{d\pi}{dt} + (\chi + h)\pi\right) = 0, \tag{5.4.21}$$

where $h = \mathbf{L}_q^0\mathbf{E}^0\mathbf{R}_q^0$ and d/dt denotes the ray derivative along the characteristics of (5.4.7), and

$$\chi = \sum_{k=0}^n \mathbf{L}_q^0\mathbf{A}_k^0\frac{\partial(\mathbf{R}_q^0)}{\partial x_k}. \tag{5.4.22}$$

Note that we have obtained the transport equation (32) of [94] except for the quadratic terms of the Burgers equation. We must now incorporate the ϵ corrections indicated in equation (5.4.14).

Since $\epsilon Z_3'$ is the $\mathbf{R}_q^0$ component of $\Sigma_{j=1}^n(\partial\theta/\partial x_j)\mathbf{B}_j^0[\mathbf{R}_q^0, \mathbf{R}_q^0]$, namely $\Gamma_{qq}^q\pi\pi_{,\xi}/\Delta_q^0$, we must add $Z_3' = (1/\epsilon)\Gamma_{qq}^q\pi\pi_{,\xi}/\Delta_q^0 = O(1)$ to the right-hand side of (5.4.21); thus we get

$$\left(\frac{d\pi}{dt} + (\chi + h)\pi\right) + \frac{\gamma}{2}\pi^2\pi_{,\xi} + \Sigma_1\pi\pi_{,\xi} = 0, \tag{5.4.23}$$

which is precisely the equation obtained in [94] taking into account the contribution due to the source term in (5.4.1), where $\Sigma_1 = \Sigma_0/(\epsilon\Delta_q^0) = \Gamma_{qq}^q/(\epsilon\Delta_q^0)$ and

$$\gamma = -\sum_{j\neq q} \frac{3\Gamma_{qq}^j\Gamma_{qj}^q}{\Delta_j^0(\lambda_j^0 - \lambda_q^0)} + \sum_{k=0}^n \mathbf{L}_q^0 \frac{\partial\theta}{\partial x_k}\mathbf{C}_k^0[\mathbf{R}_q^0, \mathbf{R}_q^0, \mathbf{R}_q^0]. \tag{5.4.24}$$

5.4.3　Comparison with an alternative approach

We compare our derivation of (5.4.23) with the derivation in [94], where for simplicity we have assumed that the system in [94] arises out of a system of conservation laws. Let us denote by $\mathbf{A}(\lambda)$ the matrix $\Sigma_{j=1}^{n}(\partial\theta/\partial x_j)\mathbf{A}_j^0 - \lambda\mathbf{I}$ and $\mathbf{a}_{\tau}(\lambda)$ its rows. Let $\mathbf{b}$ be the row vector that represents the functional $\Sigma_{j=0}^{n}(\partial\theta/\partial x_j)\mathbf{L}_q^0\mathbf{B}_j^0[\mathbf{R}_q^0,\cdot]/\Delta_q^0$. Condition (5.4.12) can be stated as vanishing of $\mathbf{b}\cdot\mathbf{R}_q^0$. The authors in [94] proceed to observe that this holds if $\mathbf{b}$ is in the row space of the matrix $\mathbf{A}(\lambda_q)$, i.e., $\mathbf{b} = \Sigma_{\tau}\beta_{\tau}\mathbf{a}_{\tau}(\lambda)$. The coefficients of the cubic Burgers equation satisfied by π, as derived in [94], involve β_{τ}, and hence there is a need to compute them explicitly. To do this, notice that $\mathbf{a}_{\tau}\mathbf{R}_p^0 = (\lambda_p^0 - \lambda)\mathbf{R}_{p,\tau}^0$, where $\mathbf{R}_{p,\tau}^0$ denotes the τ entry in the column vector $\mathbf{R}_p^0$, and similarly $\mathbf{L}_{p,\tau}^0$ denotes the τ component of the left eigenvector $\mathbf{L}_p^0$. Multiplying by β_{τ} and summing over τ we get

$$\mathbf{b}\cdot\mathbf{R}_p^0/(\lambda_p^0 - \lambda) = \Gamma_{qp}^q/\Delta_q^0(\lambda_p^0 - \lambda) = \bar{\beta}\cdot\mathbf{R}_p^0, \quad p = 1, 2, ..., n, \qquad (5.4.25)$$

where $\bar{\beta}$ denotes a row matrix, from which it follows that

$$\bar{\beta} = \sum_{l} \frac{\Gamma_{ql}^q\mathbf{L}_l^0}{\Delta_l^0\Delta_q^0(\lambda_l^0 - \lambda)},$$

and hence

$$\beta_{\tau} = \sum_{l} \frac{\Gamma_{ql}^q L_{l,\tau}^0}{\Delta_l^0\Delta_q^0(\lambda_l^0 - \lambda)}.$$

On using this in equation (32) of [94], we recover (5.4.23). It may be noticed that the summand with $l = q$ in the above formula for β_{τ} vanishes, since (5.4.10) is assumed in [94] at the level $\mathbf{A}_i^0|_{\epsilon=0}$.

5.4.4　Energy dissipated across shocks

Throughout this section we assume that $\mathbf{u}_0$ is constant, the matrices $\mathbf{A}_j^0$ are symmetric, and $\partial\mathbf{A}_j^0/\partial x_j = 0$. With the normalization $\mathbf{L}_q^0\mathbf{R}_q^0 = 1$ we get on using the relation $\partial\lambda_q/\partial\vartheta_j = \mathbf{L}_q^0\mathbf{A}_j^0\mathbf{R}_q^0$ obtained by differentiating $\mathbf{L}_q(-\lambda_q\mathbf{I} + \Sigma_{j=1}^{n}\vartheta_j\mathbf{A}_j^0)\mathbf{R}_q = 0$ with respect to ϑ_j,

$$\sum_{j=0}^{n}\mathbf{L}_q^0\frac{\partial}{\partial x_j}(\mathbf{A}_j^0\mathbf{R}_q^0\pi) = \frac{1}{2}\sum_{j=0}^{n}\pi\frac{\partial}{\partial x_j}(\mathbf{L}_q^0\mathbf{A}_j^0\mathbf{R}_q^0) + \sum_{j=0}^{n}(\mathbf{L}_q^0\mathbf{A}_j^0\mathbf{R}_q^0)\frac{\partial\pi}{\partial x_j},$$

$$= \frac{1}{2}\sum_{j=1}^{n}\frac{\partial}{\partial x_j}\frac{\partial\lambda_q}{\partial\vartheta_j}\Bigg|_{\vartheta=\nabla_x\theta}\pi + \sum_{j=0}^{n}(\mathbf{L}_q^0\mathbf{A}_j^0\mathbf{R}_q^0)\frac{\partial\pi}{\partial x_j}.$$

Numerous physical systems are governed by isotropic conservation laws, where the eigenvalues λ_q^0 have the form $\lambda_q^0(x, t, \vartheta) = |\vartheta|c_q^0(x, t)$. In particular, this is so with the example considered in ([94, 153]). For isotropic systems we have

$$\frac{1}{2} \sum_{j=1}^{n} \frac{\partial}{\partial x_j} \frac{\partial \lambda_q}{\partial \vartheta_j} \bigg|_{\vartheta = \nabla_x \theta} \pi = \frac{\pi}{2} (\hat{\mathbf{n}} \cdot \nabla_x c_q^0 + c_q^0 \chi). \tag{5.4.26}$$

Here $\hat{\mathbf{n}}$ is the unit normal vector to the wave front $\theta(x,t) = c$ and χ given by (5.4.22) is the mean curvature of the wave front. Note that due to the absence of explicit $(\mathbf{x},t)$ dependence in system (5.4.1), the eigenvalues λ_q are independent of $(\mathbf{x},t)$ except for their appearance through $\nabla_x \theta$, namely $\lambda = c_q^0 |\nabla_x \theta|$ with c_q^0 a constant; here, ∇_x is the spatial gradient operator. The PDE (5.4.23) therefore reads

$$\frac{d\pi}{ds} + \left(\frac{\gamma\pi}{2} + \Sigma_1\right) \pi \frac{\partial \pi}{\partial \xi} + (\chi + h)\pi = 0. \tag{5.4.27}$$

Here Σ_1 is the coefficient of genuine nonlinearity introduced by Lax, γ characterizes the degree of material nonlinearity and h is due to the source term in (5.4.1). The coefficient χ also has the following interpretation as the areal derivative of the wave front along rays.

Theorem 5.4.1 *Consider system (5.4.1), where (x,t) do not explicitly appear in the equations. Let s be the ray coordinate and $A(s)$ the area of the cross section of the ray tube generated by the characteristics of (5.4.7). Then*

$$\chi = \lim_{|A| \to 0} \frac{1}{2A} \frac{dA}{ds}. \tag{5.4.28}$$

Proof: The characteristic flow of (5.4.7) is given by the system of ODEs

$$\frac{dx_j}{ds} = \mathbf{L}_q^0 \mathbf{A}_j^0 \mathbf{R}_j^0 = \frac{\partial \lambda_q}{\partial x_j}, \quad \frac{d\vartheta_j}{ds} = 0, \quad \frac{d\theta}{ds} = 0, \quad 0 \le j \le n. \tag{5.4.29}$$

Let Ω_0 be an element of surface area along $\theta = c$. Assume that Ω_0 evolves to Ω_s along the bicharacteristic flow. The area $A(s)$ of the surface element Ω_s is given by

$$A(s) = \int_{\Omega_s} \frac{\theta_{,t}}{|\nabla_x \theta|} dx_1 dx_2 ... dx_n,$$

where we have assumed that $\partial\theta/\partial t \ne 0$, and $x_1, x_2, ..., x_n$ are local coordinates on the surface. Denoting by $(c_1, c_2, ..., c_n)$ the coordinates of a general point on the element Ω_0, we get

$$A(0) = \int_{\Omega_0} \frac{\theta_{,t}}{|\nabla_x \theta|} dc_1 dc_2 ... dc_n.$$

If $\mathbf{x} = \phi_s(\mathbf{c})$ denotes the bicharacteristic flow, then

$$A(s) = \int_{\Omega_0} \frac{\theta_{,t}}{|\nabla_x \theta|} \left| \frac{\partial \mathbf{x}(s)}{\partial \mathbf{c}} \right| dc_1 dc_2 ... dc_n,$$

$$\frac{dA}{ds} = \int_{\Omega_0} \frac{\theta_{,t}}{|\nabla_x \theta|} \frac{d}{ds} \left(\left| \frac{\partial \mathbf{x}(s)}{\partial \mathbf{c}} \right| \right) d\mathbf{c} = 2 \int_{\Omega_0} \frac{\theta_{,t} \chi}{|\nabla_x \theta|} \left| \frac{\partial \mathbf{x}(s)}{\partial \mathbf{c}} \right| d\mathbf{c} = 2 \int_{\Omega_s} \frac{\theta_{,t} \chi}{|\nabla_x \theta|} d\mathbf{x},$$

where we have used (5.4.29) and the variational equation for derivatives, namely,

$$\frac{d}{ds} \left(\left| \frac{\partial \mathbf{x}(s)}{\partial \mathbf{c}} \right| \right) = Trace \left(\frac{\partial}{\partial x_i} \frac{\partial \lambda_q}{\partial \vartheta_k} \right) \left| \frac{\partial \mathbf{x}(s)}{\partial \mathbf{c}} \right| = 2\chi \left| \frac{\partial \mathbf{x}(s)}{\partial \mathbf{c}} \right|.$$

Now, χ is not constant but we may write $\chi = \chi(p) + o(1)$ if p is a generic point of a sufficiently small element Ω_s so that

$$\frac{dA(s)}{ds} = (2\chi(p) + o(1)) A(s), \tag{5.4.30}$$

from which the result follows.   □

We now introduce the augmented field

$$X(t, x_1, ..., x_n, \xi) = \left(1, \mathbf{L}_q^0 \mathbf{A}_1^0 \mathbf{R}_q^0, \ ... \ , \mathbf{L}_q^0 \mathbf{A}_n^0 \mathbf{R}_q^0, \frac{2}{3}\Sigma_1 \pi + \frac{\gamma}{4}\pi^2 \right), \tag{5.4.31}$$

and show that $\pi^2/2$ is a multiplier for this vector field, giving an integral invariant along the bicharacteristic flow. It may be noted that integral invariants were introduced by Poincaré in connection with celestial mechanics [144].

Theorem 5.4.2 *(i)* $\pi^2/2$ *is a multiplier for the augmented vector field* *(5.4.31), i.e.,*

$$Div \left(\frac{\pi^2}{2} X \right) = 0.$$

(ii) The integral $\int_{\Omega_s} (\pi^2/2) d\mathbf{x} dt d\xi$ *is conserved along the flow determined by* *(5.4.31).*
(iii) At the points in (x, t) *space where the ray tube collapses, i.e.,* $|\Omega_s| \to 0$, *the amplitude becomes infinite inversely as the volume* $|\Omega_s|$.

Proof: From the calculation leading to (5.4.26) and (5.4.27) we get

$$\frac{d\pi}{ds} + \chi\pi = \frac{1}{\pi} \sum_{j=0}^{n} \frac{\partial}{\partial x_j} \left(\frac{\pi^2}{2} \mathbf{L}_q^0 \mathbf{A}_j^0 \mathbf{R}_q^0 \right), \tag{5.4.32}$$

and so the PDE for π can be recast in the divergence free form,

$$\sum_{j=0}^{n} \frac{\partial}{\partial x_j} \left(\frac{\pi^2}{2} \mathbf{L}_q^0 \mathbf{A}_j^0 \mathbf{R}_q^0 \right) + \frac{\partial}{\partial \xi} \left(\frac{\gamma}{8}\pi^4 + \frac{\Sigma_1}{3}\pi^3 \right) \equiv Div \left(\frac{\pi^2}{2} X \right) = 0, \tag{5.4.33}$$

which proves (i); the proof of (ii) follows form the classical Liouville's theorem and is similar in spirit to Theorem 5.4.1. As we approach a point where the ray tube collapses, volume $|\Omega_s|$ shrinks to zero; this implies that due to the conservation of $\int_{\Omega_s} (\pi^2/2) d\mathbf{x} dt d\xi$, π^2 blows up in a neighborhood of this point. Moreover, we see that the square of the amplitude blows up inversely as the volume $|\Omega_s|$. □

Remarks 5.4.1:

(i) The manipulations leading to (5.4.33) are valid in regions of smoothness, the boundary of which may contain a caustic point.

(ii) The energy dissipated across shocks, present in the solutions of (5.4.27), can be computed using (5.3.8) or (5.3.25).

5.4.5   Application

Here we apply the results of the foregoing sections to the basic equations that describe the propagation of sound waves in a nonuniform atmosphere, and derive the transport equations for the high frequency wave amplitude in the leading order terms in the expansion. Practically any problem of acoustics takes place in the presence of a gravitational field and as a consequence, the unperturbed state is not uniform. In problems of propagation over large distances in atmosphere or the ocean, these effects may be crucially important and produce amplifications and refractions of sound waves. The basic equations describing the propagation of sound waves through a stratified fluid may be expressed in the following form:

$$\frac{\partial \mathbf{v}}{\partial t} + (\mathbf{v} \cdot \nabla_x)\mathbf{v} + \rho^{-1}\nabla_x p = -\mathbf{g}, \tag{5.4.34}$$

$$\frac{\partial \rho}{\partial t} + (\mathbf{v} \cdot \nabla_x)\rho + \rho(\nabla_x \cdot \mathbf{v}) = 0, \tag{5.4.35}$$

$$\frac{\partial S}{\partial t} + (\mathbf{v} \cdot \nabla_x)S = 0, \tag{5.4.36}$$

where ρ is the fluid density, $\mathbf{v} = (v_1, v_2, v_3)^{tr}$ the fluid velocity vector, $p = p(\rho, S)$ the pressure, S the entropy, t the time, $\nabla_x = (\partial/\partial x_1, \partial/\partial x_2, \partial/\partial x_3)$ is the spatial gradient operator, and $\mathbf{g}$ the acceleration due to gravity. The reference state is characterized by a flow field at rest, $\mathbf{v}_0 = 0$, with a spatially varying density and entropy fields, namely $\rho_0 = \rho_0(\mathbf{x})$ and $S_0 = S_0(\mathbf{x})$ with

$$\nabla_x p_0 + \rho_0 \mathbf{g} = 0, \tag{5.4.37}$$

where the subscript zero denotes the unperturbed fluid in equilibrium. The governing system (5.4.34) – (5.4.36) can be cast into the form

$$\frac{\partial \mathbf{U}}{\partial t} + \sum_{k=1}^{3} \mathbf{A}^k(\mathbf{U}) \frac{\partial \mathbf{U}}{\partial x_k} + \mathbf{F} = 0, \tag{5.4.38}$$

representing a quasilinear hyperbolic system of equations with source term which is attributed to the influence of gravity; here $\mathbf{U}$ and $\mathbf{F}$ are column vectors defined as $\mathbf{U} = [v_1, v_2, v_3, \rho, S]^{tr}$ and $\mathbf{F} = [g_1, g_2, g_3, 0, 0]^{tr}$. The $\mathbf{A}^k(\mathbf{U})$ are 5×5 matrices with entries $A^k_{\alpha\beta}, 1 \le \alpha, \beta \le 5$, given by

$$A^k_{11} = A^k_{22} = A^k_{33} = A^k_{44} = A^k_{55} = v_k, \quad A^k_{i4} = a^2\delta_{ik}/\rho, \quad A^k_{i5} = \frac{\partial p}{\partial S}\frac{\delta_{ik}}{\rho},$$

$$A^k_{4j} = \rho\delta_{jk}, \quad A^k_{5j} = A^k_{45} = A^k_{54} = A^k_{21} = A^k_{12} = A^k_{31} = A^k_{13} = A^k_{23} = A^k_{32} = 0,$$

where $1 \le i, j, k \le 3$, δ is the Kronecker symbol, and a is the sound speed given by $a^2 = (\partial p/\partial \rho)|_S$. We look for a small amplitude high frequency asymptotic solution of (5.4.38) when the length L of the disturbed region is small in comparison with the scale height H of the stratification, defined as a typical value of $\rho_0 |\nabla_x \rho_0|^{-1}$, so that $\epsilon = L/H \ll 1$. In this limit, the perturbations caused by the wave are of size $O(\epsilon)$ and they depend significantly on the fast variable $\xi = \theta(\mathbf{x}, t)/\epsilon^2$, where θ is the phase function to be determined. The small amplitude high frequency solution to (5.4.38) that admits an asymptotic expansion of the form (5.4.8) where $\mathbf{u}_0 = [0, 0, 0, \rho_0(\mathbf{x}), S_0(\mathbf{x})]^{tr}$ is the known background state. Let π be the wave amplitude associated with the right running acoustic wave $\theta(\mathbf{x}, t) = \text{const}$, propagating with speed $a_0|\nabla_x\theta|$. The left and right eigenvectors $\mathbf{L}$ and $\mathbf{R}$ of $\Sigma^3_{k=0}(\partial\theta/\partial x_k)\mathbf{A}^k$ associated with eigenvalue $a_0|\nabla_x\theta|$ are given by

$$\mathbf{L} = \left[n_1, n_2, n_3, \frac{a_0}{\rho_0}, \frac{1}{a_0\rho_0}\left(\frac{\partial p}{\partial S}\right)_0\right]^{tr}, \qquad \mathbf{R} = \left[n_1, n_2, n_3, \frac{a_0}{\rho_0}, 0\right]^{tr},$$

where $\hat{\mathbf{n}} = (n_1, n_2, n_3)$ denotes the unit normal field $|\nabla_x\theta|^{-1}\nabla_x\theta$ to the wave front. The coefficients Σ_1, γ, χ, and h appearing in (5.4.23) and (5.4.24) can now be easily calculated; in fact, we find that $h = 0$ and

$$\Sigma_1 = \frac{|\nabla_x\theta|}{2\epsilon}\left(1 + \frac{\rho_0}{a_0}\frac{\partial a}{\partial\rho}\Big|_0\right), \quad \gamma = \frac{\rho_0^2|\nabla_x\theta|}{a_0^2}\frac{\partial}{\partial\rho}\left(\frac{1}{\rho}\left(\frac{\partial(a\rho)}{\partial\rho}\right)\right)\Big|_0,$$

$$\chi = \frac{1}{2}\left(a_0\nabla_x \cdot \hat{\mathbf{n}} - 2\left(\frac{\partial a/\partial S}{\partial p/\partial S}\right)_0(a_0^2\mathbf{n}\cdot\nabla_x\rho_0 + \rho_0\mathbf{n}.\mathbf{g})\right.$$

$$\left. - \frac{a_0}{\rho_0}\mathbf{n}\cdot\nabla_x\rho_0 - \mathbf{n}\cdot\nabla_x a_0\right). \tag{5.4.39}$$

The functions Σ_1, γ and χ characterize respectively the genuine nonlinearity coefficient of Lax, the degree of material nonlinearity, and the variation in wave amplitude due to the wavefront geometry and the varying medium ahead. It may be noticed that although $h = 0$, the gravity effects enter indirectly through its control of $\rho_0(\mathbf{x})$ as may be seen in the expression for χ. Thus the evolution equation for the amplitude π describing the propagation of acoustic waves in a stratified medium is given by

$$\frac{d\pi}{dt} + \chi\pi + \left(\frac{1}{2}\gamma\pi + \Sigma_1\right)\pi\pi_{,\xi} = 0, \tag{5.4.40}$$

with coefficients given by (5.4.39). On substituting the values of Σ_1 and γ, given by (5.4.39), one computes the energy dissipated across shocks using (5.3.8).

Remark 5.4.2: The qualitative behavior of the weak solutions of (5.1.6) is quite different from those of (5.4.40); one can check how the solutions of (5.1.6) and (5.4.40), satisfying the same periodic or compact support initial data, evolve at different times leading to a breakdown. For more details, the reader is referred to ([42] and [43]).

5.5 Singular Ray Expansions

The geometric ray expansion derived in section 5.4 breaks down when the signal variations transverse to the direction of wave propagation become significant. This is a common feature in diffraction problems, where ray theory predicts unbounded behavior near a singular ray.

Following Kluwik and Cox [94], we derive an evolution equation for a density-stratified, nonisentropic flow governed by the hyperbolic system (5.4.38) when diffraction effects are prevalent. In addition to the fast phase variable, $\xi = \phi(\mathbf{x}, t)/\epsilon^2$, we introduce other fast variables such as $\eta_n = \psi_n(\mathbf{x}, t)/\epsilon$, $n = 1, 2$, that describe the modulations of the signals in directions transverse to those of the rays. The singular rays are supposed to lie in the hypersurfaces, $\psi_n(\mathbf{x}, t), n = 1, 2$. We look for progressive wave solution of form:

$$\begin{aligned}\mathbf{U}(\mathbf{x}, t) = \ & \mathbf{U}_0(\mathbf{x}) + \epsilon \mathbf{U}^{(1)}(\xi, \eta_1, \eta_2, \mathbf{x}, t) + \epsilon^2 \mathbf{U}^{(2)}(\xi, \eta_1, \eta_2, \mathbf{x}, t) \\ & + \epsilon^3 \mathbf{U}^{(3)}(\xi, \eta_1, \eta_2, \mathbf{x}, t), \qquad \epsilon \ll 1.\end{aligned} \quad (5.5.1)$$

In terms of the new variables the system (5.4.38) can be written as

$$\mathbf{U}_{,\xi}(\phi_{,t}\mathbf{I} + \mathbf{A}^k \phi_{,k}) + \epsilon \mathbf{U}_{,\eta_n}(\psi_{n,t}\mathbf{I} + \mathbf{A}^k \psi_{n,k}) + \epsilon^2(\mathbf{U}_{,t} + \mathbf{A}^k \mathbf{U}_{,k} + \mathbf{F}) = 0. \quad (5.5.2)$$

Using (5.5.1) in (5.5.2) along with the Taylor's series expansion for $\mathbf{A}^k$ near $\mathbf{U} = \mathbf{U}_0(\mathbf{x})$, we obtain at the leading order

$$\mathbf{U}^{(1)}(\mathbf{x}, t) = \pi(\xi, \eta_1, \eta_2, \mathbf{x}, t)\mathbf{R}$$

where π is the scalar amplitude which will be determined to the next order, and $\mathbf{R}$ is the right null vector of $\mathbf{G} = \phi_{,t}\mathbf{I} + \mathbf{A}_0^k \phi_{,k}$. Equations at the next level of sequential approximations in ϵ are the following.

$$\mathbf{G}.\mathbf{U}^{(2)}_{,\xi} = -K_n \mathbf{R}\pi_{,\eta_n} - \phi_{,k}\mathbf{R}.(\nabla \mathbf{A}^k)_0 \mathbf{R}\pi\pi_{,\xi} + \Gamma \mathbf{R}\pi\pi_{,\xi}, \tag{5.5.3}$$

$$\mathbf{G}.\mathbf{U}^{(3)}_{,\xi} = -(\pi\mathbf{R})_{,t} - \mathbf{A}^k_0(\pi\mathbf{R})_{,k} - \mathbf{R}.(\nabla \mathbf{A}^k)_0 \mathbf{U}^{(0)}_{,k}\pi - \hat{\Gamma}\mathbf{R}\pi\pi_{,\xi} - K_n \mathbf{U}^{(2)}_{,\eta_n}$$
$$-\phi_{,k}\mathbf{R}.(\nabla \mathbf{A}^k)_0 \mathbf{U}^{(2)}_{,\xi}\pi - \phi_{,k}\mathbf{U}^{(2)}.(\nabla \mathbf{A}^k)_0 \mathbf{R}\pi\pi_{,\xi} - \mathbf{R}.(\nabla F)_0 \pi$$
$$-\psi_{n,k}\mathbf{R}.(\nabla \mathbf{A}^k)_0 \mathbf{R}\pi\pi_{,\eta_n} - \frac{1}{2}\phi_{,k}\mathbf{R}\mathbf{R} : (\nabla\nabla \mathbf{A}^k)_0 \mathbf{R}\pi^2\pi_{,\xi}, \tag{5.5.4}$$

where $\Gamma = \mathbf{L}(\mathbf{R}.(\nabla \mathbf{A}^k)_0)\mathbf{R}\Phi_{,k}/(\mathbf{L}\mathbf{R})$ is the quadratic nonlinearity parameter which is of order $O(\epsilon)$ with $\mathbf{L}$ as the left null vector of $\mathbf{G}$, $\hat{\Gamma} = \Gamma/\epsilon = O(1)$,

$$\mathbf{R}\mathbf{R} : (\nabla\nabla \mathbf{A}^k)_0 = \sum_{n,m=1}^{5} R_m R_n \frac{\partial^2 A^k}{\partial U_m \partial U_n}\bigg|_{U=U_0},$$

and the matrix K_n is given by

$$K_n = \mathbf{I}\psi_{n,t} + \mathbf{A}^k_0 \psi_{n,k}. \tag{5.5.5}$$

The solvability condition for $\mathbf{U}^{(2)}$ requires the satisfaction of the following:

$$\mathbf{L}K_n\mathbf{R} = 0. \tag{5.5.6}$$

The scalar product of (5.5.4) with $\mathbf{L}$ yields the following solvability condition for $\mathbf{U}^{(3)}$

$$\frac{d\pi}{dt} + \left(\hat{\Gamma} + \frac{M}{2}\pi\right)\pi\frac{\partial \pi}{\partial \xi} + \left(\mathbf{b}.\mathbf{U}^{(2)}\right)\frac{\partial \pi}{\partial \xi} + \left(\mathbf{c}.\frac{\partial \mathbf{U}^{(2)}}{\partial \xi}\right)\pi + \mathbf{e}_n.\frac{\partial \mathbf{U}^{(2)}}{\partial \eta_n}$$
$$+\frac{\mathbf{L}(\mathbf{R}.(\nabla \mathbf{A}^k)_0)\mathbf{R}\psi_{n,k}\pi\pi_{,\eta_n}}{2} + \chi\pi + \frac{\Gamma(\mathbf{L}.\mathbf{U}^{(2)}\pi)_{,\xi}}{2} = 0, \tag{5.5.7}$$

where $d/dt = \partial/\partial t + a_0\phi_{,k}\partial/\partial x_k$ denotes derivative along the ray — the trajectory of an element of the surface $\phi(\mathbf{x}, t) = $ constant, and M, $\mathbf{b}$, $\mathbf{c}$, $\mathbf{e}_n$, and χ are given by

$$\begin{aligned} M &= \{\phi_{,k}\mathbf{L}(\mathbf{R}\mathbf{R} : (\nabla\nabla \mathbf{A}^k)_0)\mathbf{R}\}/2, & \mathbf{b} &= \{\mathbf{L}\phi_{,k}(\nabla \mathbf{A}^k)_0\mathbf{R} - \Gamma\mathbf{L}\}/2, \\ \mathbf{c} &= \{\mathbf{L}(\mathbf{R}.(\nabla \mathbf{A}^k)_0)\phi_{,k} - \Gamma\mathbf{L}\}/2, & \mathbf{e}_n &= \mathbf{L}K_n/2, \tag{5.5.8} \\ \chi &= \mathbf{L}\{\mathbf{A}^k_0(\partial \mathbf{R}/\partial x_k) + \mathbf{R}.(\nabla \mathbf{A}^k)_0 \mathbf{U}^{(0)}_{,k} + \mathbf{R}.(\nabla F)_0\}/2. \end{aligned}$$

From (5.5.6), it follows that $\mathbf{e}_n$ is orthogonal to $\mathbf{R}$ and lies in the 4-dimensional row-space of $\mathbf{G}$. Thus, $\mathbf{e}_n$ can be written as

$$\mathbf{e}_n = \beta_{n\alpha}\mathbf{G}_\alpha, \qquad \alpha = 1, 2, 3, 4, \quad n = 1, 2, \tag{5.5.9}$$

where $\mathbf{G}_\alpha$ are the linearly independent rows of $\mathbf{G}$. We notice that the projection of $\partial \mathbf{U}^{(2)}/\partial \eta_n$ onto the row space of $\mathbf{G}$ will be required to construct the term $\mathbf{e}_n \cdot (\partial \mathbf{U}^{(2)}/\partial \eta_n)$ in the evolution equation of (5.5.7). Integration of (5.5.3) gives,

$$\mathbf{G}\mathbf{U}^{(2)} = -(K_n \mathbf{R}) \int^\xi \frac{\partial \pi}{\partial \eta_n} d\xi - \frac{1}{2} \left(\phi_{,k} \mathbf{R}.(\nabla \mathbf{A}^k)_0 - \Gamma \mathbf{I} \right) \mathbf{R} \pi^2, \qquad (5.5.10)$$

where we assume that $\mathbf{U}^{(2)} = 0$ when $\mathbf{U}^{(1)} = 0$. Differentiating the above equation with respect to η_n, we have,

$$\mathbf{G} \frac{\partial \mathbf{U}^{(2)}}{\partial \eta_n} = -(K_n \mathbf{R}) \int^\xi \frac{\partial^2 \pi}{\partial \eta_n \partial \eta_l} d\xi - (\phi_{,k} \mathbf{R}.(\nabla \mathbf{A}^k)_0 - \Gamma \mathbf{I}) \mathbf{R} \pi \pi_{,\eta_n}.$$

Since $\mathbf{b}$ and $\mathbf{c}$ are orthogonal to $\mathbf{R}$, we have $\mathbf{b} = \omega_\alpha \mathbf{G}_\alpha, \mathbf{c} = \delta_\alpha \mathbf{G}_\alpha$ with $\omega_i = \delta_i = n_i(\frac{\rho_0 a_{,so}}{p_{,so}})$, $i = 1, 2, 3$; $\omega_4 = \delta_4 = \rho_0^{-1}(1 + \frac{\rho_0 a_0 a_{,so}}{p_{,so}})$; thus, the terms $\mathbf{b}.\mathbf{U}^{(2)}$, $\mathbf{c}.\partial \mathbf{U}^{(2)}/\partial \xi$, and $\mathbf{e}_n.\partial \mathbf{U}^{(2)}/\partial \eta_n$ may be written as

$$\mathbf{b}.\mathbf{U}^{(2)} = -\omega \left\{ \frac{1}{2} \left(\phi_{,k} \mathbf{R}.(\nabla \mathbf{A}_1^k)_0 - \Gamma \mathbf{I}_1 \right) \mathbf{R} \pi^2 + (K_{1n} \mathbf{R}) \int^\xi \frac{\partial \pi}{\partial \eta_n} d\xi \right\},$$

$$\mathbf{c}.\frac{\partial \mathbf{U}^{(2)}}{\partial \xi} = -\delta \left\{ \left(\phi_{,k} \mathbf{R}.(\nabla \mathbf{A}_1^k)_0 - \Gamma \mathbf{I}_1 \right) \mathbf{R} \pi \pi_{,\xi} + (K_{1n} \mathbf{R}) \frac{\partial \pi}{\partial \eta_n} \right\},$$

$$\mathbf{e}_n.\frac{\partial \mathbf{U}^{(2)}}{\partial \eta_n} = -\beta_n \left\{ \left(\phi_{,k} \mathbf{R}.(\nabla \mathbf{A}_1^k)_0 - \Gamma \mathbf{I}_1 \right) \mathbf{R} \pi \frac{\partial \pi}{\partial \eta_n} + (K_{1n} \mathbf{R}) \int^\xi \frac{\partial^2 \pi}{\partial \eta_n \partial \eta_l} d\xi \right\},$$

where $\mathbf{A}_1^k$ and $\mathbf{I}_1$ are 4×5 matrices obtained from $\mathbf{A}^k$ and $\mathbf{I}$, respectively, by deleting their last row, and $K_{1n} = \mathbf{I}_1 \psi_{n,t} + \mathbf{A}_1^k \psi_{n,k}$. Substituting the above expressions in the evolution equation (5.5.7), we have,

$$\frac{d\pi}{dt} + \left(\hat{\Gamma} + \frac{\gamma}{2} \pi \right) \pi \frac{\partial \pi}{\partial \xi} + \chi \pi + T_n \pi \frac{\partial \pi}{\partial \eta_n} - W_n \frac{\partial \pi}{\partial \xi} \int^\xi \frac{\partial \pi}{\partial \eta_n} d\xi$$

$$- M_{nl} \int^\xi \frac{\partial^2 \pi}{\partial \eta_n \partial \eta_l} d\xi = 0, \qquad (5.5.11)$$

where T_n, W_n, M_{nl} and γ are defined as

$$T_n = -\delta(K_{1n} \mathbf{R}) - \beta_n \left(\phi_{,k} \mathbf{R}.(\nabla \mathbf{A}_1^k)_0 - \Gamma \mathbf{I}_1 \right) \mathbf{R} + \frac{\mathbf{L}(\mathbf{R}(\nabla \mathbf{A}^k)_0) \mathbf{R} \psi_{n,k}}{2},$$

$$W_n = \omega(K_{1n} \mathbf{R}), \quad M_{nl} = \beta_n(K_{1n} \mathbf{R}),$$

$$\gamma = \mathbf{M} - (\omega + 2\delta)(\phi_{,k} \mathbf{R}(\nabla \mathbf{A}_1^k)_0 - \Gamma \mathbf{I}_1) \mathbf{R}.$$

Equation (5.5.11) can be reduced to an equation similar to the Zabolotskaya-Khokhlov equation on differentiating with respect to ξ which is as follows:

$$\frac{\partial}{\partial \xi} \left(\frac{d\pi}{dt} + \left(\hat{\Gamma} + \frac{\gamma}{2} \pi \right) \pi \frac{\partial \pi}{\partial \xi} + \chi \pi + \sum_{n=1}^2 T_n \pi \frac{\partial \pi}{\partial \eta_n} - \sum_{n=1}^2 W_n \frac{\partial \pi}{\partial \xi} \int^\xi \frac{\partial \pi}{\partial \eta_n} d\xi \right)$$

$$= \sum_{n=1}^2 \sum_{l=1}^2 M_{nl} \frac{\partial^2 \pi}{\partial \eta_n \partial \eta_l},$$

where $\hat{\Gamma}$, the quadratic nonlinearity coefficient may be identified with the Lax's genuine nonlinearity parameter and γ, the cubic nonlinearity coefficient, signifies the degree of material nonlinearity; the linear term $\chi\pi$ brings about a variation in the wave amplitude as a result of wave interactions with the varying medium ahead and changes in the geometry of the wavefront as it moves along the rays. Each of these coefficients including the additional coefficients T_n, W_n and M_{nl} which bring about the distortion of the signal in the transverse direction, is of order $O(1)$ and is a function of the space coordinates. The nonlinearity parameters lead to distortion of the wave profile and formation of shock, while the linear term results in growth or decay of the wave amplitude depending on whether it is negative or positive. When the medium ahead is uniform and the source term is absent the evolution equation reduces to that of Kluwick and Cox [94] with the coefficients being absolute constant. The linear term vanishes too, if we consider one-dimensional wave propagation in the uniform medium and we get an evolution equation similar to that of Cramer and Sen [43].

Remarks 5.5.1:

(i) In the geometric ray theory, the intersection of rays predicts infinite amplitudes. The geometric singularity is, in fact, associated with the formation of caustics; the breakdown occurs in a thin caustic boundary layer where a local analysis is appropriate.

(ii) It is easy to incorporate the effects of a small amount of dissipation into the equations of WNGO, so that the new system is

$$\mathbf{u}_{,t} + \mathbf{A}^k(\mathbf{u})\mathbf{u}_{,k} + \mathbf{F}(\mathbf{u}) = \epsilon^2 (\mathbf{D}_{ij}(\mathbf{u})\mathbf{u}_{,j})_{,i} \,,$$

where $\mathbf{D}_{ij}$ are the $n \times n$ viscosity matrices satisfying the stability condition $\xi_i \xi_j \mathbf{L}_q^o \mathbf{D}_{ij} \mathbf{R}_q^o \geq 0$, where $\boldsymbol{\xi} \in \mathbb{R}^n$ and subscript q refers to the qth eigenmode (see Majda and Pego [114]). If one assumes that the matrices involved in the above equation have Taylor expansion about $\mathbf{u} = \mathbf{u}_0$, and repeats the weakly nonlinear analysis, it is straightforward to show that the transport equation (5.4.40) for the wave amplitude, in media exhibiting mixed nonlinearity, gets modified to the (viscous) Burgers equation with cubic nonlinearity (see Sharma et. al ([172]))

$$\frac{d\pi}{dt} + (\Sigma_1 + \frac{\gamma}{2}\pi)\pi\pi_{,\xi} + \pi\chi = \mu\pi_{,\xi\xi},$$

where $\mu = (\mathbf{L}_q^o \mathbf{D}_{ij} \mathbf{R}_q^o)/(\mathbf{L}_q^o \mathbf{R}_q^o) \geq 0$.

5.6   Resonantly Interacting Waves

A systematic approach to resonantly interacting weakly nonlinear hyperbolic waves has been proposed by Majda and Rosales [115] and Hunter et al. [77]. This approach enables one to analyze situations when many waves coexist and interact with one another resonantly. In this section we first give an outline of this method, and then use it to a hyperbolic system governing the one dimensional flow of a nonideal gas.

Let us consider a strictly hyperbolic system in one space dimension, given by (5.1.1). Let us assume that $\mathbf{A}$ and $\mathbf{b}$ admit Taylor expansion about $\mathbf{u} = \mathbf{u}_o$, which is a known constant solution of (5.1.1), such that $\mathbf{b}(\mathbf{u}_o) = 0$. Let $\lambda_1, \lambda_2, \ldots, \lambda_n$ be the n-distinct real eigenvalues of $\mathbf{A}(\mathbf{u}_o)$ with corresponding left and right eigenvectors denoted by $\mathbf{L}^{(i)}$ and $\mathbf{R}^{(j)}$, $1 \leq i, j \leq n$, respectively, satisfying the normalization condition $\mathbf{L}^{(i)}\mathbf{R}^{(j)} = \delta_{ij}$, with δ_{ij} as the Kronecker symbol.

Our object here is to construct formal asymptotic solution of (5.1.1) of the form

$$\mathbf{u} = \mathbf{u}_o + \epsilon\mathbf{u}_1(x, t, \boldsymbol{\theta}) + \epsilon^2\mathbf{u}_2(x, t, \boldsymbol{\theta}) + o(\epsilon^2), \tag{5.6.1}$$

satisfying the small amplitude rapidly oscillating initial data

$$\mathbf{u}(x, 0) = \mathbf{u}_o + \epsilon\mathbf{u}_1^o(x, x/\epsilon) + \epsilon^2\mathbf{u}_2^o(x, x/\epsilon) + o(\epsilon^2). \tag{5.6.2}$$

Here, ϵ is a small parameter defined as a $\epsilon = \tau_i/\tau \ll 1$, where τ_i is the time scale defined by the input and τ is the time scale characterizing the medium, and $\mathbf{u}_1^o$ and $\mathbf{u}_2^o$ are smooth bounded functions, with the remainder term $o(\epsilon^2)$, to be uniform in x as $\epsilon \to 0$. It is assumed that in this high frequency limit, the perturbations caused by the wave are of size $O(\epsilon)$, and they depend significantly on the fast variable $\boldsymbol{\theta} = \phi/\epsilon$, where $\boldsymbol{\theta} = (\theta_1, \theta_2, \ldots, \theta_n)$ is a vector of rapidly varying phase functions with $\theta_i = \phi_i(x, t)/\epsilon$; indeed, ϕ_i is the phase of the ith wave associated with the characteristic speed λ_i. In (5.6.1), $\mathbf{u}_1$ is a smooth bounded function of its arguments. We need to require additionally that $\mathbf{u}_2$ can have at most sublinear growth in $\boldsymbol{\theta}$, i.e., $\|\mathbf{u}_2\| \to 0$ as $\|\boldsymbol{\theta}\| \to \infty$ and $\mathbf{u}_1^o(x, x/\epsilon)$ in (5.6.2) is a periodic smooth function with a compact support and mean zero. In [115], authors show that under these conditions, the expansion (5.6.1) is a uniformly valid approximation for times of order ϵ^{-1}. Now we use (5.6.2) in (5.6.1), expand $\mathbf{A}$ and $\mathbf{b}$ in a Taylor series in powers of ϵ about $\mathbf{u} = \mathbf{u}_o$, replace the partial derivatives $\partial/\partial X$ (X being either x or t) by $\partial/\partial X + \epsilon^{-1}\sum_{i=1}^n \phi_{i,x}\partial/\partial\theta_i$, and equate to zero the coefficients of ϵ^0 and ϵ^1 in the resulting expansions, to get

$$O(\epsilon^0) : \sum_{i=1}^n (\mathbf{I}\phi_{i,t} + \mathbf{A}_o\phi_{i,x})\mathbf{u}_{1,\theta_i} = 0, \tag{5.6.3}$$

$$O(\epsilon^1) : \sum_{i=1}^n (\mathbf{I}\phi_{i,t} + \mathbf{A}_o\phi_{i,x})\mathbf{u}_{2,\theta_i} = -\mathbf{u}_{1,t} - \mathbf{A}_o\mathbf{u}_{1,x} - \nabla\mathbf{b}|_o \cdot \mathbf{u}_1$$

$$- \sum_{i=1}^n \phi_{i,x}\nabla\mathbf{A}|_0 \cdot \mathbf{u}_1\mathbf{u}_{1,\theta_i}, \tag{5.6.4}$$

where $\mathbf{I}$ is the $n \times n$ unit matrix and ∇ is the gradient operator with respect to the dependent variable $\mathbf{u}$; the subscript o refers to the evaluation at $\mathbf{u} = \mathbf{u}_o$.

A solution of (5.6.3) is

$$\mathbf{u}_1 = \sum_{i=1}^{n} \pi_i(x, t, \theta_i)\mathbf{R}^{(i)}, \qquad (5.6.5)$$

where $\theta_i = \phi_i/\epsilon$ with $\phi = x - \lambda_i(\mathbf{u}_o)t$. We suppose that the wave amplitudes π_i and their derivatives with respect to x, t and $\boldsymbol{\theta}$ are periodic functions of $\boldsymbol{\theta}$, and the wave amplitudes have 0 mean with respect to $\boldsymbol{\theta}$, i.e., $\lim_{T\to\infty}(2T)^{-1}\int_{-T}^{T}\pi_j(x, t, \theta)d\theta = 0$. We then use (5.6.5) in (5.6.4) and solve fro $\mathbf{u}_2$. To begin with, we expand $\mathbf{u}_2$ in the basis of $\mathbf{A}_o$, i.e., $\mathbf{u}_2 = \sum m_j\mathbf{R}^{(j)}$, and substitute it in (5.6.4); then on pre-multiplying the resulting equation by $L^{(i)}$, we get

$$\sum_{j=1}^{n}(\lambda_i - \lambda_j)\frac{\partial m_i}{\partial \theta_j} = -\frac{\partial \pi_i}{\partial t} - \lambda_i\frac{\partial \pi_i}{\partial x} - \mu_i\pi_i \qquad (5.6.6)$$

$$-\nu_i\pi_i\pi_i'(x, t, \theta_i) - \sum_{p}^{(i)}\mathbf{L}_i\nabla\mathbf{b}|_o\mathbf{R}^{(p)}\pi_p$$

$$-\sum_{p}^{(i)}\sum_{q}^{(i)}\mathbf{L}_i \cdot \nabla A|_o \cdot \mathbf{R}^{(p)}\mathbf{R}^{(q)}\pi_p\pi_q'(x, t, \theta_q); \quad (i\text{- unsummed}),$$

where $\pi_q'(x, t, \theta) = \partial\pi_q/\partial\theta$, $\mu_i = \mathbf{L}_i \cdot \nabla\mathbf{b}|_o\mathbf{R}^{(i)}$, $\nu_i = \mathbf{L}_i \cdot \nabla A|_o\mathbf{R}^{(i)}\mathbf{R}^{(i)}$, $\sum_{p}^{(i)}$ stands for the sum over p with $1 \leq p \leq n$ and $p \neq i$, and m_i are bounded functions of $\boldsymbol{\theta}$.

We use the standard properties of periodic functions, and appeal to the sublinearity of $\mathbf{u}_2$ in $\|\boldsymbol{\theta}\|$, which ensures that the expression (5.6.2) does not contain secular terms; the constancy of θ_i along the characteristics of the ith equation in (5.6.6), and the vanishing asymptotic means of the terms on the right-hand side of (5.6.6), imply that the wave amplitude π_i satisfy the following integro-differential equation (see [115] and [5])

$$\frac{\partial \pi_i}{\partial t} + \lambda_i\frac{\partial \pi_i}{\partial x} + \mu_i\pi_i + \nu_i\pi_i\frac{\partial \pi_i}{\partial \theta_i}$$

$$+ \sum_{i\neq j\neq k}\Lambda_{jk}^{i}\lim_{T\to\infty}\frac{1}{2T}\int_{-T}^{T}\pi_j(\theta_i + (\lambda_i - \lambda_j)s)\pi_k'(\theta_i + (\lambda_i - \lambda_k)s)ds = 0,$$

$$(5.6.7)$$

where $\Lambda_{jk}^{i} = \mathbf{L}_i \cdot \nabla\mathbf{A}|_o\mathbf{R}^{(j)}\mathbf{R}^{(k)}$ are the interaction coefficients asymmetric in the indices j and k, and represent the strength of coupling between the jth and kth wave which can produce ith wave through the nonlinear interaction. The coefficients Λ_{ii}^{i}, which we denote by ν_i, account for the nonlinear self-interaction, and are nonzero for genuinely nonlinear waves; these coefficients are indeed zero for linearly degenerate waves. The initial conditions for (5.6.7) are recovered from the initial data (5.6.2). It may be noticed that the coefficients μ_i vanish in the absence of source term in (5.1.1). If all the coupling coefficients $\Lambda^i_{jk} (i \neq j \neq k)$ are zero, or the integral averages in (5.6.7) vanish, the waves do not resonate, and (5.6.7) reduces to a system of uncoupled Burgers equations.

It may be remarked that for $n = 2$, there are no integro-differential terms in (5.6.7), i.e., for resonances, the hyperbolic system has at least three equations. The asymptotic equation (5.6.7) can be written in a conservation form, which remains valid in the weak sense after shocks form (see Cehelsky and Rosales [30]). In gasdynamics, a weakly nonlinear resonant interaction between the acoustic modes and the entropy waves occurs (see [115] and [77]); in more than one dimension, vorticity plays a role similar to that of entropy. In this resonance, each acoustic wave interacts with the entropy wave to produce the other sound wave. No new entropy is produced by interactions between the acoustic waves. The resonance results in coupling of the acoustic waves which have a dispersive nature; thus, it can turn off the nonlinear steepening of pressure waves and the shocks which are produced. The numerical calculations show that solutions without wave breaking are possible; the interested reader may consult [115] for dispersive coupling due to resonant wave interactions, and [75] for diffraction of interacting waves.

Remarks 5.6.1: It may be remarked that WNGO is not yet a complete theory; many open problems remain in this area. For instance the behavior of caustics, singular rays and the case of many phases in multidimensions still remain essentially open (see [153] and [164]).

Chapter 6

Self-Similar Solutions Involving Discontinuities

Self-similar solutions are well known and widespread. Self-similarity of the solutions of PDEs has allowed their reduction to ordinary differential equations (ODEs) which often simplifies the investigation.

Self-similarity means that a solution $u(x, t)$ at a certain instant of time t is similar to the solution $u(x, t_o)$ at a certain earlier moment t_o. In fact, similarity is connected with a change of scales, i.e., $u(x, t) = (t/t_o)^m U(x(t/t_o)^n, t_o)$, or $u(x, t) = \phi(t)\widehat{U}(x/\psi(t))$, where the dimensional scales $\phi(t)$ and $\psi(t)$ depend on time t in the manner indicated, and the dimensionless ratio u/ϕ is a function of new dimensionless variable, called the similarity variable, $\xi = x/\psi(t)$. In geometry, this type of transformation is called an affine transformation. The existence of a function $\widehat{U}(\xi)$ that does not change with time enables us to find a similarity of the distribution $u(x, t)$ at different moments. The heat conduction equation has a natural symmetry that enables us to transform the dependent variable u by a scaling factor without changing the solution; the solution of such an equation is

$$u = \frac{E}{c(2\sqrt{\pi kt})^3} \exp(-r^2/(4kt)), \qquad (6.0.1)$$

where E represents a finite amount of heat supplied initially at a point in an infinite space filled with a heat conducting medium, c is the specific heat of the medium, k its thermal diffusivity, and r the radial distance from the point at which the heat is supplied instantaneously (see Barenblatt [10]). It may be noticed from (6.0.1) that there exist temperature and length scales both depending on time, i.e., $u_o(t) = E/(c(kt)^{3/2})$ and $r_o(t) = (kt)^{1/2}$ such that

$$u/u_o = U(r/r_o), \quad U(\xi) = (8\pi^{3/2})^{-1} \exp(-\xi^2/4); \xi = r/r_o.$$

Thus, the spatial distribution of temperature, expressed in terms of the variable scales $u_o(t)$ and $r_o(t)$, is independent of time, and can be expressed as a function of the similarity variable $\xi = r/r_o$. The solution of the problem thus reduces to the solution of an ordinary ODE for $U(\xi)$. In fact, the scale invariance allows us to scale out the t-dependence by introducing a new dependent variable and new spatial variables called similarity variables; the scale invariance and similarity method may be applied to other PDEs.

Let us consider the following scaling transformations

$$u^* = \epsilon^\alpha u, \quad x^* = \epsilon^\beta x, \quad t^* = \epsilon^\gamma t,$$

where α, β, and γ are constants; in other words, we have defined

$$u^*(x^*, t^*) = \epsilon^\alpha u(\epsilon^{-\beta} x^*, \epsilon^{-\gamma} t^*).$$

In order to scale out the t-dependence, we should take $\epsilon^\gamma = t^{-1}$, so that $t^* = 1$; thus, if we introduce $U(x^*) = u^*(x^*, 1)$, and $\xi = x^*$, then

$$u(x, t) = t^{\alpha/\gamma} U(\xi), \tag{6.0.2}$$

defines the similarity transformation with the similarity variable $\xi = x/t^{\beta/r}$. Generally, it is found that at least one of α, β, and γ may be given an arbitrary value; if one takes $\gamma = 1$, then (6.0.2) simplifies to

$$u(x, t) = \phi(t) U(x/f(t)), \tag{6.0.3}$$

where $\phi(t) = t^\alpha$ and $f(t) = t^\beta$. We then substitute (6.0.2) into the given PDE and choose α and β appropriately so that we obtain a differential equation for U, which is independent of t (see McOwen [121]).

In Euler equations of gasdynamics, $u(x, t)$ is the gas velocity; if we carry out this program for this system, we find that since $\xi = x/t^\beta$, $\xi = \text{constant}$ implies $dx/dt = x\beta/t$. Thus, it is appropriate to take $u = (\beta x/t)U(\xi)$, and hence the density, pressure and sound speed can be taken as

$$\rho = x^k \Omega(\xi), \quad p = \beta^2 x^{k+2} t^{-2} P(\xi), \quad c = (\beta x/t) C(\xi). \tag{6.0.4}$$

In other words, the quantities $ut/x, ct/x$ and ρx^{-k} appear constant for an observer who moves so that $\xi = x/t^\beta$ is constant; the factor $\beta x/t$ in u and c is the velocity of such an observer.

The very idea of self-similarity is connected with the study of the structure of PDEs using the notion of symmetry group. This term is used to refer to a group of transformations that maps any solution to another solution of the system, leaving it invariant. In Lie theory, such a group depends on continuous parameters called continuous group of transformations of the differential equation/system. Lie initiated the study of such groups of transformations based on the infinitesimal properties of the group. Briefly, a Lie group is a topological group in which the group operation and taking the inverses are continuous functions, and in which the group manifold has a structure that allows calculus to be performed on it. The method of solving ODEs and PDEs, based on invariance under continuous (Lie) groups of transformations, is a natural generalization of the method of self-similar solutions. More information about the group theoretic method is to be found in [2], [16], [17], [18], [49], [50], [78], [112], [136], [137] and [138].

This chapter includes treatment of self-similar and the basic symmetry (group theoretic) methods for solving Euler equations of gasdynamics involving shocks, characteristics shocks and weak discontinuities together with their interactions. A Lie group of transformation is characterized in terms of its infinitesimal generators which form a Lie algebra; emphasis is placed to discover symmetries admitted by PDEs, and to construct solutions resulting from symmetries. The present account is largely based on our papers ([148], [168], and [170]).

6.1   Waves in Self-Similar Flows

In this section, the work of Rogers [152], which examines the closed form analytical solution for a spherically symmetric blast wave in an atmosphere of varying density, is extended to include in a unified manner the plane and cylindrical cases along with the spherical one. The flow properties behind these waves are completely characterized and certain observations are noted in respect of their contrasting behavior. Finally, the exact self-similar solution and the results of interaction theory are used to study the interaction between a weak discontinuity wave and a blast wave in plane and radially symmetric flows.

The system describing a planar ($m = 0$), cylindrically symmetric ($m = 1$) or spherically symmetric ($m = 2$) adiabatic motion of a perfect gas having velocity u, pressure p, and density ρ may be written as

$$\mathbf{u}_{,t} + \mathbf{A}\mathbf{u}_{,r} + \mathbf{b} = 0, \tag{6.1.1}$$

where $\mathbf{u} = (u, \rho, p)^{tr}, \mathbf{A} = \begin{bmatrix} u & 0 & \rho^{-1} \\ \rho & u & 0 \\ \rho a^2 & 0 & u \end{bmatrix}, \mathbf{b} = (0, m\rho u/r, m\rho a^2 u/r)^{tr}.$

Here $a^2 = \gamma p/\rho$ is the speed of sound with γ the specific heat ratio for the gas, r is the linear distance in the planar case ($m = 0$), or the distance measured from the center in the cylindrical and spherical cases ($m = 1, 2$), and t denotes the time.

At time $t = 0$, an explosion is assumed to take place at a point ($m = 2$), or line ($m = 1$), or plane ($m = 0$) driving ahead of it a blast wave (strong shock) with radius $R(t)$ and velocity $s(t) = dR/dt$, propagating into a medium the density of which decreases as the inverse power of the distance from the source of explosion, so that $\rho^* = \rho_c r^{-\alpha}$, where ρ_c is the central density before the explosion and α is a positive real number. An asterisk will be used to indicate quantities evaluated in the quiescent medium immediately ahead of the blast wave.

It should be noted that the matrix $\mathbf{A}$ in (6.1.1) has the eigenvalues

$$\lambda^{(1)} = u + a, \quad \lambda^{(2)} = u, \quad \lambda^{(3)} = u - a, \tag{6.1.2}$$

and the corresponding left and right eigenvectors

$$\mathbf{L}^{(1)} = (1, 0, 1/\rho a), \quad \mathbf{L}^{(2)} = (0, 1, -1/a^2), \quad \mathbf{L}^{(3)} = (1, 0, -1/\rho a),$$
$$\mathbf{R}^{(1)} = (1, \rho/a, \rho a)^{tr}, \quad \mathbf{R}^{(2)} = (0, 1, 0)^{tr}, \quad \mathbf{R}^{(3)} = (1, -\rho/a, -\rho a)^{tr}.$$
$$(6.1.3)$$

We now consider a C^1 discontinuity wave across the fastest characteristic determined by $dr/dt = \lambda^{(1)}$, and originating from the point (r_0, t_0) in the region swept up by the blast wave, and envisage the situation when this C^1 discontinuity wave encounters a shock wave at time $t = t_p > t_0$. In order to study the time evolution of this C^1 discontinuity wave, and to determine the initial amplitudes of the transmitted and reflected waves, if any, we need to know the solution of the system (6.1.1) in the disturbed region behind the blast wave. To achieve this, we introduce a similarity variable $\xi = r/R(t)$ and seek a solution of the form [152]

$$u = sf(\xi), \quad \rho = \rho^*(R)h(\xi), \quad p = \rho^* s^2 g(\xi)/\gamma, \tag{6.1.4}$$

where $\rho^* = \rho_c R^{-\alpha}, s = dR/dt$, and

$$f(1) = \frac{2}{\gamma + 1}, \quad h(1) = \frac{\gamma + 1}{\gamma - 1}, \quad g(1) = \frac{2\gamma}{\gamma + 1}. \tag{6.1.5}$$

Since the total energy inside a blast wave remains constant under adiabatic conditions for all times, it turns out that $\alpha \le m + 3$, and the radius of the blast wave is given by

$$R(t) = \begin{cases} Kt^{2/(3+m-\alpha)}, & \text{when } \alpha < m + 3 \\ R_0 \exp(Kt), & \text{when } \alpha = m + 3 \end{cases}, \tag{6.1.6}$$

where K is a constant related to the total amount of energy generated by the explosion, and R_0 is the blast wave radius at $t = 0$.

6.1.1   Self-similar solutions and their asymptotic behavior

In view of equations (6.1.4) and (6.1.6)$_1$, the basic set of equations (6.1.1) can now be expressed in the following matrix form

$$\begin{bmatrix} \xi - f & -1/\gamma h & 0 \\ -\gamma & (\xi - f)/g & 0 \\ -1 & 0 & (\xi - f)/h \end{bmatrix} \begin{bmatrix} f' \\ g' \\ h' \end{bmatrix} = \begin{bmatrix} f(\alpha - m - 1)/2 \\ (m\gamma f/\xi) - (m + 1) \\ (mf/\xi) - \alpha \end{bmatrix},$$
$$(6.1.7)$$

where a prime denotes differentiation with respect to ξ. The energy conservation equation provides a first order differential equation which on integration yields

$$2(\gamma f - \xi)g = \gamma(\gamma - 1)(\xi - f)hf^2, \tag{6.1.8}$$

while equations (6.1.7)$_{2,3}$ can be combined to yield

$$h = C_1 g^{\frac{1+m-\alpha}{(m+1)(\gamma-1)}} ((\xi - f)\xi^m)^{\frac{1+m-\alpha}{(m+1)(\gamma-1)}}, \tag{6.1.9}$$

where the integration constant C_1 can be determined by using (6.1.5), as has been done in writing (6.1.8). The compatibility condition for f can be obtained from $(6.1.7)_1$ on using (6.1.8) and (6.1.9) and it has the following form

$$(\xi/f)\, f' = \frac{m\gamma\,(1-\gamma)\,(f/\xi)^2 + [(m+1)\,(2\gamma-1)-\alpha\gamma]\,(f/\xi) - (m+1)+\alpha}{\gamma\,(\gamma+1)\,(f/\xi)^2 - 2\,(\gamma+1)\,(f/\xi) + 2}.$$

(6.1.10)

It is clear that the system (6.1.7) together with (6.1.5) admits single valued nonsingular solutions provided the matrix on the left of (6.1.7) is nonsingular in the interval (0,1). That is, the functions

$$\phi(\xi) = \xi - f \text{ and } \Psi(\xi) = (\xi - f) - (g/h)^2$$

do not vanish at any point, ξ, lying in the interval (0,1). Equation (6.1.8) shows that $\Psi(\xi)$ vanishes nowhere on (0,1); for if it does for some ξ (say, ξ_1), then $f(\xi_1)$ is given by the quadratic equation,

$$\gamma\,(\gamma+1)\,f^2\,(\xi_1) - 2\xi_1\,(\gamma+1)\,f\,(\xi_1) + 2\xi_1^2 = 0,$$

but this does not yield a real expression for $f(\xi_1)$, and so the assertion is justified. However, the function $\phi(\xi)$ may vanish for some $\xi_1 \in (0,1)$, thereby indicating a singularity in the solution at $\xi = \xi_1$; this situation depends on the values of α. Indeed, it turns out that there exists a critical value α_c of α given by

$$\alpha_c = 1 + (m\,(3-\gamma)\,/\,(\gamma+1))\,,$$

such that no singularity in the solution occurs for $\alpha \leq \alpha_c$; possible singularities, if any, can occur only for $\alpha > \alpha_c$. On integration, (6.1.10) yields

$$f = C_2\xi^{k_1}\,(\gamma f - \xi)^{\omega_1}\,((3+m-\alpha)\,\xi/\,\{m\,(\gamma-1)+\gamma+1\} - f)^{s_1}, \quad (6.1.11)$$

where

$$
\begin{aligned}
k_1 &= m\,(\gamma-1)\,(3+m-\alpha)\,/\,(2\,(m-1-m\gamma-\gamma))\,, \\
\omega_1 &= (\gamma-1)\,(\alpha-m-3)\,/\,(2\,(m-1+2\gamma-\alpha\gamma))\,, \\
s_1 &= \frac{\left(m^2 + \alpha^2 + 2m - 4\alpha + 5\right)\gamma^2 + \left(\alpha^2 - 3m^2 + 2m - 2\alpha - 2\alpha m + 1\right]\gamma}{2\,[m\,(1-\gamma)-(\gamma+1))\,(m-1+(2-\alpha)\gamma)} \\
&\quad + \frac{\left(4m^2 - 2\alpha m + 2\alpha - 4\right)}{2\,(m\,(1-\gamma)-(\gamma+1))\,(m-1+(2-\alpha)\gamma)}\,,
\end{aligned}
$$

and C_2 is determined by the boundary conditions (6.1.5). The forms of ω_1 and s_1 suggest that the solution (6.1.11) is valid as long as $\alpha \neq 2 + (m-1)/\gamma$. For $\alpha = 2 + (m-1)/\gamma > \alpha_c$, integration of (6.1.10) gives

$$f = C_3\xi^{\frac{m(1-\gamma)}{2\gamma}}\,(\gamma\,f-\xi)^{\frac{\gamma-1}{\gamma}}\,\exp\left(\frac{(\gamma-1)\,\xi}{2\gamma\,(\gamma\,f-\xi)}\right), \quad (6.1.12)$$

where C_3 is determined by (6.1.5). It should be noted that the corresponding solution for the spherical ($m = 2$) case obtained by Rogers (see equation (65) of [152]) is in error, as can easily be verified; however, the correct solution should be as follows

$$f = K_6 x^{(1-\gamma)/\gamma}(\gamma f - x)^{(\gamma-1)/2\gamma} \exp\left(\frac{(\gamma - 1)\,x}{2\gamma\,(\gamma f - x)}\right),$$

which is recovered from (6.1.12) with $m = 2$, since $\xi = x$ and $C_3 = K_6$. The asymptotic behavior of the solutions of (6.1.11) and (6.1.12) near the origin are as follows.

Case I. When $\alpha = \alpha_c = 1 + (m\,(3 - \gamma)\,/\,(\gamma + 1))$, the velocity, density and pressure behind plane, cylindrical and spherical waves are given by

$$f(\xi) = \frac{2\xi}{\gamma + 1}, \quad h(\xi) = \frac{\gamma + 1}{\gamma - 1}\xi^{m-1}, \quad g(\xi) = \frac{2\gamma}{\gamma + 1}\xi^{m+1}. \tag{6.1.13}$$

Equation $(6.1.13)_1$ follows from (6.1.10), while equation $(6.1.13)_{2,3}$ follow from (6.1.8) and (6.1.9). The solution behind spherical waves found by Kopal et al. [95] and Rogers [152] is recovered from (6.1.13) with $m = 2$. It should be noted that, unlike the case of spherical waves, in the case of plane and cylindrical waves α_c does not vanish for γ. The density behind plane waves, unlike cylindrical and spherical waves, exhibits singular behavior at the origin; indeed, it increases (respectively; decreases) from the shock towards the center in the case of plane (respectively; spherical) shocks, while it remains constant behind cylindrical waves. Numerical solutions of equations (6.1.7) satisfying (6.1.5) are exhibited in Figs. 6.1.1 – 6.1.3; the situation under consideration here is illustrated by the curve VI in Fig. 6.1.1.

Case II. When $\alpha < \alpha_c$, it follows from (6.1.11) that near $\xi = 0$,

$$f(\xi) \sim \frac{\xi}{\gamma} + \frac{\gamma + 1}{(\gamma - 1)\gamma}\xi^\theta,$$

where $\theta = ((3 - \alpha)\,\gamma + m - 2)\,/\,(\gamma - 1)$. Thus the velocity for plane, cylindrical and spherical waves is nearly a linear function of ξ. The density and pressure distributions near $\xi = 0$ can be obtained from (6.1.8) and (6.1.9) in the following form

$$h(\xi) = O\left(\xi^\delta\right), \quad g(\xi) = O(1) \quad \text{as } \xi \to 0,$$

where $\delta = (1 + m - \alpha\gamma)/(\gamma - 1)$. Thus, depending on the value of α, we have the following cases: (i) when $\alpha \le (2 + m - \gamma)/\gamma$, the density vanishes at the center. Curves I and II in Fig. 6.1.1 are representative of the corresponding situations. (ii) When $(m + 2 - \gamma)/\gamma < \alpha < (m + 1)/\gamma$, the density vanishes at the origin but the density gradient becomes unbounded there as shown by curve III in Fig. 6.1.1. (iii) When α is equal to (respectively; greater than) the value $(m+1)/\gamma$, the density at the origin is finite (respectively; infinite). Thus, the density distribution behind plane, cylindrical and spherical waves is regular on $(0,1)$ except, perhaps, at the origin, where the behavior could be singular; the pressure, however, remains finite at the origin in this case. The respective situations are delineated by curves IV and V in Fig. 6.1.1. When $\alpha = (m+1)\gamma$, we have the following interesting situation. Equation (6.1.9) yields $gh^{-\gamma} = $ constant, while equations $(6.1.4)_{2,3}$ together with (6.1.5) imply

$$p\rho^{-\gamma} = 4\rho_c^{(1-\gamma)}K^{(3+m-\alpha)}R^{(\alpha\gamma-1-m)}gh^{-\gamma}\left(\gamma(3+m-\alpha)^2\right)^{-1},$$

which, for $\alpha = (m+1)/\gamma$, yields $p\rho^{-\gamma} = $ constant, and thus in this case the flow behind a plane, cylindrical or spherical shock is isentropic.

Case III. When $\alpha > \alpha_c$, as remarked earlier, a singularity in the solution appears at $\xi = \xi_1 \in (0,1)$, where $\phi(\xi_1) = 0$. In this case, it turns out that the velocity behind cylindrical and spherical waves tends to infinity near the center, whereas it remains finite there for plane waves. Indeed, in spherical waves the rise near the center is sharper when compared to the cylindrical wave case. Following a standard procedure, we find from (6.1.11) that

$$f(\xi) = O\left(\xi^{-m(\gamma-1)/(\gamma+1)}\right) \quad \text{as} \quad \xi \to 0. \tag{6.1.14}$$

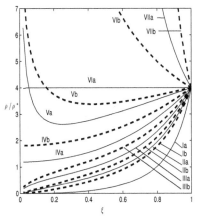

Figure 6.1.1: Density distribution behind plane ($m = 0$, - - - -) and cylindrical ($m = 1$,———) shock fronts for $\gamma = 5/3..$ Ia: $\alpha =0.0$, Ib : $\alpha =0.0$, IIa: $\alpha =0.8$, IIb : $\alpha =0.2$, IIIa: $\alpha =1.0$, IIIb : $\alpha =0.4$, IVa: $\alpha =1.2$, IVb : $\alpha =0.6$, Va: $\alpha =1.4$, Vb : $\alpha =0.75$, VIa: $\alpha =1.5$, VIb : $\alpha =1.0$, VIIa: $\alpha =2.0$, VIIb : $\alpha =2.0$.

It follows from (6.1.8) and (6.1.9) that for $\alpha < 2(m+1)/(\gamma+1)$, and thus for plane and cylindrical waves, $\alpha < \alpha_c$ and so the possibility of a singularity in the solution on $(0,1)$ is ruled out. However, for spherical waves this situation includes the possibility that $\alpha > \alpha_c$, thereby indicating the possibility of a singularity in the solution at $\xi_1 \in (0,1)$, where the density vanishes but the density gradient becomes unbounded. It should be noted that for $\alpha = 2(m+1)/(\gamma+1)$ there is no singularity in the density, in the sense that both the density and its gradient are finite at $\xi = \xi_1$. However, when $\alpha > 2(m+1)/(\gamma+1)$, the density becomes infinite at ξ_1 while its gradient changes sign indicating that the gas is further compressed after the passage of the shock. The pressure distribution behind plane, cylindrical and spherical waves can be discussed in a similar manner by exploiting the result obtained from (6.1.8) and (6.1.9) after eliminating the function $h(\xi)$. It turns out that for $\alpha \le 2\,(m+1)\,/\,(\gamma+1)$, as shown by curves I and II in Fig. 6.1.2, and by curves IV and V in Fig. 6.1.3, the pressure distribution behind plane and cylindrical waves is regular on $(0,1)$, while for spherical waves the solution exhibits a singularity at a point $\xi_1 \in (0,1)$, where the gas pressure and its gradient are finite.

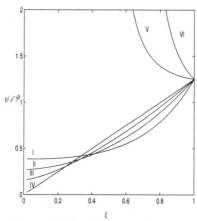

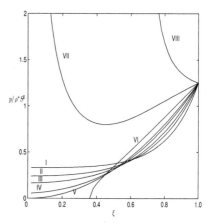

Figure 6.1.2: Pressure distribution behind plane shock front $(m=0)$ for $\gamma = 5/3$. I: $\alpha =0.4$, II: $\alpha =0.75$, III: $\alpha=0.9$,   IV: $\alpha =1.0$, V: $\alpha =1.5$,   VI: $\alpha =2.0$.

Figure 6.1.3:   Pressure distribution behind cylindrical shock front $(m=1)$ for $\gamma = 5/3$. I: $\alpha =0.6$, II: $\alpha =1.0$, III: $\alpha =1.2$, IV: $\alpha =1.4$, V: $\alpha =1.5$, VI: $\alpha =1.7$, VII: $\alpha =2.0$, VIII: $\alpha =2.5$.

When $\alpha > 2(1+m)/(\gamma+1)$, as shown by curves III, IV and V in Fig. 6.1.2, and by curves VI, VII and VIII in Fig. 6.1.3, the pressure at ξ_1 is zero, finite or infinite according as α is, respectively, lees than, equal to, or greater than $m+1$; in these cases the pressure gradient is, respectively, infinite, finite or negatively infinite at ξ_1, showing in effect that when $\alpha = m+1$, there is no singularity in pressure.

Case IV. When $\alpha = m + 3$, equation (6.1.6) shows that the system (6.1.1) admits a similarity solution of the type given in (6.1.4) in a region headed by a strong shock of radius $R = R_0 \exp(Kt)$, propagating into a medium the density of which varies according to $\rho^* = \rho_c R^{-(m+3)}$. The solution for $f(\xi)$ in this case can be found from (6.1.10), for $\alpha = m + 3$, in the form

$$f = K_1 \left(\xi^m \left(\gamma f - \xi\right)\right)^{-(\gamma-1)/2} \exp(-\xi/f), \qquad (6.1.15)$$

where K_1 is determined by (6.1.5)$_1$. The density follows from (6.1.8) and (6.1.9). The asymptotic behavior of the velocity near the origin is the same as that given by (6.1.14); that is, unlike plane waves where the velocity remains finite near the center, the velocity in cylindrical and spherical waves tends to ∞; however, the rise near the center is more rapid in spherical waves than in cylindrical waves. Again, a singularity in the solution occurs at the point $\xi_1 \in (0,1)$, which from (6.1.15) is given by

$$\xi_1 = \left\{ 2 \exp((\gamma - 1)/2)/(\gamma + 1)^{(\gamma+1)/2} \right\}^{2/[\gamma+1+m(\gamma-1)]},$$

which shows that for a given $\gamma > 1$, an increase in m causes ξ_1 to increase. That is, an increase in the wave front curvature causes the singular point to move toward the wave front. However, an increase in γ has the reverse effect, so that for a given geometry of the wave front, an increase in γ causes ξ_1 to move towards the wave center.

6.1.2 Collision of a C^1-wave with a blast wave

We now consider system (6.1.1). Let $\mathbf{\Lambda}_1$ denote the jump in $\mathbf{u}_{,r}$ across the C^1 discontinuity propagating along the curve determined by $dr/dt = \lambda^{(1)}$, originating from the point (r_0, t_0) in the region behind the blast wave, considered in the preceding sections. In view of equations (6.1.4), (6.1.6) and (6.1.13), the medium ahead of this C^1 discontinuity wave is characterized by the solution vector $\mathbf{u}_0$ of the form

$$\mathbf{u}_0 = \left\{ \frac{2}{\gamma+1} \frac{r}{\lambda t}, \; \frac{\gamma+1}{\gamma-1} \rho_c R^{-\alpha} \left(\frac{r}{R} \right)^{m-1}, \; \frac{2}{\gamma+1} \rho_c R^{-\alpha} \left(\frac{R}{\lambda t} \right)^2 \left(\frac{r}{R} \right)^{m+1} \right\}^{t_r},$$

$$(6.1.16)$$

where $\lambda = (3 + m - \alpha)/2, \alpha = 1 + \{m(3 - \gamma)/(\gamma + 1)\}$ and $R = Kt^{1/\lambda}$. Let $\alpha^{(1)}(t)$ be the amplitude of this C^1 discontinuity; then $\mathbf{\Lambda}_1 = \alpha^{(1)} \mathbf{R}_0^{(1)}$, where $\mathbf{R}_0^{(1)}$ is the vector $\mathbf{R}^{(1)}$ in (6.1.3) evaluated in the state $\mathbf{u} = \mathbf{u}_0$ described by (6.1.16). In view of (6.1.1), (6.1.2), (6.1.3) and (6.1.16), system (4.3.8) reduces to a single Bernoulli type equation in $\alpha^{(1)}$, which on integration yields

$$\alpha^{(1)}(t) = [t/t_0]^{-\nu} \alpha_0^{(1)}(t_0) \left\{ 1 + \left(\alpha_0^{(1)}(t_0)/Z \right) \left(1 - (t_0/t)^{\nu-1} \right) \right\}^{-1}, \quad (6.1.17)$$

where $\alpha_0^{(1)}$ is the initial wave amplitude at $t = t_0$

$$Z = 2(\gamma - 1)\{(\gamma + 1)t_0\}^{-1} > 0, \; \nu = \{(m+1)(\gamma + 2\Gamma) - m + 5\}/(2\theta),$$

with $\Gamma^2 = 2\gamma(\gamma - 1)$ and $\theta = (\gamma + 1) + m(\gamma - 1)$. It should be noted that for a plane ($m = 0$), or a radially symmetric ($m = 1$ or 2) flow, $\nu > 1$ for values of $\gamma > 1$. Thus the occurrence of a secondary shock, in the region behind the blast wave, can take place at a time $t = \tau > t_0$, given by $\tau = t_0\{1 - Z/|\alpha_0^{(1)}|\}^{1/(\nu-1)}$, provided the initial weak discontinuity wave is compressive ($\alpha_0^{(1)} < 0$) with its initial amplitude satisfying the inequality $\alpha_0^{(1)} < -Z < 0$. It is clear that when the initial weak discontinuity is an expansion wave ($\alpha_0^{(1)} > 0$), or a compression wave ($\alpha_0^{(1)} < 0$) with $-Z \leq \alpha_0^{(1)} < 0$, there is no secondary shock formation in the self-similar region behind the blast wave, and the C^1 wave will impinge upon the blast wave at some $t > t_0$. However, a compressive C^1 wave with $\alpha_0^{(1)} < -Z < 0$ can meet the blast at, say, $t = t_p \equiv \hat{t}$, provided $t_0 < \hat{t} < \tau$; an equivalent condition for this to occur follows from the foregoing relations and the fact that $r(\hat{t}) = R(\hat{t})$ in the form

$$t_0 \left| \alpha_0^{(1)} \right| < 2(\nu - 1)/(\gamma + 1) \left\{ 1 - [R_0/r_0]^{-(\nu-1)\theta/(\Gamma-\gamma+1)} \right\}.$$

The eigenvalues, the corresponding eigenvectors and the shock velocity $s = dR/dt = (\gamma + 1)R/\theta t$, when evaluated at the collision time $t = \hat{t}$, yield

$$\hat{\lambda}^{(1)} = \lambda^{(1)}(\hat{t}) = \hat{s}(2 + \Gamma)/(\gamma + 1), \quad \hat{\lambda}^{(2)} = 2\hat{s}/(\gamma + 1), \quad \hat{\lambda}^{(3)} = \hat{s}(2 - \Gamma)/(\gamma + 1),$$

where $\hat{s} \equiv s(\hat{t}) = (\gamma + 1)\hat{R}/(\theta\hat{t})$ with $\hat{R} = R(\hat{t})$. Also, since the pressure ahead of the blast wave (where the variables are designated by an asterisk) is very small when compared with the pressure behind, it follows immediately that the Lax evolutionary conditions (4.3.22) for a "physical shock" are satisfied for the index $\ell = 2$, so that

$$\hat{\lambda}^{(3)} < \hat{\lambda}^{(2)} < \hat{s} < \hat{\lambda}^{(2)}, \quad \hat{\lambda}^{*(3)} < \hat{\lambda}^{*(2)} < \hat{\lambda}^{*(1)} < \hat{s}.$$

In effect, this asserts that when the incident wave with velocity $\hat{\lambda}^{(1)}$ encounters the blast wave, it gives rise to two reflected waves with velocities $\hat{\lambda}^{(2)}$ and $\hat{\lambda}^{(3)}$, but no transmitted wave. The reflection coefficients $\alpha^{(2)}, \alpha^{(3)}$ and the jump in shock acceleration $[\![s]\!]$ at the collision time $t = \hat{t}$ can be determined from the algebraic system (4.3.23), which can be written in the following form

$$[\![\dot{s}]\!](\mathbf{G} - \mathbf{G}^*)_0 + (\nabla\mathbf{G})_0 \mathbf{R}_0^{(2)} \alpha^{(2)} \left(s - \lambda^{(2)}\right)^2 + (\nabla\mathbf{G})_0 \mathbf{R}_0^{(3)} \alpha^{(3)} \left(s - \lambda^{(3)}\right)^2$$
$$= -(\nabla\mathbf{G})_0 \mathbf{R}_0^{(1)} \alpha^{(1)} \left(s - \lambda^{(1)}\right)^2, \tag{6.1.18}$$

where all the quantities are evaluated in the state $\mathbf{u}_0$ at $t = \hat{t}$, so that

$$(\mathbf{G} - \mathbf{G}^*)_0 = \left\{ 2\hat{\rho}(\gamma - 1)^{-1}(\gamma + 1)\left(\hat{R}/\theta\hat{t}\right), 1, 2(\gamma + 1)\left(\hat{R}/\theta\hat{t}\right)^2 \right\}^{t_r},$$

$$(\nabla\mathbf{G})_0 \, \mathbf{R}_0^{(2)} = \left\{ 2\left(\hat{R}/\theta\hat{t}\right), 1, 2\left(\hat{R}/\theta\hat{t}\right)^2 \right\}^{t_r},$$

$$(\nabla\mathbf{G})_0 \, \mathbf{R}_0^{(3)} = \Psi \left\{ (\Gamma - 2), -\left(\hat{R}/\theta\hat{t}\right)^{-1}, -2\left(\hat{R}/\theta\hat{t}\right)((\gamma + 1 - \Gamma)\right\}^{t_r},$$

$$(\nabla\mathbf{G})_0 \, \mathbf{R}_0^{(1)} = \Psi \left\{ (\Gamma + 2), \left(\hat{R}/\theta\hat{t}\right)^{-1}, 2\left(\hat{R}/\theta\hat{t}\right)((\gamma + 1 - \Gamma)\right\}^{t_r},$$

where $\Psi = \hat{\rho}(\gamma + 1)\left\{\Gamma\left(\gamma - 1\right)\right\}^{-1}$ and $\hat{\rho} = \hat{\rho}_c \hat{R}^{-\alpha}$.

The system (6.1.18) yields the following solution at $t = \hat{t}$

$$[[s]] = -\frac{\Gamma(\gamma + 1)\left\{3\gamma - 2\Gamma - 1\right\}(\hat{R}/\theta\hat{t})[\hat{t}/t_0]^{-\nu}\alpha_0^{(1)}}{(\Gamma + 2)\left\{1 + \left[\alpha_0^{(1)}/Z\right]\left[1 - [t_0/\hat{t}]^{\nu-1}\right]\right\}},$$

$$\alpha^{(2)} = -\frac{4(\gamma + 1)\left\{3\gamma - 2\Gamma - 1\right\}\hat{\rho}\left[\hat{t}/t_0\right]^{-\nu}\alpha_0^{(1)}}{(\Gamma + 2)\Gamma(\gamma - 1)\left(\hat{R}/\theta\hat{t}\right)\left\{1 + \left[\alpha_0^{(1)}/Z\right]\left[1 - [t_0/\hat{t}]^{\nu-1}\right]\right\}}, \qquad (6.1.19)$$

$$\alpha^{(3)} = -\frac{(\Gamma - 2)\left(3\gamma - 2\Gamma - 1\right)^2}{(\Gamma + 2)\left(\Gamma + \gamma - 1\right)^2}\frac{[\hat{t}/t_0]^{-\nu}}{\left\{1 + \left[\alpha_0^{(1)}/Z\right]\left[1 - [t_0/\hat{t}]^{\nu-1}\right]\right\}}\alpha_0^{(1)}.$$

The coefficients $\alpha^{(2)}$ and $\alpha^{(3)}$ determine the amplitude vectors

$$\mathbf{\Lambda}_2^{(R)} = \alpha^{(2)}(\hat{t})\mathbf{R}^{(2)}, \quad \mathbf{\Lambda}_3^{(R)} = \alpha^{(3)}(\hat{t})\mathbf{R}^{(3)}$$

of the reflected waves along the characteristic lines with the velocities $\lambda^{(2)}$ and $\lambda^{(3)}$, respectively. Equations (6.1.19) demonstrate, as would be expected, that in the absence of the incident wave, i.e.,

$$\alpha_0^{(1)} = 0,$$

the jump in the shock acceleration vanishes and there are no reflected waves. Moreover an increase in the magnitude of the initial discontinuity of the incident wave causes the reflection coefficients and the jump in shock acceleration to increase in magnitude. It is clear from (6.1.19)$_1$ that for γ lying in the range $1 < \gamma < 2$, the coefficient $\alpha_0^{(1)}$ is always negative, thereby indicating that after the impact the shock will either accelerate or decelerate, depending on whether the incident wave is compressive or expansive. This conclusion is in agreement with the observations made by Friedrichs [41], that if the shock front is overtaken by a compression (respectively; expansion) wave, it is accelerated (respectively; decelerated) and consequently the strength of the shock increases (respectively; decreases). Moreover, since the coefficient $\alpha_0^{(1)}$ in (6.1.19)$_3$ is negative for values of γ in the range $1 < \gamma < 2$, this is indicative of the situation that the reflected wave along the characteristic $dr/dt = u-a$ is a compression wave whenever the incident wave is expansive, and conversely. This behavior may, perhaps, be attributed to the nonlinear effects. The present analysis can really only hint at the manner in which the incident wave interacts with the shock and the reflected waves evolve. However, a more elaborate treatment offering a simultaneous account of the solution in the general wave region behind the incident and reflected waves together with the evolutionary behavior of the reflected waves which seem to be responsible for some of the interesting features of the wave motion that finally develop, offer substantial hope for progress in the understanding of this complex problem.

6.2   Imploding Shocks in a Relaxing Gas

It is well known that in contrast to the self-similar solutions of the first type, where the self-similarity of the solution and its representation can be obtained from dimensional analysis, the nature of the similarity in respect of self-similar solutions of the second type is revealed neither by dimensional analysis nor by other group properties of the medium, but only by actually solving the equations as a nonlinear eigenvalue problem. The collapse of an imploding shock wave, is one of the examples of a class of self-similar solutions of the second type. A theoretical investigation of imploding shock waves in a perfect gas has been presented by Guderley [66], Zeldovich and Raizer [214], and Lazarus [103]; Van Dyke and Guttmann [202] as well as Hafner [68] have derived solutions to the convergent shock problem by means of alternative methods. Assuming the shock to be infinitely strong, Logan and Perez [112] discussed the self-similar solutions to a problem concerning plane flow of a reacting gas with an arbitrary form of the reaction rate.

In flows with imploding shocks, conditions of very high temperature and pressure can be produced near the center (axis) of implosion on account of the self-amplifying nature of imploding shocks. As a result of high temperatures attained by gases in motion, the effects of nonequilibrium thermodynamics on the dynamic motion of a converging shock wave can be important. Ever since the nonequilibrium effects were recognized to be important in high temperature gasdynamics, considerable efforts have been devoted to the restudy of the classical problem in the light of these new considerations. Here, we use the method of Lie group invariance under infinitesimal point transformations ([112], [17]) to determine the class of self-similar solutions for unsteady planar and radially symmetric flows of an inviscid relaxing gas involving shocks, and discuss the qualitative features of the possible self-similar solutions in a unified manner. The departure from equilibrium is due to vibrational relaxation of a diatomic gas; the rotational and translational modes are assumed to be in local thermodynamical equilibrium throughout. The group theoretic method enables us to characterize completely the state dependent form of the relaxation rate for which the problem is invariant and admits self-similar solutions. The arbitrary constants, occurring in the expressions for the generators of the local Lie group of transformations, give rise to different cases of possible solutions with a power law, exponential or logarithmic shock paths. A particular case of the collapse of an imploding shock is worked out in detail for radially symmetric flows. For typical values of the flow parameters, the value of the self-similar exponent is uniquely determined from the condition that the solution of the system of ordinary differential equations describing the self-similar motion is regular on a regular characteristic passing through the center (axis) of implosion. Indeed, the system of ordinary differential equations is integrated iteratively with the similarity exponent so adjusted that the system is made regular. Numerical calculations have been performed to determine the values of the self-similarity exponent and the profiles of the flow variables behind the shock; for the sake of comparison the similarity exponent is also computed using Whitham's rule. This work brings out some interesting features of the interaction between the relaxing mode and the fluid flow. In the absence of vibrational relaxation, our results are in good agreement with the values of the similarity parameter quoted in [66], [68] and [159].

6.2.1 Basic equations

Assuming that the gas molecules have only one lagging internal mode (i.e., vibrational relaxation) and the various transport effects are negligible, the basic equations for an unsteady one dimensional planar $(m = 0)$, cylindrically $(m = 1)$ or spherically $(m = 2)$ symmetric motion of a relaxing gas are [37]

$$\rho_{,t} + \rho u_{,x} + u\rho_{,x} + m\rho u/x = 0, \tag{6.2.1}$$
$$u_{,t} + uu_{,x} + \rho^{-1}p_{,x} = 0, \tag{6.2.2}$$
$$p_{,t} + up_{,x} + \rho a^2(u_{,x} + mu/x) + (\gamma - 1)\rho Q = 0, \tag{6.2.3}$$
$$\sigma_{,t} + u\sigma_{,x} = Q, \tag{6.2.4}$$

where x is the spatial coordinate being either axial in flows with planar geometry or radial in cylindrically and spherically symmetric flows, t the time, u the particle velocity, ρ the density, p the pressure, σ the vibrational energy and $a = (\gamma p/\rho)^{1/2}$ the frozen speed of sound with γ being the frozen specific heat ratio of the gas. The quantity Q, which is a known function of p, ρ and σ, denotes the rate of change of vibrational energy, and is given by [163]

$$Q = \{\bar{\sigma}(p,\rho) - \sigma\}/\tau(p,\rho), \tag{6.2.5}$$

where $\bar{\sigma}$ is the equilibrium value of σ defined as $\bar{\sigma} = \bar{\sigma}_0(x) + c(p,\rho)R(T - T_0(x))$; here T is the translational temperature, R is the specific gas constant, suffix 0 refers to the initial rest conditions, and the quantities τ and c are respectively the relaxation time and the ratio of vibrational specific heat to the specific gas constant. The equation of state is taken to be of the form

$$p = \rho RT. \tag{6.2.6}$$

Now, we consider the motion of a shock front, $x = X(t)$, propagating into an inhomogeneous medium specified by

$$u_0 \equiv 0, \quad p_0 \equiv \text{constant}, \quad \rho_0 = \rho_0(x), \quad \sigma_0 = \sigma_0(x). \tag{6.2.7}$$

Let V be the shock velocity; then the usual Rankine Hugoniot jump conditions across the shock front may be written as

$$u(X(t),t) = \frac{2}{\gamma+1}\frac{\rho_0(X(t))V^2 - \gamma p_0}{\rho_0(X(t))V}, \quad \rho(X(t),t) = \frac{(\gamma+1)\rho_0^2(X(t))V^2}{(\gamma-1)\rho_0(X(t))V^2 + 2\gamma p_0},$$
$$p(X(t),t) = \frac{2\rho_0(X(t))V^2 - (\gamma-1)p_0}{\gamma+1}, \quad \sigma(X(t),t) = \sigma_0(X(t)).$$

$$\tag{6.2.8}$$

6.2.2   Similarity analysis by invariance groups

Since the system (6.2.1) – (6.2.4) is a set of quasilinear hyperbolic partial differential equations, it is hard, in general, to determine a solution without approximations. Here, we assume that there exists a solution of (6.2.1) – (6.2.4) subject to (6.2.8) along a family of curves, called similarity curves, for which the set of partial differential equations reduce to a set of ordinary differential equations, and also assume that the shock trajectory is embedded in the family of similarity curves; this type of solution is called a similarity solution.

In order to determine a similarity solution and the similarity curves, we seek a one parameter infinitesimal group of transformations

$$
\begin{aligned}
t^* &= t + \epsilon\psi(x, t, \rho, u, p, \sigma), & x^* &= x + \epsilon\chi(x, t, \rho, u, p, \sigma), \\
\rho^* &= \rho + \epsilon S(x, t, \rho, u, p, \sigma), & u^* &= u + \epsilon U(x, t, \rho, u, p, \sigma), \quad (6.2.9) \\
p^* &= p + \epsilon P(x, t, \rho, u, p, \sigma), & \sigma^* &= \sigma + \epsilon E(x, t, \rho, u, p, \sigma),
\end{aligned}
$$

where the generators ψ, χ, S, U, P and E are to be determined in such a way that the partial differential equations (6.2.1) – (6.2.4), together with the conditions (6.2.7) and (6.2.8), are invariant with respect to the transformations (6.2.9); the entity ϵ is so small that its square and higher powers may be neglected. The existence of such a group allows the number of independent variables in the problem to be reduced by one, and thereby allowing the system (6.2.1) – (6.2.4) to be replaced by a system of ordinary differential equations.

In the sequel, we shall use summation convention and, therefore, introduce the notation

$$
x_1 = t, \quad x_2 = x, \quad u_1 = \rho, \quad u_2 = u, \quad u_3 = p, \quad u_4 = \sigma, \quad p_j^i = \partial u_i / \partial x_j,
$$

where $i = 1, 2, 3, 4$ and $j = 1, 2$. The system (6.2.1) – (6.2.4), which can be represented as

$$
F_k(x_j, u_i, p_j^i) = 0, \quad k = 1, 2, 3, 4,
$$

is said to be constantly conformally invariant under the infinitesimal group (6.2.9), if there exist constants $\alpha_{rs}(r, s = 1, 2, 3, 4)$ such that for all smooth surfaces, $u_i = u_i(x_j)$, we have

$$
\mathcal{L}F_k = \alpha_{kr}F_r, \tag{6.2.10}
$$

where $\mathcal{L}$ is the Lie derivative in the direction of the extended vector field

$$
\mathcal{L} = \xi_x^j \frac{\partial}{\partial x_j} + \xi_u^i \frac{\partial}{\partial u_i} + \xi_{p_j}^i \frac{\partial}{\partial p_j^i},
$$

with $\xi_x^1 = \psi, \ \xi_x^2 = \chi, \ \xi_u^1 = S, \ \xi_u^2 = U, \ \xi_u^3 = P, \ \xi_u^4 = E$, and

$$
\xi_{p_j}^i = \frac{\partial\xi_u^i}{\partial x_j} + \frac{\partial\xi_u^i}{\partial u_k}p_j^k - \frac{\partial\xi_x^\ell}{\partial x_j}p_\ell^i - \frac{\partial\xi_x^\ell}{\partial u_n}p_\ell^i p_j^n, \quad \ell = 1, 2, \quad n = 1, 2, 3, 4 \quad (6.2.11)
$$

being the generators of the derivative transformation.

Equation (6.2.10) implies

$$\frac{\partial F_k}{\partial x_j}\xi_x^j + \frac{\partial F_k}{\partial u_i}\xi_u^i + \frac{\partial F_k}{\partial p_j^i}\xi_{p_j}^i = \alpha_{ks}F_s, \quad k = 1, 2, 3, 4.$$

Substitution of $\xi_{p_j}^i$ from (6.2.11) into the last equation gives a polynomial in the p_j^i. Setting the coefficients of p_j^i and $p_j^i p_l^k$ to zero gives a system of first order, linear partial differential equations in the generators ψ, χ, S, U, P and E. This system, which is called the system of determining equations of the group, can be solved to find the invariance group (6.2.9).

If the above program is carried out for the the system of partial differential equations (6.2.1) – (6.2.4), we obtain the most general group under which the system is invariant; indeed, the invariance of the continuity equation (6.2.1) yields the following system of determining equations

$$S_{,t} + uS_{,x} + \rho U_{,x} + \frac{m}{x}\left(\rho U + uS - \frac{\rho u\chi}{x}\right) = \alpha_{13}\left\{(\gamma - 1)\rho Q + \frac{m\gamma pu}{x}\right\}$$
$$+ \alpha_{11}\frac{m\rho u}{x} - \alpha_{14}Q,$$

$$S_{,\rho} - \psi_{,t} - u\psi_{,x} = \alpha_{11}, \qquad S_{,u} - \rho\psi_{,x} = \alpha_{12}, \qquad S_{,p} = \alpha_{13},$$
$$S_{,\sigma} = \alpha_{14}, \qquad U - \chi_{,t} + uS_{,\rho} - u\chi_{,x} + \rho U_{,\rho} = \alpha_{11}u, \tag{6.2.12}$$
$$S + uS_{,u} + \rho(U_{,u} - \chi_{,x}) = \alpha_{11}\rho + \alpha_{12}u + \alpha_{13}\gamma p,$$
$$\rho U_{,p} + uS_{,p} = \alpha_{12}/\rho + \alpha_{13}u, \qquad \rho U_{,\sigma} + uS_{,\sigma} = \alpha_{14}u.$$

In a similar way, invariance of the momentum equation (6.2.2) yields

$$U_{,t} + uU_{,x} + \rho^{-1}P_{,x} = \alpha_{21}m\rho u/x + \alpha_{23}\left\{(\gamma - 1)\rho Q + m\gamma pu/x\right\} - \alpha_{24}Q,$$
$$U_{,\rho} = \alpha_{21}, \quad U_{,u} - \psi_{,t} - u\psi_{,x} = \alpha_{22}, \quad U_{,p} - \rho^{-1}\psi_{,x} = \alpha_{23}, \quad U_{,\sigma} = \alpha_{24},$$
$$U - \chi_{,t} + u(U_{,u} - \chi_{,x}) + \rho^{-1}P_{,u} = \alpha_{21}\rho + \alpha_{22}u + \alpha_{23}\gamma p, \tag{6.2.13}$$
$$uU_{,\rho} + \rho^{-1}P_{,\rho} = \alpha_{21}u, \quad -S/\rho^2 + uU_{,p} + \rho^{-1}(P_{,p} - \chi_{,x}) = \alpha_{22}\rho^{-1} + \alpha_{23}u,$$
$$uU_{,\sigma} + \rho^{-1}P_{,\sigma} = \alpha_{24}u.$$

Next, invariance of the energy equation (6.2.3) yields

$$P_{,t} + uP_{,x} + \gamma pU_{,x} + m\gamma(uP + pU - pu\chi/x)/x + (\gamma - 1)SQ$$
$$+ (\gamma - 1)\left\{SQ_{,\rho} + PQ_{,p} + EQ_{,\sigma}\right\}\rho = \alpha_{31}m\rho u/x$$
$$+ \alpha_{33}\left\{(\gamma - 1)\rho Q + m\gamma pu/x\right\} - \alpha_{34}Q,$$
$$P_{,\rho} = \alpha_{31}, \quad P_{,u} - \gamma p\psi_{,x} = \alpha_{32}, \quad P_{,p} - \psi_{,t} - u\psi_{,x} = \alpha_{33},$$
$$P_{,\sigma} = \alpha_{34}, \qquad uP_{,\rho} + \gamma pU_{,\rho} = \alpha_{31}u, \tag{6.2.14}$$
$$\gamma P + uP_{,u} + \gamma p(U_{,u} - \chi_{,x}) = \alpha_{31}\rho + \alpha_{32}u + \alpha_{33}\gamma p,$$
$$U - \chi_{,t} + uP_{,p} - u\chi_{,x} + \gamma pU_{,p} = \alpha_{32}/\rho + \alpha_{33}u, \quad uP_{,\sigma} + \gamma pU_{,\sigma} = \alpha_{34}u.$$

Finally, invariance of the rate equation (6.2.4) yields

$$E_{,t} + uE_{,x} - (SQ_{,\rho} + PQ_{,p} + EQ_{,\sigma}) = \alpha_{41}m\rho u/x$$
$$+ \alpha_{43}\left\{(\gamma - 1)\rho Q + m\gamma pu/x\right\} - \alpha_{44}Q,$$
$$E_{,\rho} = \alpha_{41}, \qquad E_{,u} = \alpha_{42}, \qquad E_{,p} = \alpha_{43},$$
$$E_{,\sigma} - u\psi_{,x} - \psi_{,t} = \alpha_{44}, \qquad uE_{,\rho} = \alpha_{41}u, \qquad (6.2.15)$$
$$uE_{,u} = \alpha_{41}\rho + \alpha_{42}u + \alpha_{43}\gamma p, \qquad uE_{,p} = \alpha_{42}/\rho + \alpha_{43}u,$$
$$U - \chi_{,t} + uE_{,\sigma} - u\chi_{,x} = \alpha_{44}u, \qquad \psi = \psi(x,t), \quad \chi = \chi(x,t).$$

From $(6.2.12)_{5,9}$, $(6.2.13)_{2,5,9}$ and $(6.2.14)_{2,5,6}$ we have $U_{,\sigma} = P_{,\sigma} = U_{,\rho} = P_{,\rho} = \alpha_{21} = \alpha_{31} = \alpha_{24} = \alpha_{34} = 0$; consequently, derivatives of $(6.2.13)_4$ and $(6.2.13)_6$ with respect to ρ and p yield $U_{,\rho\rho} = \psi_{,x} = U_{,p} = \alpha_{23} = 0$. Thus, we have from $(6.2.12)_{4,8}$ and $(6.2.15)_{4,8}$ that $S_{,u} = E_{,u} = \alpha_{12} = \alpha_{42} = 0$.

Next, differentiating $(6.2.12)_7$ with respect to p and σ and using $(6.2.12)_4$, we get $S_{,p} = S_{,\sigma} = \alpha_{13} = \alpha_{14} = 0$. Similarly, using $(6.2.14)_7$ and $(6.2.15)_7$, we have $S = S(x,t,\rho), U = U(x,t,u), P = P(x,t,p), E = E(x,t,\sigma)$ and $\alpha_{ij} = 0$ for $i \neq j$. Finally, equations $(6.2.12)_2$, $(6.2.13)_3$, $(6.2.14)_4$ and $(6.2.15)_5$ lead to

$$S_{,\rho\rho} = U_{,uu} = P_{,pp} = E_{,\sigma\sigma} = 0,$$

which, in turn, imply

$$S = (\alpha_{11} + \psi_{,t})\rho + S_1(x,t), \quad U = (\alpha_{22} + \psi_{,t})u + U_1(x,t),$$
$$P = (\alpha_{33} + \psi_{,t})p + P_1(x,t), \quad E = (\alpha_{44} + \psi_{,t})\sigma + E_1(x,t), \quad \psi = \psi(t),$$
$$(6.2.16)$$

where S_1, U_1, P_1 and E_1 are arbitrary functions of x and t only. Substitution of (6.2.16) into $(6.2.12)_7$, $(6.2.13)_{1,6}$, $(6.2.14)_{1,7}$, and $(6.2.15)_1$ yields on simplification

$$S_1 \equiv P_1 \equiv 0, \quad E_1 \equiv d, \quad U_1 \equiv \begin{cases} 0, & \text{if } m = 1,2 \\ k_1, & \text{if } m = 0 \end{cases},$$
$$\psi = at + b, \quad \chi = \begin{cases} (2a + \alpha_{22})x, & \text{if } \quad m = 1,2 \\ (2a + \alpha_{22})x + k_1 t + c, & \text{if } \quad m = 0 \end{cases}, \quad (6.2.17)$$
$$SQ_{,\rho} + PQ_{,p} + EQ_{,\sigma} = \alpha_{44}Q, \quad \alpha_{33} = 2\alpha_{22} + \alpha_{11} + 2a, \quad \alpha_{44} = 2\alpha_{22} + a,$$

where a, b, c, d and k_1 are integration constants. Using equations $(6.2.17)_{1,2,3,4,8}$ in equations (6.2.16) and $(6.2.17)_6$, we obtain

$$S = (\alpha_{11} + a)\rho, \qquad U = \begin{cases} (\alpha_{22} + a)u, & \text{if } m = 1,2 \\ (\alpha_{22} + a)u + k_1, & \text{if } m = 0 \end{cases},$$
$$P = (2\alpha_{22} + \alpha_{11} + 3a)p, \qquad E = 2(\alpha_{22} + a)\sigma + d, \qquad (6.2.18)$$
$$\psi = at + b, \qquad \chi = \begin{cases} (\alpha_{22} + 2a)x, & \text{if } \quad m = 1,2 \\ (\alpha_{22} + 2a)x + k_1 t + c, & \text{if } \quad m = 0 \end{cases},$$
$$SQ_{,\rho} + PQ_{,p} + EQ_{,\sigma} = (2\alpha_{22} + a)Q.$$

Thus, the generators of the local Lie group of transformations, involving arbitrary constants $a, b, c, d, k_1, \alpha_{11}$ and α_{22}, are known.

6.2.3 Self-similar solutions and constraints

The arbitrary constants, occurring in the expressions for the generators of the local Lie group of transformations, give rise to several cases of possible solutions, which we discuss below.

Case I. When $a \neq 0$ and $\alpha_{22} + 2a \neq 0$, the change of variables from (x, t) to $(\tilde{x}, \tilde{t})$ defined as

$$\tilde{x} = \begin{cases} x, & \text{if } m = 1, 2 \\ x + c\,(\alpha_{22} + 2a)^{-1}, & \text{if } m = 0 \end{cases} \quad ; \quad \tilde{t} = t + b/a,$$

does not alter the equations (6.2.1) – (6.2.4); thus, rewriting the set of equations (6.2.18) in terms of the new variables $\tilde{x}$ and $\tilde{t}$, and then suppressing the tilde sign, we have

$$S = (\alpha_{11} + a)\rho, \qquad U = \begin{cases} (\alpha_{22} + a)u, & \text{if } m = 1, 2 \\ (\alpha_{22} + a)u + k_1, & \text{if } m = 0 \end{cases},$$

$$P = (2\alpha_{22} + \alpha_{11} + 3a)p, \quad E = 2(\alpha_{22} + a)\sigma + d, \quad \psi = at, \qquad (6.2.19)$$

$$\chi = \begin{cases} (\alpha_{22} + 2a)x, & \text{if } m = 1, 2 \\ (\alpha_{22} + 2a)x + k_1 t, & \text{if } m = 0 \end{cases},$$

$$SQ_{,\rho} + PQ_{,p} + EQ_{,\sigma} = (2\alpha_{22} + a)Q.$$

The similarity variable and the form of similarity solutions for ρ, u, p, σ and Q readily follow from the invariant surface condition, $u_i(x^*, t^*) = u_i^*(x, t)$, which yields

$$\psi\rho_{,t} + \chi\rho_{,x} = S, \qquad \psi u_{,t} + \chi u_{,x} = U,$$

$$\psi p_{,t} + \chi p_{,x} = P, \qquad \psi\sigma_{,t} + \chi\sigma_{,x} = E.$$

The last set of equations together with $(6.2.19)_7$ yields on integration the following forms of the flow variables

$$\rho = t^{(1+\alpha_{11}/a)}\hat{S}(\xi), \quad u = \begin{cases} t^{(\delta-1)}\hat{U}(\xi), & \text{if } m = 1, 2 \\ t^{(\delta-1)}\hat{U}(\xi) - k^*, & \text{if } m = 0 \end{cases},$$

$$p = t^{(2\delta-1+\alpha_{11}/a)}\hat{P}(\xi), \quad \sigma = t^{2(\delta-1)}\hat{E}(\xi) - d^*, \quad Q = p^{\frac{(2\delta-3)a}{\alpha_{11}+(2\delta-1)a}}q(\eta, \zeta),$$

$$(6.2.20)$$

where $k^* = k_1/(\delta - 1)a, d^* = d/2(\delta - 1)a$ and q is an arbitrary function of η and ζ

$$\eta = \rho p^{-\frac{\alpha_{11}+a}{\alpha_{11}+(2\delta-1)a}}, \quad \zeta = p\,(\sigma + d^*)^{-\frac{\alpha_{11}+(2\delta-1)a}{2(\delta-1)a}}, \quad \delta = (\alpha_{22} + 2a)/a. \quad (6.2.21)$$

The functions $\hat{S}, \hat{U}, \hat{P}$ and $\hat{E}$ depend only on dimensionless form of the similarity variable ξ, which is determined as

$$\xi = \begin{cases} x/(At^\delta), & \text{if } m = 1, 2 \\ x/(At^\delta) + k^* t^{(1-\delta)}/A, & \text{if } m = 0, \end{cases} \qquad (6.2.22)$$

where A is a dimensional constant, whose dimensions are obtained by the similarity exponent δ. Since the shock must be a similarity curve, it may be normalized to be at $\xi = 1$. The shock path and the shock velocity V are, then, given by

$$X = \begin{cases} At^\delta, & \text{if} \quad m = 1, 2, \\ At\left\{t^{(\delta-1)} - k^*/A\right\}. & \text{if} \quad m = 0. \end{cases} \tag{6.2.23}$$

$$V = \begin{cases} \delta X/t, & \text{if} \quad m = 1, 2 \\ \left\{A\delta t^{(\delta-1)} - k^*\right\}, & \text{if} \quad m = 0. \end{cases} \tag{6.2.24}$$

Equation $(6.2.20)_5$, together with $(6.2.21)$, yields the general form of Q for which the self-similar solutions exist. At the shock, we have the following conditions on the functions $\hat{S}, \hat{U}, \hat{P}$ and $\hat{E}$

$$\rho|_{\xi=1} = t^{(1+\alpha_{11}/a)}\hat{S}(1), \quad u|_{\xi=1} = \begin{cases} t^{(\delta-1)}\hat{U}(1), & \text{if} \quad m = 1, 2 \\ t^{(\delta-1)}\hat{U}(1) - k^*, & \text{if} \quad m = 0 \end{cases}$$

$$p|_{\xi=1} = t^{(2\delta-1+\alpha_{11}/a)}\hat{P}(1), \quad \sigma|_{\xi=1} = t^{2(\delta-1)}\hat{E}(1) - d^*. \tag{6.2.25}$$

Equations $(6.2.25)$, in view of the invariance of jump conditions $(6.2.8)$, suggest the following forms of $\rho_0(x), \sigma_0(x)$

$$\rho_0(x) \equiv \rho_c(x/x_0)^\theta, \quad \overline{\sigma}_0(x) \equiv \sigma_0(x) \equiv \sigma_c(x/x_0)^\Gamma + \sigma_{c0}, \tag{6.2.26}$$

and the following conditions on the functions $\hat{S}, \hat{U}, \hat{P}$ and $\hat{E}$ at the shock

$$\hat{U}(1) = \frac{2\delta A}{\gamma+1}\frac{\rho_0 V^2 - \gamma p_0}{\rho_0 V^2}, \quad \hat{E}(1) = \begin{cases} \sigma_c(A/x_0)^\Gamma, & \text{if } \sigma_0 \text{ is varying} \\ 0, & \text{if } \sigma_0 \text{ is constant} \end{cases}$$

$$\hat{P}(1) = \frac{\left(2\rho_0 V^2 - (\gamma-1)p_0\right)\rho_c\delta^2 A^{2+\theta}}{(\gamma+1)\rho_0 V^2 x_0^\theta}, \quad \hat{S}(1) = \frac{(\gamma+1)\rho_0 V^2 \rho_c A^\theta}{\{(\gamma-1)\rho_0 V^2 + 2\gamma p_0\}x_0^\theta}, \tag{6.2.27}$$

together with

$$k^* = 0, \quad d^* = \begin{cases} -(\sigma_c + \sigma_{c0}), & \text{if } \sigma_0 \text{ is constant} \\ -\sigma_{c0}, & \text{if } \sigma_0 \text{ is varying} \end{cases} \tag{6.2.28}$$

where $\theta = (\alpha_{11} + a)/\delta a$,

$$\Gamma = \begin{cases} 0, & \text{if } \sigma_0 \text{ is constant} \\ 2(\delta - 1)/\delta, & \text{if } \sigma_0 \text{ is varying} \end{cases} \tag{6.2.29}$$

and ρ_c, x_0, σ_c and σ_{c0} are some reference constants associated with the medium.

Equations $(6.2.27)$ show that the necessary condition for the existence of a similarity solution for shocks of arbitrary strength is that $\rho_0(X(t))V^2$ must be a constant, which implies that θ and δ are not independent, but rather

$$\theta\delta + 2(\delta - 1) = 0. \tag{6.2.30}$$

However, for strong shocks, $V \gg a_0$, constancy of $\rho_0 V^2$ or equivalently the condition (6.2.30) is not required, and thus, we have

$$\hat{S}(1) = \frac{(\gamma+1)\rho_c A^\theta}{(\gamma-1)x_0^\theta}, \quad \hat{U}(1) = \frac{2\delta A}{\gamma+1}, \quad \hat{P}(1) = \frac{2\delta^2 \rho_c A^{\theta+2}}{(\gamma+1)x_0^\theta},$$

$$\hat{E}(1) = \begin{cases} \sigma_c (A/x_0)^\Gamma, & \text{if } \sigma_0 \text{ is varying} \\ 0, & \text{if } \sigma_0 \text{ is constant} \end{cases}, \tag{6.2.31}$$

together with the equations defined in (6.2.26), (6.2.28) and (6.2.29).

Using (6.2.27) or (6.2.31), we rewrite the equations (6.2.20), (6.2.21), (6.2.22), (6.2.23) and (6.2.24) as

$$\begin{aligned}
&\rho = \rho_0(X(t))S^*(\xi), \quad u = VU^*(\xi), \\
&p = \rho_0(X(t))V^2 P^*(\xi), \quad \sigma = V^2 E^*(\xi) - d^*, \\
&Q = p^{\frac{2\delta-3}{2\delta-2}} q(\eta, \zeta), \quad \eta = \rho p^{\frac{-\theta\delta}{(2+\theta)\delta-2}}, \quad \zeta = p(\sigma + d^*)^{-\frac{(2+\theta)\delta-2}{2(\delta-1)}}, \\
&\xi = x/At^\delta, \quad X = At^\delta, \quad V = \delta X/t,
\end{aligned} \tag{6.2.32}$$

where $S^*(\xi) = x_0^\theta \hat{S}(\xi)/\rho_c A^\theta, U^*(\xi) = \hat{U}(\xi)/(\delta A), P^*(\xi) = x_0^\theta \hat{P}(\xi)/(\rho_c \delta^2 A^{\theta+2})$ and $E^* = \hat{E}(\xi)/\delta^2 A^2$.

We may note that this case leads to a class of similarity solutions where the shock path is given by a power law $(6.2.32)_9$. The form of Q, given by (6.2.5), should be consistent with equation $(6.2.32)_5$ in order to have a similarity solution; thus, if the quantities τ and c are assumed to be of the form

$$c = c^* p^{\beta_1} \rho^{\beta_2}, \quad \tau = \tau^* p^{\beta_3} \rho^{\beta_4}, \tag{6.2.33}$$

with $\beta_1, \beta_2, \beta_3, \beta_4, c^*$ and τ^* being constants associated with the reference medium, it is required that the following relations hold

$$\beta_1 \{\theta\delta + 2(\delta-1)\} = -\beta_2\theta\delta, \quad \beta_3 \{\theta\delta + 2(\delta-1)\} = 1 - \beta_4\theta\delta, \tag{6.2.34}$$

We, thus find that the requirement of a self-similar flow pattern poses the similarity conditions (6.2.29) and (6.2.34) on the parameters $\beta_1, \beta_2, \beta_3, \beta_4, \theta, \Gamma$ and δ. It may, however, be noted that there are no constraints, like (6.2.34), in the absence of relaxation. Substituting (6.2.32) in the governing equations (6.2.1) – (6.2.4) and using (6.2.5), (6.2.26), (6.2.28), (6.2.29), (6.2.33) and (6.2.34) we obtain the following system of ordinary differential equations in S^*, U^*, P^* and E^* which on suppressing the asterisk sign becomes,

$$(U - \xi)S' + SU' + \theta S + \frac{mSU}{\xi} = 0,$$

$$(U - \xi)U' + \frac{1}{S} P' + \frac{(\delta-1)U}{\delta} = 0, \tag{6.2.35}$$

$$(U - \xi)P' + \gamma PU' + \theta P + \frac{2(\delta-1)P}{\delta} + \frac{m\gamma PU}{\xi} + (\gamma-1)SQ_* = 0,$$

$$(U - \xi)E' + 2(\delta-1)E/\delta - Q_* = 0.$$

where a prime denotes differentiation with respect to the independent variable ξ and

$$
Q_* =
\begin{cases}
\left(\left(e_0 \delta^{2\beta_1} P^{\beta_1} S^{\beta_2} \left(\dfrac{P}{S} - \dfrac{p_0}{\rho_0 V^2} \right) - E \right) \Big/ \left(\tau_0 \delta^{2\beta_3+1} P^{\beta_3} S^{\beta_4} \right), \\
\qquad\qquad\qquad\qquad\qquad\qquad\qquad\qquad \text{if } \sigma_0 \text{ is constant} \\[2ex]
\left(e_0 \delta^{2\beta_1} P^{\beta_1} S^{\beta_2} \left(\dfrac{P}{S} - \dfrac{p_0}{\rho_0 V^2} \right) - E + \dfrac{\hat\sigma_c}{\delta^2} \right) \Big/ \left(\tau_0 \delta^{2\beta_3+1} P^{\beta_3} S^{\beta_4} \right), \\
\qquad\qquad\qquad\qquad\qquad\qquad\qquad\qquad \text{if } \sigma_0 \text{ is varying}
\end{cases}
$$

with $\hat\sigma_c, \tau_0$ and e_0 as dimensionless parameters defined as

$$
\hat\sigma_c = \sigma_c A^{\Gamma-2} x_0^{-\Gamma}, \quad \tau_0 = \tau^* x_0^{-\theta(\beta_3+\beta_4)} A^{2\beta_3+\theta(\beta_3+\beta_4)} \rho_c^{(\beta_3+\beta_4)},
$$
$$
e_0 = c^* x_0^{-\theta(\beta_1+\beta_2)} A^{2\beta_1+\theta(\beta_1+\beta_2)} \rho_c^{\beta_1+\beta_2}.
$$

In view of (6.2.30), the term $p_0/\rho_0 V^2$ in the expression of Q_*, is either a constant or can be neglected, depending on whether the shock is of arbitrary strength or of infinite strength respectively. Indeed, for a shock of arbitrary strength, the jump conditions become

$$
U(1) = \tfrac{2}{\gamma+1} \tfrac{\rho_0 V^2 - \gamma p_0}{\rho_0 V^2}, \quad S(1) = \tfrac{(\gamma+1)\rho_0 V^2}{(\gamma-1)\rho_0 V^2 + 2\gamma p_0}, \quad \delta\theta + 2(\delta-1) = 0,
$$
$$
P(1) = \tfrac{(2\rho_0 V^2 - (\gamma-1)p_0)}{(\gamma+1)\rho_0 V^2}, \quad E(1) = \begin{cases} \hat\sigma_c/\delta^2, & \text{if } \sigma_0 \text{ is varying} \\ 0, & \text{if } \sigma_0 \text{ is constant} \end{cases}.
$$

$$\tag{6.2.36}$$

However, for an infinitely strong shock, the jump conditions are

$$
S(1) = \tfrac{(\gamma+1)}{(\gamma-1)}, \quad U(1) = \tfrac{2}{\gamma+1}, \quad P(1) = \tfrac{2}{\gamma+1},
$$
$$
E(1) = \begin{cases} \hat\sigma_c/\delta^2, & \text{if } \sigma_0 \text{ is varying} \\ 0, & \text{if } \sigma_0 \text{ is constant.} \end{cases}
$$

$$\tag{6.2.37}$$

Thus, the system (6.2.35) is to be solved subjected to the jump conditions (6.2.36) and (6.2.37) depending on whether the shock is of arbitrary strength or an infinite strength.

Case II. When $a = 0$ and $\alpha_{22} \neq 0$, the change of variables from (x, t) to $(\tilde{x}, \tilde{t})$ defined as

$$
\tilde{x} = x + c/\alpha_{22}, \quad \tilde{t} = t,
$$

leaves the basic equations (6.2.1) – (6.2.4) unchanged. The similarity variable and the form of similarity solutions for the flow variables readily follow from (6.2.18), and can be expressed in the following forms on suppressing the tilde sign

$$\rho = \rho_0(X(t))S(\xi), \quad u = VU(\xi), \quad p = \rho_0 V^2 P(\xi), \quad \sigma = V^2 E(\xi) - d^*,$$
$$Q = p^{\frac{2+\theta}{2}} q(\eta,\zeta), \quad \eta = \rho p^{-\frac{\theta}{2+\theta}}, \quad \zeta = p\,(\sigma + d^*)^{-\frac{2+\theta}{2}},$$
$$\rho_0(x) = \rho_c(x/x_0)^\theta, \quad \sigma_0(x) = \sigma_c(x/x_0)^\Gamma + \sigma_{c0},$$
$$\Gamma = \begin{cases} 0, & \text{if } \sigma_0 \text{ is constant} \\ 2, & \text{if } \sigma_0 \text{ is varying} \end{cases}, \quad d^* = \begin{cases} -(\sigma_c + \sigma_{c0}), & \text{if } \sigma_0 \text{ is constant} \\ -\sigma_{c0}, & \text{if } \sigma_0 \text{ is varying} \end{cases},$$
$$\tag{6.2.38}$$

where the dimensionless similarity variable ξ, the shock location X and the shock velocity V are given by

$$\xi = (x/x_0)\exp(-\delta t/A), \quad X = x_0 \exp(\delta t/A), \quad V = (\delta x_0/A)\exp(\delta t/A), \tag{6.2.39}$$

with A as a dimensional constant. Here, we note that this case leads to a class of similarity solutions with an exponential shock path given by $(6.2.39)_2$. If the relaxation rate is given by (6.2.5) and (6.2.33), the requirement of a self-similar flow pattern in this case poses the following similarity conditions

$$\beta_1 = -\frac{\beta_2 \theta}{2 + \theta}, \quad \beta_3 = -\frac{\beta_4 \theta}{2 + \theta}. \tag{6.2.40}$$

Thus, substituting (6.2.38) in the system of equations (6.2.1) – (6.2.4) and using (6.2.5), (6.2.33), (6.2.39) and (6.2.40), we obtain the following set of ordinary differential equations

$$
\begin{aligned}
&(U - \xi)S' + SU' + \theta S + \frac{mSU}{\xi} = 0, \\
&(U - \xi)U' + \frac{1}{S} P' + \delta U = 0, \\
&(U - \xi)P' + \gamma PU' + (\theta + 2)P + \frac{m\gamma PU}{\xi} + (\gamma - 1)SQ_* = 0, \\
&(U - \xi)E' + 2E - Q_* = 0,
\end{aligned}
\tag{6.2.41}
$$

where Q_* is the same as in (6.2.35) with
$$\hat{\sigma}_c = \sigma_c A^2/x_0^2, \quad \tau_0 = \tau^* \rho_c^{(\beta_3 + \beta_4)} x_0^{2\beta_3} A^{-(2\beta_3 + 1)}, \quad e_0 = c^* x_0^{2\beta_1} \rho_c^{\beta_1 + \beta_2} A^{-2\beta_1}.$$
The system (6.2.41) is to be solved subject to the constraints (6.2.40) and conditions for $S(\xi), U(\xi), P(\xi)$ and $E(\xi)$ at $\xi = 1$, which are given by $(6.2.36)_{1,2,4,5}$ together with $\theta = -2$ for a shock of arbitrary strength; however for a shock of infinite strength, the corresponding conditions are given by (6.2.37).

Case III. Here, we consider a situation when $a \neq 0$ and $\alpha_{22} + 2a = 0$. The analysis reveals that this situation can not arise in a radially symmetric $(m = 1, 2)$ flow as it does not allow for the existence of a similarity solution in such a flow configuration. However, this situation can occur in a plane $(m = 0)$ flow where the change of variables from (x, t) to $(\tilde{x}, \tilde{t})$ defined as

$$\tilde{x} = x, \quad \tilde{t} = t + b/a,$$

does not alter the basic equations (6.2.1) – (6.2.4); consequently, equations

(6.2.18) imply the following forms of the flow and similarity variables.

$$\rho = \rho_0(X(t))S(\xi), \qquad u = VU(\xi),$$
$$p = \rho_0 V^2 P(\xi), \qquad \sigma = V^2 E(\xi) + d^*,$$
$$Q = p^{-\frac{3}{\theta\delta-2}} q(\eta, \zeta), \qquad \eta = \rho p^{\frac{\delta\theta}{2-\delta\theta}}, \qquad \zeta = p(\sigma - d^*)^{\frac{\theta\delta-2}{2}}, \qquad (6.2.42)$$
$$\xi = (x - x_0\delta ln(t/A))/x_0, \qquad X = x_0\delta ln(t/A), \qquad V = \delta x_0/t,$$
$$\rho_0(x) = \rho_c exp(\theta x/x_0), \qquad \sigma_0(x) = \sigma_c exp(\Gamma x/x_0) + \sigma_{c0},$$

$$\text{where } \Gamma = \begin{cases} 0 & \text{if } \sigma_0 \text{ is constant} \\ -2/\delta & \text{if } \sigma_0 \text{ is varying} \end{cases}, \quad d^* = \begin{cases} \sigma_c + \sigma_{c0}, & \text{if } \sigma_0 \text{ is constant} \\ \sigma_{c0}, & \text{if } \sigma_0 \text{ is varying} \end{cases}$$

and A is a dimensional constant. It may be noted that this case leads to a class of similarity solutions with logarithmic shock path given by $(6.2.42)_9$. In addition, for the relaxation rate given by (6.2.5) and (6.2.33), the similarity considerations imply the following constraints

$$\beta_1 = \frac{\beta_2\theta\delta}{2 - \delta\theta}, \qquad \beta_3 = \frac{\beta_4\theta\delta - 1}{2 - \theta\delta}. \qquad (6.2.43)$$

Substituting (6.2.42) in equations (6.2.1) – (6.2.4) and using (6.2.5), (6.2.33) and (6.2.43), we obtain the following set of ordinary differential equations

$$\begin{aligned}
&(U - 1)S' + SU' + \delta\theta S = 0, \\
&(U - 1)U' + SP' - U/\delta = 0, \\
&(U - 1)P' + \gamma PU' + (\theta - 2/\delta)P + (\gamma - 1)SQ_* = 0, \\
&(U - 1)\xi E' - 2E/\delta - Q_* = 0,
\end{aligned} \qquad (6.2.44)$$

where Q_* is the same as in (6.2.35) with $\hat{\sigma}_c, \tau_0$ and e_0 given by

$$\hat{\sigma}_c = \sigma_c A^2/x_0^2, \tau_0 = \tau^* \rho_c^{(\beta_3+\beta_4)} A^{-\theta\delta(\beta_3+\beta_4)} x_0^{2\beta_3}, e_0 = c^* \rho_c^{\beta_1+\beta_2} A^{-\theta\delta(\beta_1+\beta_2)} x_0^{2\beta_1}.$$

It is interesting to note that for a nonrelaxing gas ($Q = 0$) with $\theta = 1$, the similarity variable, the shock path and the form of similarity solutions given by (6.2.42) together with similarity equations (6.2.44) are precisely the same as those employed by Hayes [69] in analyzing the self-similar propagation of a shock wave in an exponential medium.

For a shock of arbitrary strength, the jump conditions at $\xi = 0$ are

$$U(0) = \frac{2}{\gamma+1}\frac{\rho_0 V^2 - \gamma p_0}{\rho_0 V^2}, \quad S(0) = \frac{(\gamma+1)\rho_0 V^2}{(\gamma-1)\rho_0 V^2 + 2\gamma p_0}, \quad \theta\delta = 2,$$
$$P(0) = \frac{(2\rho_0 V^2 - (\gamma-1)p_0)}{(\gamma+1)\rho_0 V^2}, \quad E(0) = \begin{cases} \hat{\sigma}_c/\delta^2, & \text{if } \sigma_0 \text{ is varying} \\ 0, & \text{if } \sigma_0 \text{ is constant.} \end{cases}$$
$$(6.2.45)$$

However, for an infinitely strong shock, the jump conditions at $\xi = 0$, become

$$S(0) = \frac{(\gamma+1)}{(\gamma-1)}, \qquad U(0) = \frac{2}{\gamma+1}, \qquad P(0) = \frac{2}{\gamma+1},$$
$$E(0) = \begin{cases} \hat{\sigma}_c/\delta^2, & \text{if } \sigma_0 \text{ is varying} \\ 0, & \text{if } \sigma_0 \text{ is constant.} \end{cases} \qquad (6.2.46)$$

The system (6.2.44) is to be solved subject to the constraints (6.2.43) and the conditions (6.2.45) or (6.2.46) depending on whether the shock is of arbitrary strength or of infinite strength.

Case IV. When $a = \alpha_{22} = 0$, the situation is similar to the preceding case in the sense that it does not allow for the existence of a self-similar solution in a radially symmetric flow. However, the plane flow of a relaxing gas involving a shock wave moving at constant speed admits a self-similar solution; and the corresponding similarity variable and the flow variables have the following form

$$\rho = \rho_0 S(\xi), \quad u = VU(\xi), \quad p = \rho_0 V^2 P(\xi), \quad \sigma = V^2 E(\xi) + d^* t,$$
$$Q = q(\eta, \zeta), \quad \xi = (x - \delta x_0 t/A)/x_0, \quad X = x_0(1 + \delta t/A),$$
$$V = \delta x_0/A, \quad \rho_0(x) = \rho_c \exp(\theta(x - x_0)/x_0),$$

$$\sigma_0(x) = \begin{cases} \sigma_c(x - x_0)/x_0 + \sigma_{c0}, & \text{if } \sigma_0 \text{ is varying} \\ \sigma_{c0}, & \text{if } \sigma_0 \text{ is constant} \end{cases},$$

(6.2.47)

where $\eta = \rho/p$, $\zeta = p^d \exp(-\sigma)$ and $d^* = \begin{cases} (\sigma_c \delta x_0/A), & \text{if } \sigma_0 \text{ is varying,} \\ 0, & \text{if } \sigma_0 \text{ is constant.} \end{cases}$

We note that the shock velocity in this case is constant and the density and the vibrational energy ahead of the shock are varying according to the law (6.2.47)$_{9,10}$. For the relaxation rate given by (6.2.5) and (6.2.33), the similarity constraints in this case for $\theta \neq 0$ turn out to be

$$\beta_1 + \beta_2 = 0, \text{ and } \beta_3 + \beta_4 = 0. \tag{6.2.48}$$

It may, however, be noted that when $\theta = 0$, there are no constraints like (6.2.48).

Substituting (6.2.47) in the basic set of equations (6.2.1) – (6.2.4) for $m = 0$, and using (6.2.5), (6.2.33) and (6.2.48), we obtain

$$\begin{aligned} &(U - 1)S' + SU' + \theta S = 0, \\ &(U - 1)U' + P'/S = 0, \\ &(U - 1)P' + \gamma PU' + \theta P + (\gamma - 1)SQ_* = 0, \\ &(U - 1)E' - Q_* = 0, \end{aligned} \tag{6.2.49}$$

where Q_* is the same as in (6.2.35) with

$$\hat{\sigma}_c = \sigma_{c0} A^2/x_0^2, e_0 = c^* \rho_c^{\beta_1 + \beta_2} x_0^{2\beta_1} A^{-2\beta_1}, \tau_0 = \tau^* \rho_c^{\beta_3 + \beta_4} x_0^{2\beta_3} A^{-(2\beta_3 + 1)}.$$

The condition for E at $\xi = 1$ is given by $E(1) = \hat{\sigma}_c/\delta^2$, while the functions U, S and P at $\xi = 1$ are given by (6.2.36)$_{1,2,4}$ or (6.2.37)$_{1,2,3}$ depending on whether the shock is of arbitrary strength with a constant density ahead or of infinite strength, respectively. In the absence of relaxation, $(Q = 0)$, and for a constant ambient density, $(\theta = 0)$ it follows from equations (6.2.49) that the flow variables, namely the velocity, pressure and density behind the shock are constant.

6.2.4   Imploding shocks

Here, we consider the problem of an imploding shock for which $V \gg a_0$ in the neighbourhood of implosion, and assume σ_0 to be a constant. For the problem of a converging shock collapsing to the center (axis), the origin of time t is taken to be the instant at which the shock reaches the center (axis) so that $t \leq 0$ in (6.2.35). In this regard, we modify slightly the definition of the similarity variable by setting

$$X = A(-t)^\delta, \quad \xi = x/A(-t)^\delta, \tag{6.2.50}$$

so that the intervals of the variables are $-\infty < t \leq 0, X \leq x < \infty$ and $1 \leq \xi < \infty$. Thus, the system (6.2.35) is to be solved subject to the similarity conditions (6.2.34) and the jump conditions at the shock

$$S(1) \;=\; (\gamma+1)/(\gamma-1), \; U(1) = 2/(\gamma+1), \; P(1) = 2/(\gamma+1), \; E(1) = 0,$$

$$Q_* \;=\; \{e_0\delta^{2\beta_1}P^{1+\beta_1}S^{1-\beta_2} - E\}/\{\tau_0\delta^{2\beta_3+1}P^{\beta_3}S^{\beta_4}\},$$

$$\tag{6.2.51}$$

and the boundary conditions at ∞. At the instant of collapse $t = 0$, the gas velocity, pressure, density and the sound speed at any finite radius x are bounded. But with $t = 0$ and finite $x, \xi = \infty$. In order for the quantities $u = (\delta X/t)U(\xi)$, $p = \rho_0(X)V^2P(\xi)$, $\rho = \rho_0(X)S(\xi)$, $\sigma = V^2E(\xi)$ and $c^2 = \gamma(\delta X/t)^2(P(\xi)/S(\xi))$ to be bounded when $t = 0$ and x is finite, the entities $U, P/S$ and E must vanish. Thus, we have the following boundary conditions at $\xi = \infty$,

$$U(\infty) = 0, \quad \gamma P(\infty)/S(\infty) = 0, \quad E(\infty) = 0. \tag{6.2.52}$$

In the matrix notation the equations (6.2.35) can be written as

$$CW' = B, \tag{6.2.53}$$

where $W = (U, S, P, E)^{tr}$, and the matrix C and the column vector B can be read off by inspection of equations (6.2.35). It may be noted that the system (6.2.35) has an unknown parameter δ, which is not obtainable from an energy balance or the dimensional considerations as is the case of diverging shock waves driven by release of energy at a plane or an axis or a point of implosion; indeed, it is computed only by solving a nonlinear eigenvalue problem for a system of ordinary differential equations. For the implosion problem, the range of similarity variable is $1 \leq \xi < \infty$. System (6.2.53) can be solved for the derivatives U', S', P' and E' in the following form:

$$U' = \Delta_1/\Delta, \quad S' = \Delta_2/\Delta, \quad P' = \Delta_3/\Delta, \quad E' = \Delta_4/\Delta, \tag{6.2.54}$$

where Δ, which is the determinant of the system, is given by

$$\Delta = (U - \xi)^2 \left\{ (U - \xi)^2 - \gamma P/S \right\},$$

and Δ_k, $k = 1, 2, 3, 4$ are the determinants obtained from Δ by replacing the kth column by the column vector B, and are given by

$$\Delta_1 = \left\{ (\xi - U)U\frac{(\delta - 1)}{\delta} + \left(\theta + \frac{2(\delta - 1)}{\delta}\right)\frac{P}{S} + \frac{m\gamma PU}{S\xi} \right.$$
$$\left. + (\gamma - 1)\frac{e_0\delta^{2\beta_1}P^{\beta_1+1}S^{\beta_2-1} - E}{\tau_0\delta^{(2\beta_3+1)}P^{\beta_3}S^{\beta_4}} \right\}(U - \xi)^2,$$

$$\Delta_2 = -\left\{ S\Delta_1 + (\theta S + mSU/\xi)\Delta \right\} / \left\{ (U - \xi) \right\},$$

$$\Delta_3 = -\left\{ \gamma P\Delta_1 + \theta P + 2(\delta - 1)P/\delta + m\gamma PU/\xi \right.$$
$$\left. + (\gamma - 1)\frac{e_0\delta^{2\beta_1}P^{\beta_1+1}S^{\beta_2-1} - E}{\tau_0\delta^{(2\beta_3+1)}P^{\beta_3}S^{\beta_4-1}} \right\} / (U - \xi),$$

$$\Delta_4 = -\Delta\left\{ 2(\delta - 1)E - \frac{e_0\delta^{2\beta_1}P^{\beta_1+1}S^{\beta_2-1} - E}{\tau_0\delta^{2\beta_3}P^{\beta_3}S^{\beta_4}} \right\}\frac{1}{\delta(U - \xi)}.$$

It may be noted that $U < \xi$ in the interval $[1, \infty)$, whilst Δ is positive at $\xi = 1$ and negative at $\xi = \infty$ indicating thereby that there exists a $\xi \in [1, \infty)$ at which Δ vanishes, and consequently the solutions become singular. In order to get a nonsingular solution of (6.2.35) in the interval $[1, \infty)$, we choose the exponent δ such that Δ vanishes only at the points where the determinant Δ_1 is zero too; it can be checked that at points where Δ and Δ_1 vanish, the determinants Δ_2, Δ_3 and Δ_4 also vanish simultaneously. To find the exponent δ in such a manner, we introduce the variable Z,

$$Z(\xi) = (U(\xi) - \xi)^2 - \gamma P(\xi)/S(\xi), \qquad (6.2.55)$$

which, in view of (6.2.54), implies

$$Z' = \left\{ 2(U - \xi)(\Delta_1 - \Delta) - \gamma\Delta_3/S + \gamma P\Delta_2/S^2 \right\}/\Delta. \qquad (6.2.56)$$

Equations (6.2.54), in view of (6.2.56), become

$$\frac{dU}{dZ} = \frac{\Delta_1}{\Delta_5}, \quad \frac{dS}{dZ} = \frac{\Delta_2}{\Delta_5}, \quad \frac{dP}{dZ} = \frac{\Delta_3}{\Delta_5}, \quad \frac{dE}{dZ} = \frac{\Delta_4}{\Delta_5}, \qquad (6.2.57)$$

where

$$\Delta_5 = 2(U - \xi)(\Delta_1 - \Delta) - \gamma\Delta_3/S + \gamma P\Delta_2/S^2$$

with ξ given by (6.2.55) in the form

$$\xi = U + \{Z + \gamma P/S\}^{1/2}.$$

6.2.5 Numerical results and discussion

We integrate the equations (6.2.57) from the shock $Z = Z(1)$ to the singular point $Z = 0$, by choosing a trial value of δ, and compute the values of

U, S, P, E and Δ_1 at $Z = 0$; the value of δ is corrected by successive approximations in such a way that for this value, the determinant Δ_1 vanishes at $Z = 0$. The computation has been performed by choosing $e_0 = 1$, $\tau_0 = 0.1$, $\beta_2 = 0$, $\beta_3 = -1$ and $\gamma = 1.4$. The corresponding values of β_1 and β_4 are computed from (6.2.34). The values of δ, obtained from the numerical calculations and CCW approximation, [210], for different values of θ, τ_0, γ and m are given in Table 1. The relaxation effects enter through the parameter τ_0 while the effects of density stratification become important through the parameter θ. The fact that δ is always less than 1 shows that the shock wave is continuously accelerated; indeed, the shock velocity V becomes infinite as $X \to 0$, but less rapidly than X^{-1}. We find that an increase in any of the parameter γ, θ, m or τ_0^{-1}, causes the similarity exponent δ to decrease, and consequently an increase in the shock velocity, as the shock approaches the center (axis) of implosion.

Analytical expressions of the dimensionless flow variables at the instant of collapse of the shock wave at $t = 0, X = 0$ ($t = 0, x \neq 0$ correspond to $\xi = \infty$), where conditions (6.2.52) hold, can easily be obtained from the equations (6.2.54) in the following form.

$$U \sim \xi^{\frac{\delta-1}{\delta}}, \quad S \sim \xi^{\theta}, \quad P \sim \xi^{\theta+2(\delta-1)/\delta}, \quad E \sim \xi^{2(\delta-1)/\delta}.$$

The above relations imply that both the velocity U and the vibrational energy E tend to zero, while the gas pressure P remains bounded (respectively; unbounded) at the instant of collapse of the shock wave if $\delta\theta + 2(\delta - 1)$ is negative (respectively; positive). However, the density S at the time of collapse is infinite; this is in contrast to the corresponding situation of uniform initial density, where S remains bounded. Numerical integration of equations (6.2.57) for $1 \leq \xi < \infty$ has been carried out, and the values of the flow variables before collapse and at the instant of collapse are depicted in Figs. (6.2.1 – 6.2.3). The numerical solutions in the neighborhood of $\xi = \infty$ are in conformity with the above mentioned asymptotic results. The typical flow profiles show that the pressure and density increase behind the shock wave, while the velocity decreases; this is because a gas particle passing through the shock is subjected to a shock compression; indeed, this increase in pressure and density behind the shock may also be attributed to the geometrical convergence or area contraction of the shock wave.

This increase in pressure and density, and decrease in particle velocity are further reinforced by an increase in the ambient density exponent θ. In contrast to the density profiles, which exhibit monotonic variations, the pressure profiles in cylindrically and spherically symmetric flow configurations exhibit nonmonotonic variation for values of $\theta < \theta_c$, where $\theta_c = 2(1 - \delta)/\delta$. Indeed, for $\theta \leq \theta_c$, the pressure profiles attain a maximum value, which appears to move towards the point (axis) of implosion as the shock collapses. However for $\theta > \theta_c$, the pressure variations are monotonic like density variations, and the profiles steepen as the shock converges; the steepening rate of pressure and density profiles enhances with an increase in θ. However, a decrease in τ_0

counteracts, to some extent, the steepening rate and results in a falling off of the particle velocity, density and pressure relative to what they would be in the absence of relaxation.

As might be expected, the differences between planar and nonplanar flow configurations are considerable. For plane waves, the flow distribution is relatively less influenced by the interaction between the relaxation and gasdynamic phenomena as compared to cylindrical and spherical waves; however, for imploding shocks, where the changes in flow variables may be attributed to the geometrical convergence or area contraction of the shock waves, the variations that result from relaxing gasdynamics are noteworthy. A study of the vibrational energy and temperature profiles behind the shock indicates sizeable reductions in these quantities. It is noted that a decrease in τ_0 causes an increase in temperature in the region behind the shock as compared to what it would be in the absence of relaxation.

Table 1 : Similarity exponent δ for plane and radially symmetric flow of a relaxing gas with $e_0 = 1.$, $\beta_1 = \beta_2 = 0$, $\beta_3 = -1$ and for different values of the ambient density exponent θ, the specific heat gas constant γ and relaxation time τ_0.

m	γ	θ	τ_0	computed delta δ	Whitham's rule	β_4
2	1.4	0.5	0.1	0.6648	0.6659	1.9918
2	1.4	1.0	0.1	0.6158	0.6213	1.3762
2	1.4	2.0	0.1	0.5446	0.5480	1.0818
1	1.4	0.5	0.1	0.7670	0.7665	2.3926
1	1.4	1.0	0.1	0.7044	0.7080	1.5804
1	1.4	2.0	0.1	0.6121	0.6144	1.1831
0	1.4	0.5	0.1	0.9042	0.9028	2.7880
0	1.4	1.5	0.1	0.7525	0.7559	1.4474
0	1.4	2.0	0.1	0.6980	0.6991	1.2836
2	5/3	1.0	0.1	0.5575	0.5928	1.2063
1	5/3	1.0	0.1	0.6505	0.6842	1.4627
0	5/3	1.0	0.1	0.7730	0.8090	1.7064
2	1.4	1.0	1.0	0.6257	0.6214	1.4019
1	1.4	1.0	1.0	0.7140	0.7081	1.5995
0	1.4	1.0	1.0	0.8309	0.8229	1.7965
2	5/3	1.0	1.0	0.5919	0.5928	1.3106
1	5/3	1.0	1.0	0.6855	0.6842	1.5412
0	5/3	1.0	1.0	0.8141	0.8090	1.7717

In an implosion problem, the self-similar region decreases with time in proportion to the radius of the shock front; the effective boundary of the self-similar region is then considered to be at some constant value of $\xi = \xi_1$. Thus, the mass, M, per unit area (m=0), or length (m=1) or volume (m=2) contained in this region is proportional to

$$M \propto \int_X^{x_1} \rho x^m \, dx = \rho_0 X^{(1+m)} \int_1^{\xi_1} S(\xi)\xi^m \, d\xi.$$

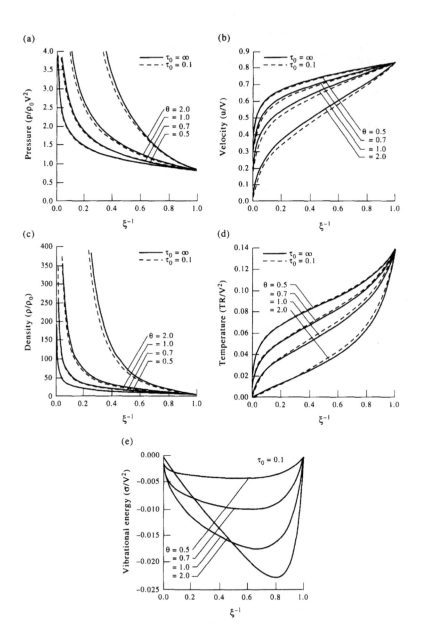

Figure 6.2.1: (a) Pressure, (b) velocity, (c) density, (d) temperature, (e) vibrational energy. Flow pattern; $m = 0$, $\gamma = 1.4$, $\beta_2 = 0$, $\beta_3 = -l$, $\Gamma = 0$ and $e_0 = 1$.

In nonrelaxing gases (i.e., $\tau_0 = \infty$), where there are no constraints such as (6.2.37), we have obtained the exponent δ for different values of γ and θ for plane and radially symmetric shocks, in a quiescent perfect gas with initial density which is either constant or decreasing towards the axis of implosion according to a power law; the results which are given in Tables 2, 3, 4 compare well with those obtained by Whitham [210], Guderley [66], Hafner [68], and Sakurai [159].

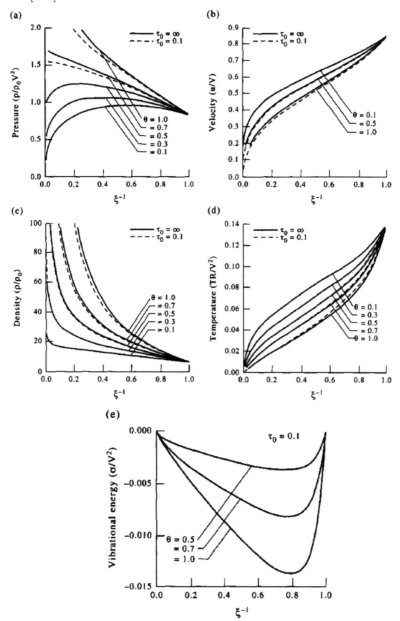

Figure 6.2.2: (a) Pressure, (b) velocity, (c) density, (d) temperature, (e) vibrational energy. Flow pattern; $m = 1$, $\gamma = 1.4$, $\beta_2 = 0$, $\beta_3 = -l$, $\Gamma = 0$ and $e_0 = 1$.

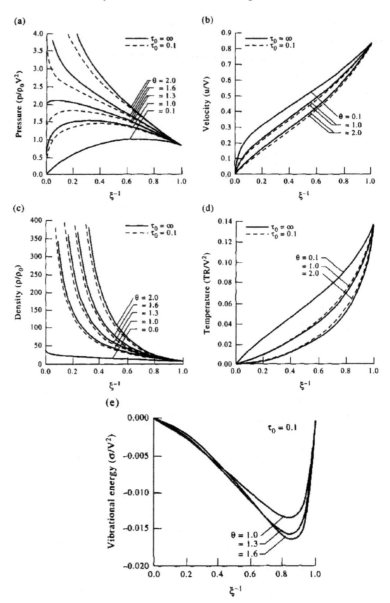

Figure 6.2.3: (a) Pressure, (b) velocity, (c) density, (d) temperature, (e) vibrational energy. Flow pattern; $m = 2$, $\gamma = 1.4$, $\beta_2 = 0$, $\beta_3 = -l$, $\Gamma = 0$ and $e_0 = 1$.

As the above integral with respect to ξ from 1 to ξ_1 is a constant, the mass content in this region with a finite radius is finite, and tends to zero as $X \to 0$, i.e., $M \sim X^{1+m+\theta} \to 0$ as $X \to 0$. Thus, the density variation ahead of the shock causes the total mass to decrease more rapidly with time as compared to the corresponding case in which the initial density is uniform.

Table 2 : Similarity exponent δ for a spherically symmetric flow of an ideal gas for various values of the heat exponent γ and the ambient density exponent θ.

θ	γ	computed delta δ	Whitham's rule	Hafner	Guderley
0.0	1.2	0.7571	0.7540	0.7571	0.7571
0.0	1.4	0.7172	0.7173	0.7172	0.7172
0.0	5/3	0.6884	0.6893	0.6884	0.6883
0.0	2.0	0.6670	0.6667	0.6670	
0.0	3.0	0.6364	0.6295	0.6364	
0.1	1.4	0.7067	0.7064		
0.5	1.2	0.7122	0.7054		
0.5	1.4	0.6683	0.6659		
0.5	5/3	0.6373	0.6374		
1.0	1.2	0.6730	0.6626		
1.0	5/3	0.5947	0.5928		
2.0	1.2	0.6073	0.5909	0.6073	
2.0	1.4	0.5588	0.5481	0.5588	
2.0	5/3	0.5264	0.5200	0.5264	
2.0	2.0	0.5034	0.5000	0.5034	
2.0	3.0	0.4711	0.4707	0.4711	
1.0	1.4	0.6267	0.6214		
1.2	1.4	0.6112	0.6052		
1.5	1.4	0.5906	0.5824		

Table 3 : Similarity exponent δ for a cylindrically symmetric flow of an ideal gas for various values of the heat exponent γ and the ambient density exponent θ.

γ	θ	computed delta δ	Whitham's rule	Hafner	Guderley
1.2	0.	0.8612	0.8598	0.8621	0.8612
1.4	0.	0.8353	0.8354	0.8353	
5/3	0.	0.8156	0.8160	0.8156	0.8157
2.0	0.	0.8001	0.8000	0.8001	
3.0	0.	0.7757	0.7727	0.7757	
5/3	.5	0.7454	0.7443		
2.0	.5	0.7273	0.7273		
5/3	1.	0.6884	0.6842		
2.0	1.	0.6689	0.6667		
1.1	2.	0.7178	0.6983	0.7178	
1.2	2.	0.6718	0.6540	0.6718	
1.4	2.	0.6283	0.6144	0.6283	
5/3	2.	0.5994	0.5891	0.5994	
2.0	2.	0.5790	0.5714	0.5790	
3.0	2.	0.5500	0.5464	0.5502	
1.4	.3	0.7941	0.7926		
1.4	.5	0.7694	0.7665		
1.4	.8	0.7356	0.7303		

Table 4 : Similarity exponent δ for a plane flow of an ideal gas for various values
of the heat exponent γ and the ambient density exponent θ.

θ	γ	computed delta δ	Whitham's rule	Hafner	Sakurai
.3	1.4	0.9408	0.9393		
.4	1.4	0.9231	0.9207		
.5	1.4	0.9062	0.9028	0.9062	0.9062
.6	1.4	0.8901	0.8856		
.7	1.4	0.8746	0.8691		
.8	1.4	0.8598	0.8531		
.9	1.4	0.8455	0.8377		
1	1.4	0.8318	0.8229	0.8319	0.8318
.5	1.1	0.9335	0.9305	0.9335	
.5	1.2	0.9195	0.9162	0.9196	0.9195
.5	5/3	0.8976	0.8944	0.8976	0.8976
.5	2.0	0.8918	0.8889	0.8918	
.5	3.0	0.8844	0.8819	0.8844	
1	1.1	0.8783	0.8700	0.8783	
1	1.2	0.8544	0.8453	0.8544	0.8544
1	5/3	0.8174	0.8090	0.8174	0.8174
1	2.0	0.8078	0.8000	0.8078	
1	3.0	0.7954	0.7887	0.7954	
2	1.1	0.7881	0.7699	0.7881	
2	1.2	0.7514	0.7321	0.7514	0.7514
2	1.4	0.7177	0.6991	0.7177	0.7177
2	5/3	0.6966	0.6793	0.6966	0.6966
2	2.0	0.6826	0.6667	0.6826	
2	3.0	0.6648	0.6511	0.6648	

6.3 Exact Solutions of Euler Equations via Lie Group Analysis

For nonlinear systems involving discontinuities such as shocks, we do not
normally have the luxury of complete exact solutions, and for analytical work
have to rely on some approximate analytical or numerical methods which may
be useful to set the scene and provide useful information towards our understanding of the complex physical phenomenon involved. One of the most pow-
erful methods used to determine particular solutions to PDEs is based upon
the study of their invariance with respect to one parameter Lie group of point
transformations (see [2], [17], [18], [78], [136], [138], and [151]). Indeed, with the
help of symmetry generators of these equations, one can construct similarity
variables which can reduce these equations to ordinary differential equations
(ODEs); in some cases, it is possible to solve these ODEs exactly. Besides these

similarity solutions, the symmetries admitted by given PDEs enable us to look for appropriate canonical variables which transform the original system to an equivalent one whose simple solutions provide nontrivial solutions of the original system (see Refs. [48], [49], [134], and [135]). Using this procedure, Ames and Donato [1] obtained solutions for the problem of elastic-plastic deformation generated by a torque, and analyzed the evolution of weak discontinuity in a state characterized by invariant solutions. Donato and Ruggeri [50] used this procedure to study similarity solutions for the system of a monoatomic gas, within the context of the theory of extended thermodynamics, assuming spherical symmetry. In this section, we use this approach to characterize a class of solutions of the basic equations governing the one dimensional planar and radially symmetric flows of an adiabatic gas involving shock waves. Since the system involves only two independent variables, we need two commuting Lie vector fields, which are constructed by taking a linear combination of the infinitesimal operators of the Lie point symmetries admitted by the system at hand. It is interesting to note that one of the special exact solutions obtained in this manner is the well known solution to the blast wave problem studied in the theory of explosion in the gaseous media, studied in section 6.1 (see [1], [50], [96], [129], and [148]).

6.3.1   Symmetry group analysis

As in [48], [49], [134], and [135], let us assume that the system of N nonlinear partial differential equations

$$F_R\left(x, t, \bar{u}, \frac{\partial \bar{u}}{\partial x}, \frac{\partial \bar{u}}{\partial t}\right) = 0, \quad R = 1, 2, \cdots, N, \tag{6.3.1}$$

involving two independent variables x, t and the unknown vector $\bar{u}(x, t)$, where $\bar{u}(x, t) = (u_1(x, t), u_2(x, t), \cdots, u_N(x, t)) \in R^N$, admits $s-$ parameter Lie group of transformations with infinitesimal operators

$$\zeta_i = X_i(x, t, \bar{u})\frac{\partial}{\partial x} + T_i(x, t, \bar{u})\frac{\partial}{\partial t} + \sum_{j=1}^{N} U_{ij}(x, t, \bar{u})\frac{\partial}{\partial u_j}, \quad i = 1, 2, \cdots, N$$

$$\tag{6.3.2}$$

such that there exist $r(\leq s)$ infinitesimal generators $\zeta_1, \zeta_2, \cdots, \zeta_r$ that form a solvable Lie algebra. Let us now construct generators

$$Y_1 = \sum_{k=1}^{r} \alpha_k \zeta_k, \quad Y_2 = \sum_{k=1}^{r} \beta_k \zeta_k$$

where α_k and β_k are constants, to be determined, such that $[Y_1, Y_2] = 0$.

Now we introduce the canonical variables τ, ξ and $\bar{\nu} = (\nu_1, \nu_2, ..., \nu_N) \in R^N$ related to the infinitesimal generator Y_1, defined by $Y_1\tau = 1$, $Y_1\xi = 0$, and $Y_1\nu_i = 0$, $i = 1, 2, \cdots, N$. In terms of these canonical variables, the infinitesimal operator Y_1 reduces to $\tilde{Y}_1 = \partial/\partial\tau$, i.e., it corresponds to a translation in the variable τ only; consequently, owing to the invariance, the system (6.3.1) must assume the form

$$\tilde{F}_R\left(\xi, \bar{\nu}, \frac{\partial \bar{\nu}}{\partial \xi}, \frac{\partial \bar{\nu}}{\partial \tau}\right) = 0, \quad R = 1, 2, \cdots, N. \tag{6.3.3}$$

In terms of these new variables, the operator Y_2 can be written as

$$\tilde{Y}_2 = (Y_2\xi)\frac{\partial}{\partial \xi} + (Y_2\tau)\frac{\partial}{\partial \tau} + \sum_{j=1}^{N}(Y_2\nu_j)\frac{\partial}{\partial \nu_j}, \tag{6.3.4}$$

where we require that the condition $Y_2\xi \neq 0$ holds; thus, it is possible to introduce new canonical variables η, τ^* and $\bar{w} = (w_1, w_2, \cdots, w_N) \in R^N$ defined by $\tilde{Y}_2\eta = 1$, $\tilde{Y}_2\tau^* = 0$, and $\tilde{Y}_2 w_i = 0$, $i = 1, 2, \cdots, N$ which transform (6.3.3) to the form

$$\hat{F}_R\left(\bar{w}, \frac{\partial \bar{w}}{\partial \eta}, \frac{\partial \bar{w}}{\partial \tau^*}\right) = 0, \quad R = 1, 2, \cdots, N. \tag{6.3.5}$$

The resulting system (6.3.5) is an autonomous system associated with (6.3.1); in order to illustrate the method, outlined as above, we consider the system of Euler equations of ideal gasdynamics in the next section.

6.3.2   Euler equations of ideal gas dynamics

The equations governing the one dimensional unsteady planar and radially symmetric flows of an adiabatic index γ in the absence of viscosity, heat conduction and body forces can be written in the form [210]

$$\rho_t + u\rho_x + \rho u_x + \frac{m\rho u}{x} = 0,$$
$$u_t + uu_x + \rho^{-1}p_x = 0, \tag{6.3.6}$$
$$p_t + up_x + \gamma p_x \frac{m\gamma p u}{x} = 0,$$

where t is the time, x the spatial coordinate being either axial in flows with planar ($m = 0$) geometry or radial in cylindrically ($m = 1$) and spherically ($m = 2$) symmetric flows. The state variable u denotes the gas velocity, p the pressure, and ρ the density. By a straightforward analysis, it is found that the Lie groups of point transformations that leave the system (6.3.6) invariant constitute a 4-dimensional Lie algebra generated by the following infinitesimal operators:

$$\zeta_1 = \rho\frac{\partial}{\partial \rho} + p\frac{\partial}{\partial p}, \qquad\qquad \zeta_2 = t\frac{\partial}{\partial t} - u\frac{\partial}{\partial u} - 2p\frac{\partial}{\partial p},$$
$$\zeta_3 = x\frac{\partial}{\partial x} + u\frac{\partial}{\partial u} + 2p\frac{\partial}{\partial p}, \qquad \zeta_4 = \frac{\partial}{\partial t}.$$

In order to construct generators Y_1, Y_2 such that $[Y_1, Y_2] = 0$, let

$$Y_1 = \alpha_1\zeta_1 + \alpha_2\zeta_2 + \alpha_3\zeta_3 + \alpha_4\zeta_4,$$
$$= (\alpha_2 t + \alpha_4)\frac{\partial}{\partial t} + \alpha_3 x\frac{\partial}{\partial x} + (\alpha_3 - \alpha_2)u\frac{\partial}{\partial u}$$
$$+ (\alpha_1 - 2\alpha_2 + 2\alpha_3)p\frac{\partial}{\partial p},$$
$$Y_2 = \beta_1\zeta_1 + \beta_2\zeta_2 + \beta_3\zeta_3 + \beta_4\zeta_4,$$
$$= (\beta_2 t + \beta_4)\frac{\partial}{\partial t} + \beta_3 x\frac{\partial}{\partial x} + (\beta_3 - \beta_2)u\frac{\partial}{\partial u}$$
$$+ (\beta_1 - 2\beta_2 + 2\beta_3)p\frac{\partial}{\partial p},$$

where $\alpha_2\beta_4 - \alpha_4\beta_2 = 0$ and α_1, α_3, β_1, β_3 are arbitrary constants. Since the system is invariant under the group generated by the generator Y_1, we introduce canonical variables $\bar{\tau}$, $\bar{\xi}$, $\bar{R}$, $\bar{U}$ and $\bar{P}$ such that $Y_1\bar{\tau} = 1$, $Y_1\bar{\xi} = 0$, $Y_1\bar{R} = 0$, $Y_1\bar{U} = 0$ and $Y_1\bar{P} = 0$. This implies that when α_2, $\alpha_3 \neq 0$, we have

$$\bar{\tau} = (1/\alpha_2)\log(\alpha_2 t + \alpha_4), \quad \bar{\xi} = (\alpha_2 t + \alpha_4)x^{-\alpha_2/\alpha_3}, \qquad (6.3.7)$$
$$\bar{R} = \rho x^{-\alpha_1/\alpha_3}, \quad \bar{U} = ux^{(\alpha_2 - \alpha_3)/\alpha_3}, \quad \bar{P} = px^{2\alpha_2 - \alpha_1 - 2\alpha_3)/\alpha_3}.$$

In terms of these new variables, Y_2 becomes

$$\bar{Y}_2 = \frac{\beta_2}{\alpha_2}\frac{\partial}{\partial\bar{\tau}} + \frac{\beta_2\alpha_3 - \beta_3\alpha_2}{\alpha_3}\bar{\xi}\frac{\partial}{\partial\bar{\xi}} + \frac{\beta_1\alpha_3 - \beta_3\alpha_1}{\alpha_3}\bar{R}\frac{\partial}{\partial\bar{R}} + \frac{\alpha_2\beta_3 - \alpha_3\beta_2}{\alpha_3}\bar{U}\frac{\partial}{\partial\bar{U}}$$
$$+ \frac{(\alpha_3\beta_1 - \alpha_1\beta_3) + 2(\alpha_2\beta_3 - \beta_2\alpha_3)}{\alpha_3}\bar{P}\frac{\partial}{\partial\bar{P}}.$$

Now, we introduce canonical variables τ, ξ, R, U and P such that $\tilde{Y}_2\tau = 0$, $\tilde{Y}_2\xi = 1$, $\tilde{Y}_2R = 0$, $\tilde{Y}_2U = 0$ and $\tilde{Y}_2P = 0$; thus, the corresponding characteristic conditions yield

$$\xi = (\alpha_3/A)\log(\bar{\xi}), \quad \tau = \bar{\tau} - (\beta_2/\alpha_2)\xi,$$
$$R = \bar{R}\bar{\xi}^{(\alpha_1\beta_3 - \alpha_3\beta_1)/A}, \quad U = \bar{U}\bar{\xi}, \quad P = \bar{P}\bar{\xi}^{2 + ((\alpha_1\beta_3 - \alpha_3\beta_1)/A)}, \qquad (6.3.8)$$

where $A = \alpha_3\beta_2 - \alpha_2\beta_3 \neq 0$. In view of (6.3.7) and (6.3.8), we are led to the following transformations

$$\tau = (-\beta_3/A)\log\{(\alpha_2 t + \alpha_4)x^{-\beta_2/\beta_3}\}, \quad \xi = (\alpha_3/A)\log\{(\alpha_2 t + \alpha_4)x^{-\alpha_2/\alpha_3}\},$$
$$\rho = R(\xi, \tau)x^L(\alpha_2 t + \alpha_4)^K, \quad u = U(\xi, \tau)x(\alpha_2 t + \alpha_4)^{-1}, \qquad (6.3.9)$$
$$p = P(\xi, \tau)x^{L+2}(\alpha_2 t + \alpha_4)^{K-2},$$

where $L = (\alpha_1\beta_2 - \alpha_2\beta_1)/A$, $K = (\alpha_3\beta_1 - \alpha_1\beta_3)/A$ with $\beta_3 \neq 0$, and R, U, P are arbitrary functions of ξ and τ. Using (6.3.9) in (6.3.6) we get

$$(\alpha_2\alpha_3 - U\alpha_2)\frac{\partial R}{\partial \xi} + (\beta_2 U - \beta_3\alpha_2)\frac{\partial R}{\partial \tau} + \beta_2 R\frac{\partial R}{\partial \tau} - \alpha_2 R\frac{\partial U}{\partial \xi}$$
$$+ (K\alpha_2 + LU + U + mU)AR = 0,$$

$$(\alpha_2\alpha_3 - U\alpha_2)\frac{\partial U}{\partial \xi} + (\beta_2 U - \beta_3\alpha_2)\frac{\partial U}{\partial \tau} + \frac{\beta_2}{R}\frac{\partial P}{\partial \tau} - \frac{\alpha_2}{R}\frac{\partial P}{\partial \xi}$$
$$+ (U^2 - \alpha_2 U)A + (L+2)APR^{-1} = 0, \qquad (6.3.10)$$

$$(\alpha_2\alpha_3 - U\alpha_2)\frac{\partial P}{\partial \xi} + (\beta_2 U - \beta_3\alpha_2)\frac{\partial P}{\partial \tau} + \beta_2\gamma P\frac{\partial U}{\partial \tau} - \alpha_2\gamma P\frac{\partial U}{\partial \xi}$$
$$+ (K\alpha_2 - 2\alpha_2 + LU + 2U + m\gamma U + \gamma U)AP = 0,$$

where A is same as in (6.3.8). The above equations can be solved completely (6.3.10)$_{1,3}$ have the closed form solutions as

$$R(\xi, \tau) = R_1(\eta)\exp\{-((K\alpha_2 + (L+1+m)U)/(\alpha_2(\alpha_3 - U)))A\xi\}, \qquad (6.3.11)$$
$$P(\xi, \tau) = P_1(\eta)\exp\{-((K\alpha_2 - 2\alpha_2 + (L+2+\gamma+\gamma m)U)/(\alpha_2(\alpha_3 - U)))A\xi\},$$

where

$$\eta = \tau - \frac{\beta_2 U - \alpha_2\beta_3}{\alpha_2(\alpha_3 - U)}\xi = \frac{1}{\alpha_3 - U}\log\left(x(\alpha_2 t + \alpha_4)^{-U/\alpha_2}\right),$$

and $R_1(\eta)$ is an arbitrary function of η. Using (6.3.11) into (6.3.10)$_2$, we get the compatibility conditions for U and $P_1(\eta)$ as

$$U = 2\alpha_2/(\gamma + 1 + m\gamma - m) \text{ or } U = \alpha_2, \qquad (6.3.12)$$

and $P_1'(\eta) + (\alpha_1 + 2\alpha_3 + mU - U)P_1(\eta) + (\alpha_3 - U)U(U - \alpha_2)R_1(\eta) = 0$. Thus, in view of (6.3.9), (6.3.11) and (6.3.12), the solution of the system (6.3.6) can be expressed as follows.

Case-Ia. When $U = \alpha_2 \neq \alpha_3$, the solution of the system (6.3.6) takes the form

$$\rho = R_1(\eta)x^{(\alpha_1+(m+1)\alpha_2)/(\alpha_3-\alpha_2)}(\alpha_2 t + \alpha_4)^{-(\alpha_1+(m+1)\alpha_3)/(\alpha_3-\alpha_2)}, \qquad (6.3.13)$$
$$u = \alpha_2 x/(\alpha_2 t + \alpha_4), \quad p = C(\alpha_2 t + \alpha_4)^{-(m+1)\gamma},$$

where C is an arbitrary constant, $R_1(\eta)$ is an arbitrary function of η, and

$$\eta = (1/\alpha_3 - \alpha_2)\log(x/(\alpha_2 t + \alpha_4)).$$

Case-Ib. When $U = 2\alpha_2/\Gamma \neq \alpha_3$, where $\Gamma = \gamma + 1 + m\gamma - m \neq 0$, the solution of the system (6.3.6) takes the form

$$\rho = R_1(\eta)x^{A_1}(\alpha_2 t + \alpha_4)^{-A_2}, \quad u = (2\alpha_2/\Gamma)x/(\alpha_2 t + \alpha_4),$$
$$p = \{C - A_3\int R_1(\eta)\exp(A_4\eta)d\eta\}(\alpha_2 t + \alpha_4)^{-2(m+1)\gamma/\Gamma}, \qquad (6.3.14)$$

where C is an arbitrary constant, $R_1(\eta)$ is an arbitrary function of η, and

$$\eta = \frac{\Gamma}{\Gamma\alpha_3 - 2\alpha_2} \log\left(x(\alpha_2 t + \alpha_4)^{-2/\Gamma}\right),$$

$$A_1 = \frac{\Gamma\alpha_1 + 2(m+1)\alpha_2}{\Gamma\alpha_3 - 2\alpha_2}, \quad A_2 = \frac{2(\alpha_1 + (m+1)\alpha_3)}{\Gamma\alpha_3 - 2\alpha_2},$$

$$A_3 = \frac{2\alpha_2^2(m+1)(1-\gamma)(\Gamma\alpha_3 - 2\alpha_2)}{\Gamma^3}, \quad A_4 = \frac{\Gamma(\alpha_1 + 2\alpha_3) + 2(m-1)\alpha_2}{\Gamma}.$$

Case-II. Let $U \equiv \alpha_3$. Then $(6.3.10)_{1,3}$ imply that

$$R(\xi, \eta) = R_1(\xi) \exp\{-(K\alpha_2 + (L+1+m)U)\tau\},$$
$$P(\xi, \eta) = P_1(\xi) \exp\{-(K\alpha_2 - 2\alpha_2 + (L+2+m\gamma+\gamma)U)\tau\}, \qquad (6.3.15)$$

where τ and ξ are the same as defined in (6.3.9), and $R_1(\xi)$ and $P_1(\xi)$ are arbitrary functions of ξ. Moreover, on using (6.3.15) into $(6.3.10)_2$, we get the compatibility conditions for U and $P_1(\xi)$ as

$$U = \alpha_3 = 2\alpha_2/\Gamma, \qquad (6.3.16)$$

$$P_1'(\xi) + \left(\beta_1 + 2\beta_3 + \frac{2(m-1)\beta_2}{\Gamma}\right) P_1(\xi)$$

$$+ \frac{2\alpha_2(\gamma-1)(m+1)(\alpha_3\beta_2 - \alpha_2\beta_3)}{\Gamma^2} R_1(\xi) = 0.$$

Thus, in view of the equations (6.3.9), (6.3.15) and (6.3.16), the solution of the system (6.3.6) can be written as

$$\rho = R_1(\xi)x^{B_1}(\alpha_2 t + \alpha_4)^{-B_2} \quad u = (2\alpha_2/\Gamma)x/(\alpha_2 t + \alpha_4), \quad (6.3.17)$$

$$p = \left\{C - B_3 \int R_1(\xi) \exp(B_4\xi)d\xi\right\}(\alpha_2 t + \alpha_4)^{-2(m+1)\gamma/\Gamma},$$

where C is an arbitrary constant, $R_1(\xi)$ is an arbitrary function of ξ, and

$$\xi = \frac{\Gamma}{\Gamma\beta_3 - 2\beta_2} \log\left(x(\alpha_2 t + \alpha_4)^{-2/\Gamma}\right),$$

$$B_1 = \frac{\Gamma\beta_1 + 2(m+1)\beta_2}{\Gamma\beta_3 - 2\beta_2}, \quad B_2 = \frac{2(\beta_1 + (m+1)\beta_3)}{\Gamma\beta_3 - 2\beta_2},$$

$$B_3 = \frac{2\alpha_2^2(m+1)(1-\gamma)(\Gamma\beta_3 - 2\beta_2)}{\Gamma^3}, \quad B_4 = \frac{\Gamma(\beta_1 + 2\beta_3) + 2(m-1)\beta_2}{\Gamma}.$$

It may be remarked that the solution (6.3.13), (respectively, (6.3.14)) involves arbitrary parameters, α_1, α_2 and α_3 with $\alpha_2 \neq \alpha_3$ (respectively, $\alpha_3 \neq 2\alpha_2/\Gamma$), whereas the solution (6.3.17) depends on the parameters, α_2, β_1, β_2 and β_3. In fact, solution (6.3.17) is exactly the same as (6.3.14) i.e., the constants B_1, B_2, B_3 and B_4 and ξ are exactly the same as A_1, A_2, A_3, A_4 and η when α_1, α_2 and α_3 are replaced by β_1, β_2 and β_3.

6.3.3 Solution with shocks

As is well known, a shock wave may be initiated in the flow region, and once it is formed, it will propagate by separating the portions of the continuous region. At shock, the correct generalized solution satisfies the Rankine-Hugoniot jump conditions. Let $x = X(t)$ be the shock location in the $x - t$ plane propagating in to the medium where $\rho = \rho_0(x)$, $u = \equiv 0$ and $p = p_0 = $ constant. If the shock speed $V = dX/dt$ is very large compared with the sound speed $a_0 = \sqrt{\gamma p_0/\rho_0(x)}$, and the medium behind the shock is given by the solution (6.3.13) or (6.3.14), then at the shock front the following relations hold [210]:

$$\rho = \frac{\gamma + 1}{\gamma - 1}\rho_0(X(t)), \quad u = \frac{2}{\gamma + 1}V, \quad p = \frac{2}{\gamma + 1}\rho_0(X(t))V^2. \tag{6.3.18}$$

I. Let the medium behind the shock be represented by the solution (6.3.13). Then equations (6.3.18) imply

$$R_1(\eta_s)X(t)^{\left[\frac{\alpha_1+(m+1)\alpha_2}{\alpha_3-\alpha_2}\right]}(\alpha_2 t + \alpha_4)^{\left[\frac{\alpha_1+(m+1)\alpha_3}{\alpha_3-\alpha_2}\right]} = \frac{\gamma+1}{\gamma-1}\rho_0(X(t)),$$
$$\frac{\alpha_2 X(t)}{\alpha_2 t + \alpha_4} = \frac{2}{\gamma+1}V, \quad C(\alpha_2 t + \alpha_4)^{-(m+1)\gamma} = \frac{2}{\gamma+1}\rho_0(X(t))V^2, \tag{6.3.19}$$

where C is an arbitrary constant and $\eta_s = (\alpha_3 - \alpha_2)^{-1}\log\left(X(t)(\alpha_2 t + \alpha_4)^{-1}\right)$. From $(6.3.19)_2$, the shock speed $V(t)$ can be written as $V = ((\gamma + 1)\alpha_2/2)X(t)/(\alpha_2 t + \alpha_4)$, implying that

$$X(t) = X_0(T/T_0)^{(\gamma+1)/2}, \tag{6.3.20}$$

where $T = \alpha_2 t + \alpha_4$ and $T_0 = \alpha_2 t_0 + \alpha_4$, with X_0 and t_0 being related with the position and time of the shock. Thus, on using (6.3.20) in equations $(6.3.19)_{1,3}$ we find that $R_1(\eta_s)$ and $\rho(X(t))$ must have the following forms:

$$R_1(\eta_s) = \frac{2C\hat{R}_{10}}{(\gamma-1)\alpha_2^2}\left(\frac{T}{T_0}\right)^{\frac{(1-\gamma)(\alpha_1-(m+3)\alpha_2+2(m+2)\alpha_3)}{2(\alpha_3-\alpha_2)}},$$
$$\rho_0(X(t)) = \rho_c(X(t)/X_0)^{2(1-2\gamma-m\gamma)/(\gamma+1)}, \tag{6.3.21}$$

where $\rho_c = \{2CT_0^{2-m\gamma-\gamma}\}/\{(\gamma + 1)\alpha_2^2 X_0^2\}$ and $\hat{R}_{10} = X_0^{-B_1}T_0^{B_2}$ with

$$B_1 = \{\alpha_1 + (m - 1)\alpha_2 + 2\alpha_3\}/\{\alpha_3 - \alpha_2\},$$
$$B_2 = \{\alpha_1 - (2 - m\gamma - \gamma)\alpha_2 + (3 - m\gamma - \gamma\alpha_3)\}/\{\alpha_3 - \alpha_2\}.$$

In view of (6.3.20), η_s can be written as

$$\eta_s = \frac{1}{\alpha_3 - \alpha_2}\log\left(\frac{X_0}{T_0}\left(\frac{T}{T_0}\right)^{(\gamma-1)/2}\right),$$

and hence $R_1(\eta_s) = R_{10}\exp\left[-\alpha_1 + (m + 3)\alpha_2 - 2(m + 2)\alpha_3\right]\eta_s$, where

$$R_{10} = \frac{2C}{(\gamma - 1)\alpha_2^2}X_0^{2(m+1)}T_0^{-(m+1)(\gamma+1)}.$$

Thus, for a shock, $X(t) = X_0(T/T_0)^{(\gamma+1)/2}$, propagating into a nonuniform region $\rho(x) = \rho_c(x/X_0)^{2(1-2\gamma-m\gamma)/(\gamma+1)}$, $u = 0$, $p = p_0$, the downstream flow given by (6.3.13) takes the form

$$\rho = \tfrac{\gamma+1}{\gamma-1}\rho_c \left(\tfrac{x}{X_0}\right)^{-2(m+2)} \left(\tfrac{T}{T_0}\right)^{m+3}, \quad u = \tfrac{\alpha_2 x}{T},$$

$$p = \tfrac{\gamma+1}{2}\rho_c \left(\tfrac{\alpha_2 X_0}{T_0}\right)^2 \left(\tfrac{T}{T_0}\right)^{-(m+1)\gamma}.$$

(6.3.22)

II. Similarly, when the medium behind the shock is represented by (6.3.14), conditions (6.3.18) imply that the speed of such a shock is given by

$$X(t) = X_0 \left(\frac{\alpha_2 t + \alpha_4}{\alpha_2 t_0 + \alpha_4}\right)^\delta,$$

(6.3.23)

where $\delta = (\gamma+1)/\Gamma$, and the following conditions for $R_1(\eta_s)$ and $\rho_0(x)$ must hold:

$$A_3 \int R_1(\eta_s) \exp(A_4 \eta_s) d\eta_s + A_5 R_1(\eta_s)(\alpha_2 t + \alpha_4)^{A_6} - C = 0,$$

$$\rho_0(X(t)) = \frac{\gamma-1}{\gamma+1} X_0^{A_1} (\alpha_2 t_0 + \alpha_4)^{\delta A_1 - A_2} \left(\frac{\alpha_2 t + \alpha_4}{\alpha_2 t_0 + \alpha_4}\right)^{\delta A_1 - A_2},$$

where A_3 is same as in (6.3.14), $\eta_s = \frac{\Gamma}{\Gamma \alpha_3 - 2\alpha_2} \log \left(X_0(\alpha_2 t + \alpha_4)^{(\gamma-1)/\Gamma}\right)$ and

$$A_5 = \frac{2(\gamma-1)\alpha_2^2 X_0^{2+A_1}}{\Gamma^2(\alpha_2 t_0 + \alpha_4)^{(2+A_1)(\delta A_1 - A - 2)}}, \quad A_6 = \delta A_1 - A_2 + \frac{2(m+\gamma)}{\Gamma}.$$

The solution of the above integral equation exits when $C = 0$, and it can be expressed as

$$R_1(\eta) \propto \exp\{-(\alpha_1 + (1-m)\alpha_3 + (4m/\Gamma)\alpha_2)\eta\}.$$

(6.3.24)

Thus, in view of (6.3.24), equations (6.3.14), which describe the flow downstream from the shock $x = X(t)$, yield

$$\rho = \frac{\gamma+1}{\gamma-1}\rho_c \left(\frac{x}{X_0}\right)^{m-1} \left(\frac{\alpha_2 t + \alpha_4}{\alpha_2 t_0 + \alpha_4}\right)^{-4m/\Gamma}, \quad u = \frac{2\alpha_2}{\Gamma}\frac{x}{\alpha_2 t + \alpha_4},$$

$$p = \left(\frac{2(\gamma+1)\rho_c \alpha_2 X_0}{\Gamma^2(\alpha_2 t_0 + \alpha_4)}\right)^2 \left(\frac{x}{X_0}\right)^{m+1} \left(\frac{\alpha_2 t + \alpha_4}{\alpha_2 t_0 + \alpha_4}\right)^{-2(m+1)(\gamma+1)/\Gamma},$$

(6.3.25)

$$\rho_0(X(t)) = \rho_c \left(\frac{\alpha_2 t + \alpha_4}{\alpha_2 t_0 + \alpha_4}\right)^{(m\gamma-3m-\gamma-1)/\Gamma}, \quad X(t) = X_0 \left(\frac{\alpha_2 t + \alpha_4}{\alpha_2 t_0 + \alpha_4}\right)^{(\gamma+1)/\Gamma},$$

where Γ is same as in (6.3.14). It is interesting to note that the above solution (6.3.25) is exactly the same as the one obtained in the literature, using different approaches, for describing a blast wave (strong shock) propagating into a medium the density of which varies according to a power law of the distance measured from the source of explosion (see Subsection 6.1.1, [129], and Chapter 2 in [96]).

Chapter 7

Kinematics of a Shock of Arbitrary Strength

The study of hyperbolic system of equations and the associated problem of determining the motion of a shock of arbitrary strength has received considerable attention in the literature. The determination of the shock motion requires the calculation of the flow in the region behind the shock. If the shock is weak, Friedrich's theory [55] offers a solution to this problem. Other methods by which the rise of entropy across the shock can be accounted for approximately have been given by Pillow [143], Meyer [124], and Lighthill [109]. Approximate analytical solutions to the problem of decay of a plane shock wave have been proposed by Ardavan-Rhad [6] and Sharma et. al [179]. Other methods which are generally employed for solving this problem in case of strong shock waves belong to the so-called shock expansion theory; the reader is referred to the work carried out by Sirovich and his co-workers ([33] and [187]). A simple rule proposed by Whitham [210], called the characteristic rule or the CCW approximation, determines the motion of a shock without explicitly calculating the flow behind for a large, though restricted class of problems, with good accuracy. Although these methods have been developed in an ad hoc manner, they yield remarkably accurate results; see for example, the papers on shock propagation problems involving diffraction [40], refraction [29], focusing ([71], [28]), and stability [161].

The work presented in this chapter derives motivation from the study relating to an intrinsic description of shock wave propagation carried out by Nunziato and Walsh [131], Chen [32], Wright [212], and Sharma and Radha [169]. A rigorous mathematical approach which can be used for describing kinematics of a shock of arbitrary strength has been proposed by Maslov [119], within the context of a weak shock, by using the theory of generalized functions to derive an infinite system of identities that hold on the shock front. Ravindran and Prasad [150], following Maslov [119], have derived a pair of compatibility conditions for the derivatives of flow variables that hold on the shock front propagating into a nonisentropic gas region. In an attempt to describe wave motion, Best [15] also derived a sequence of transport equations, which hold on the shock front. However, his approach is based on CCW approximation and admits the fact that the application of a characteristic equation at the shock is somewhat ad hoc.

Different methods for studying shock kinematics have been proposed by

Grinfeld [65], Anile and Russo [4], and Fu and Scott [56] who use the theory of bicharacteristics (rays) to derive identity relationships that hold along the rays. Using a procedure based on the kinematics of one dimensional motion, Sharma and Radha [169] studied the behavior of 1-dimensional shock waves in inhomogeneous atmosphere by considering an infinite system of transport equations for the variation of coupling terms $\Pi_{(k)}$ and providing a natural closure on the infinite hierarchy by setting $\Pi_{(k)}$ equal to zero for the largest k retained. This truncation criterion yields a good approximation of the infinite hierarchy; (see [4], [11], and [169]).

The present Chapter, which is largely based on our papers [147] and [174], makes use of the singular surface theory to study the evolutionary behavior of shocks in an unsteady flow of an ideal gas, and of bores in shallow water equations by considering a sequence of transport equations for the variation of jumps in flow variables and their space derivatives across the wave front.

7.1 Shock Wave through an Ideal Gas in 3-Space Dimensions

In this section, we consider a shock wave propagating into an ideal gas region with varying density, velocity and pressure fields, and derive a transport equation for the shock strength. This equation shows that the evolution of the shock depends on coupling to the rearward flow; as a matter of fact, this single transport equation is inadequate to reveal the complete evolutionary behavior of a shock of arbitrary strength because the coupling term in this equation remains undetermined. We, therefore, need to examine certain aspects in more detail in subsequent sections, where the shock strength equation turns out to be just one of the infinite sequence of compatibility conditions that hold on a shock front. This enables us to determine the coupling term approximately using a truncation criterion, the accuracy of which has been tested by comparing the results with the Guderley's similarity solution.

As pointed out by Wright [212], it is found that the shock front can develop discontinuities in slope and Mach number (or, shock strength), which are referred to as shock-shocks [210]; intuitively, it appears that shock-shocks might only persist in the case of intersecting shocks, and that otherwise the underlying shock surface would tend to modify its curvature in such a way that shock-shocks are eliminated. Here, we specialize the results for the two-dimensional shock motion and notice that our exact equations bear a structural resemblance to those of geometrical shock dynamics [210]; indeed, it is shown that the shock ray tube area and the shock Mach number are related according to a differential relation which arises in a natural way from the analysis. It is interesting to note that our relations, in the weak shock limit, coincide exactly with the corresponding equations of geometrical shock dynamics. The implications of the lowest order and first order truncation approximations are discussed when the shock strength varies from the acoustic limit to the limit of geometrical shock dynamics. The lowest order truncation approximation yields the results of geometrical acoustics. The nonlinear breaking of wave motion for disturbances propagating on weak shocks is discussed and the time taken for the onset of a shock-shock is determined. In the strong shock limit, the results for the implosion problem are compared with Guderley's similarity solution. The first order truncation approximation in the weak shock limit leads to the governing equation for the amplitude of an acceleration wave, which is in full agreement with the results obtained in [205]. In the weak shock limit, the asymptotic decay formulae for the shock and rearward precursor disturbances are found to be in full agreement with the earlier results. In passing, we remark that one could have derived an equivalent system of transport equations for the variation of the jumps in density or velocity and their respective derivatives, as they have the required dependence through basic equations and jump conditions. In fact, the process of converting transport equations for the shock strength and coupling terms, defined respectively in terms of the jump in pressure and its derivatives, into an equivalent system of equations involving the jumps in density or velocity and their respective derivatives is quite straightforward. We have carried out the computation of the exponent for the implosion problem using the second order truncation approximation, and the approximate values are found to be close to the numerical results. However, there are still open questions relating to (i) the dependence of truncation on the choice of the forms of shock strength and coupling terms, and the search for an appropriate choice which could provide desired accuracy with a minimum number of ordinary differential equations obtained after employing the truncation criterion and (ii) the rate of convergence of the underlying truncation approximation and the estimates of error bounds of the solution at a particular level of approximation. To tackle these problems some serious efforts are presently underway; if the underlying approach turns out to be sufficiently general, it will be very useful in dealing with still unresolved problems of shock propagation (e.g., Von - Neumann paradox in the field of Mach reflection). Finally, we refer to an alternative approach proposed by Maslov [119], and show that the compatibility conditions derived by using this approach coincide exactly with those obtained by using the singular surface theory.

We use a fixed cartesian coordinate system x_i ($i = 1, 2, 3$) so that in regions adjacent to the shock wave, the conservation laws for mass, momentum and energy may be expressed in the form

$$\partial_t \rho + u^i \rho_{,i} + \rho u^i_{,i} = 0; \quad \partial_t u^i + u^j u^i_{,j} + \rho^{-1} p_{,i} = 0; \quad \partial_t p + u^i p_{,i} + \gamma p u^i_{,i} = 0,$$
$$(7.1.1)$$

where the range of Latin indices is 1,2,3 and the summation convention on repeated indices is implied; ∂_t and a comma followed by an index i denote, respectively, the partial differentiation with respect to time t and space coordinate x_i ; ρ, u^i, p and γ are respectively the gas density, velocity, pressure and specific heat ratio. The shock is propagating through a given state $u^i = u_0^i(t, \vec{x})$, $p = p_0(t, \vec{x})$, $\rho = \rho_0(t, \vec{x})$ where $\vec{x} = (x_1, x_2, x_3)$. Let the shock surface at any time t and position $\vec{x}$ be given by $S(t, \vec{x}) = 0$. Across a shock surface, the jump in a physical quantity is denoted by $[f] = f^- - f^+$, where f^+ and f^- are values of f immediately ahead of and just behind the shock surface respectively. If G is the normal speed of the shock surface with $\vec{n} = (n_1, n_2, n_3)$ as its unit normal, then

$$G = - (\partial_t S)(S_{,j} S_{,j})^{-1/2} \quad ; \quad n_i = (S_{,i})(S_{,j} S_{,j})^{-1/2} . \tag{7.1.2}$$

Across the shock wave, the conservation laws of mass, momentum and energy yield the following jump relations

$$[u^i] = \{z_s (a_0^+)^2 / \gamma U\} n_i \ , \quad [\rho] = 2\rho_0^+ z_s \{2\gamma + (\gamma - 1)z_s\}^{-1} \ , \quad z_s = [p]/p_0 \ , \tag{7.1.3}$$

where $U = G - u_0^{i+} n_i$, is the normal shock speed relative to the fluid, and it is given by

$$U^2 = (a_0^+)^2 \{1 + (\gamma + 1)z_s/2\gamma\} , \tag{7.1.4}$$

with $a_0^+ = (\gamma p_0^+ / \rho_0^+)^{1/2}$ as the equilibrium sound speed. In addition, we have the following compatibility relations [201]

$$[\partial_t f] = d[f]/dt - G[n_i f_{,i}] \ , \quad [f_{,i}] = n_i [n_j f_{,j}] + \tilde{\partial}_i[f] \ , \tag{7.1.5}$$

where d/dt and $\tilde{\partial}_i$ denote respectively the temporal rate of change following the shock and the spatial rate of change along the shock surface defined as

$$df/dt = \partial_t f + (U + u_0^{i+} n_i) n_j f_{,j} \ , \quad \tilde{\partial}_i f = f_{,i} - n_i n_j f_{,j} . \tag{7.1.6}$$

It immediately follows from $(7.1.6)_2$ that

$$n_i \tilde{\partial}_i f = 0, \quad \tilde{\partial}_i n_i = n_{i,i}. \tag{7.1.7}$$

Taking jump in equations (7.1.1) across the shock and using (7.1.5) and (7.1.7) and the fact that jump in u^i is parallel to n_i, we get

$$\partial[\rho]/\partial\tau + \{[u^i]n_i - U\}[n_k \rho_{,k}] + ([\rho] + \rho_0^+)\{n_i[n_j u^i_{,j}] + \tilde{\partial}_i[u^i]\}$$
$$+ [u^i](\rho_{0,i})^+ + [\rho](u^i_{0,i})^+ = 0, \tag{7.1.8}$$

$$\partial[u^i]/\partial\tau + \{[u^j]n_j - U\}[n_k u^i_{,k}] + ([\rho] + \rho_0^+)^{-1}\{n_i[n_j p_{,j}] + \tilde{\partial}_i[p]$$
$$- [\rho](p_{0,i})^+/\rho_0^+\} + [u^j](u^i_{0,j})^+ = 0, \tag{7.1.9}$$

$$\partial[p]/\partial\tau + \{[u^i]n_i - U\}[n_k p_{,k}] + \gamma([p] + p_0^+)\{n_i[n_j u^i_{,j}] + \tilde{\partial}_i[u^i]\}$$
$$+ [u^i](p_{0,i})^+ + \gamma[p](u^i_{0,i})^+ = 0, \tag{7.1.10}$$

where $\partial/\partial\tau = d/dt + \tilde{u}_0^{i+}\tilde{\partial}_i$; $\tilde{u}_0^i = u_0^i - n_i n_j u_0^j$.

Eliminating $[n_j u^i_{,j}]$ between (7.1.8) and (7.1.9) and then using (7.1.3) and (7.1.5), we obtain the following nonlinear equation for the shock strength z_s:

$$\partial z_s / \partial \tau + k_1 \, n_{i,i} = I - k_2 \, [n_i p_{,i}] , \qquad (7.1.11)$$

where

$$k_1 = \{2\gamma U z_s (1 + z_s)(2\gamma + (\gamma - 1)z_s)\} / \Theta,$$
$$k_2 = \{(\gamma + 1)U z_s (2\gamma + (\gamma - 1)z_s)\} / \Theta p_0^+,$$
$$I = k_3(\partial p_0^+ / \partial \tau) + k_4(\partial p_0^+ / \partial \tau) + k_5 n_i (p_{0,i})^+ + k_6 n_i n_j (u^i_{0,j})^+ + k_7 (u^i_{0,i})^+,$$
$$k_3 = \{2\gamma U^2 z_s (1 + z_s)\} / \Theta p_0^+,$$
$$k_4 = -z_s \{(3\gamma + (2\gamma - 1)z_s)(2\gamma + (\gamma + 1)z_s)\} / \Theta p_0^+,$$
$$k_5 = \{2(\gamma + 1)z_s^2 U\} / \Theta p_0^+,$$
$$k_6 = - \{2\gamma(1 + z_s)z_s(2\gamma + (\gamma + 1)z_s)\} / \Theta,$$
$$k_7 = - \{\gamma z_s (2\gamma + (\gamma - 1)z_s)(2\gamma + (\gamma + 1)z_s)\} / \Theta,$$
$$\Theta = 8\gamma^2 + (9\gamma^2 + \gamma)z_s + (2\gamma^2 + \gamma - 1)z_s^2.$$

The growth and decay behavior of a shock wave propagating into an inhomogeneous medium is governed by the shock strength equation (7.1.11). The term $n_{i,i}$ in (7.1.11) is equal to the twice the mean curvature of the shock surface. The inhomogeneous term I in (7.1.11) is a known function of $z_s, \vec{n}, \vec{x}$ and t; it may be noted that this inhomogeneous term vanishes for the homogeneous and rest conditions ahead of the shock. If we differentiate equations (7.1.2), we obtain on using the definitions of $\partial / \partial \tau$ and $\tilde{\partial}_i$ the following kinematic relation

$$\partial n_i / \partial \tau = -n_j (\tilde{\partial}_i u_0^{j+}) - \tilde{\partial}_i U. \qquad (7.1.12)$$

Equation (7.1.12), on using (7.1.4), can be rewritten

$$\frac{\partial n_i}{\partial \tau} + \frac{(\gamma + 1)(a_0^+)^2}{4\gamma U} \tilde{\partial}_i z_s = -\frac{U}{2p_0^+} \tilde{\partial}_i p_0^+ + \frac{U}{2\rho_0^+} \tilde{\partial}_i \rho_0^+ - n_j \tilde{\partial}_i u_0^{j+}, \qquad (7.1.13)$$

where the right-hand side is a known function of $z_s, \vec{n}, \vec{x}$ and t. From the definition of $\partial / \partial \tau$, it follows that the instantaneous motion of the shock surface is described by the equation

$$\partial x_i / \partial \tau = u_0^{i+} + U n_i . \qquad (7.1.14)$$

It may be noted that the shock rays are described by equations (7.1.13) and (7.1.14). If the shock is propagating into a region at rest, equation (7.1.14) shows that the shock rays are also normal and they form a family of orthogonal trajectories to the successive positions of the shock surface. However, equation (7.1.13) shows that the shock rays bend around in response to the gradients in z_s, p_0^+, ρ_0^+ and u_0^{i+}.

Equation (7.1.11), (7.1.13) and (7.1.14), which form a coupled system of partial differential equations in the unknowns z_s, $\vec{x}$ and $\vec{n}$ can be used to calculate successive positions of the shock front, its orientation, and the distribution of shock strength z_s on it, provided the term $[n_i p_{,i}]$ in (7.1.11) is made known. This term, which is the jump in the pressure gradient along the normal to the shock surface, provides a coupling of the shock with the rearward flow; indeed, it represents the effect of disturbances or induced discontinuities that catch-up with the shock from behind. This coupling term rather depends on the past history of the flow. Thus, the equations (7.1.11), (7.1.13) and (7.1.14) show that the evolutionary behavior of the shock strength at any time t depends not only on the distribution of shock strength z_s on the shock surface, but it also depends on the inhomogenities present in the medium ahead of the shock, the normal and the mean curvature of the shock surface, and the jump in pressure gradient along normal to the shock wave at time t. This coupling term can be estimated using a natural truncation criterion; however, we assume for the moment that this coupling term is known exactly.

7.1.1   Wave propagation on the shock

Let θ and ϕ be the angles which the normal $\vec{n}$ to the shock surface makes with the x_1 and x_3 axes respectively. Then the unit normal n_i and its divergence can be expressed as

$$\vec{n} = (\cos\theta\,\sin\phi, \sin\theta\,\sin\phi, \cos\phi), \qquad n_{i,i} = \sin\phi\,(\partial\theta/\partial\eta_2) - (\partial\phi/\partial\eta_1),$$
$$(7.1.15)$$

where

$$\partial/\partial\eta_1 = -\cos\theta\,\cos\phi\,\partial/\partial x_1 - \sin\theta\,\cos\phi\,\partial/\partial x_2 + \sin\phi\,\partial/\partial x_3, \qquad (7.1.16)$$
$$\partial/\partial\eta_2 = -\sin\theta\,\partial/\partial x_1 + \cos\theta\,\partial/\partial x_2,$$

represent spatial rates of change along two mutually orthogonal directions tangential to the shock surface. In view of (7.1.15) and (7.1.16), the shock strength equation (7.1.11) and the shock ray equations (7.1.13) and (7.1.14) become

$$\frac{\partial z_s}{\partial\tau} + k_1\left(\sin\phi\,\frac{\partial\theta}{\partial\eta_2} - \frac{\partial\phi}{\partial\eta_1}\right) = F_1(z_s, \theta, \phi, \vec{x}(\vec{\eta}), \tau),$$

$$\frac{\partial\theta}{\partial\tau} + \frac{\Gamma}{\sin\phi}\frac{\partial z_s}{\partial\eta_2} = F_2(z_s, \theta, \phi, \vec{x}(\vec{\eta}), \tau),$$

$$\frac{\partial\phi}{\partial\tau} - \Gamma\frac{\partial z_s}{\partial\eta_1} = F_3(z_s, \theta, \phi, \vec{x}(\vec{\eta}), \tau), \qquad (7.1.17)$$

$$\frac{\partial x_1}{\partial\tau} = u_0^{1+} + U\cos\theta\,\sin\phi,$$

$$\frac{\partial x_2}{\partial\tau} = u_0^{2+} + U\sin\theta\,\sin\phi,$$

$$\frac{\partial x_3}{\partial\tau} = u_0^{3+} + U\cos\phi,$$

where

$$F_1 = I - k_1[n_i p_{,i}] \quad ; \quad \Gamma = (\gamma + 1)(a_0^+)^2/4\gamma U,$$

$$F_2 = -\frac{U}{2\sin\phi}\left(\frac{1}{p_0^+}\frac{\partial p_0^+}{\partial \eta_2} - \frac{1}{\rho_0^+}\frac{\partial \rho_0^+}{\partial \eta_2}\right) - \cos\theta \frac{\partial u_0^{1+}}{\partial \eta_2}$$

$$- \sin\theta \frac{\partial u_0^{2+}}{\partial \eta_2} - \cot\phi \frac{\partial u_0^{3+}}{\partial \eta_2},$$

$$F_3 = \frac{U}{2}\left(\frac{1}{p_0^+}\frac{\partial p_0^+}{\partial \eta_1} - \frac{1}{\rho_0^+}\frac{\partial \rho_0^+}{\partial \eta_1}\right)\cos\theta\sin\phi \frac{\partial u_0^{1+}}{\partial \eta_1}$$

$$+ \sin\theta\sin\phi \frac{\partial u_0^{2+}}{\partial \eta_1} + \cos\phi \frac{\partial u_0^{3+}}{\partial \eta_1}.$$

It is interesting to note that equations (7.1.17) form a hyperbolic system of six coupled partial differential equations in the unknowns $z_s, \vec{n}$ and $\vec{x}$, and therefore represent a wave motion for disturbances propagating on the shock. Indeed, equations (7.1.17) together with the initial values of θ, ϕ, z_s and $\vec{x}$, specified as functions of η_1 and η_2 at $\tau = 0$, allow us to determine the successive positions x_i of the shock front, the inclinations θ and ϕ of its normal, and the distribution of strength z_s on it. The wave motion on the shock surface itself is brought out by studying the characteristic surfaces of equations (7.1.17), which can be rewritten in the form

$$\frac{\partial W_I}{\partial \tau} + A_{IJ}^\mu \frac{\partial W_J}{\partial \eta_\mu} = B_I, \quad I, J = 1, 2, \dots, 6; \quad \mu = 1, 2, \tag{7.1.18}$$

where $W = (z_s, \theta, \phi, x_1, x_2, x_3)^{tr}$, and the matrices A_{IJ}^μ and B_I, which are known functions of η_1, η_2, and τ, can be read off by inspection of (7.1.17); here the summation on J and μ is implied.

Let $\psi(\eta_1, \eta_2, \tau) = 0$ be the equation for the characteristic surface of (7.1.17). If ξ_μ denotes the unit normal to the characteristic surface in its direction of propagation, and if $V(>0)$ denotes its speed then

$$\xi_\mu = -V(\partial\psi/\partial\eta_\mu)/(\partial\psi/\partial\tau), \tag{7.1.19}$$

and it can be easily shown that all possible speeds of propagation satisfy the characteristic condition

$$\det|A_{IJ}^\mu \xi_\mu - V\delta_{IJ}| = 0, \tag{7.1.20}$$

where δ_{IJ} is the Kronecker function.

Equation (7.1.20) has six roots; the root $V=0$, which is of multiplicity four, corresponds to the requirement that the derivatives of z_s, θ, ϕ, etc. are continuous on the shock surface. The other two nonzero characteristic roots, $V = \pm c^{1/2}$, where

$$c = \frac{\gamma(\gamma + 1)z_s(1 + z_s)\left\{1 + (\gamma - 1)z_s/2\gamma\right\}(a_0^+)^2}{8\gamma^2 + (9\gamma^2 + \gamma)z_s + (2\gamma^2 + \gamma - 1)z_s^2}, \tag{7.1.21}$$

correspond to wave propagation on the shock surface. It is evident from (7.1.21) that the speed V of the wave propagation is independent of the direction ξ_μ.

7.1.2   Shock-shocks

Let us consider the characteristic surface $\psi(\eta_1, \eta_2, \tau) = 0$, which is propagating with the speed $V = c^{1/2}$, say. Then across $\psi = 0$, the variables z_s, θ, ϕ, etc. are continuous but discontinuities in their derivatives are permitted. The jump in an entity f across $\psi = 0$ is denoted by $[\![f]\!]$, where the brackets stand for the quantity enclosed immediately behind the wave front $\psi = 0$ minus its value just ahead of $\psi = 0$, which we shall denote by f_*. The first and second order geometric and kinematic compatibility conditions, which hold across $\psi = 0$, are

$$\left[\!\!\left[\frac{\partial f}{\partial \tau}\right]\!\!\right] = -V\hat{f}, \quad \left[\!\!\left[\frac{\partial f}{\partial \eta_\mu}\right]\!\!\right] = \xi_\mu \hat{f},$$

$$\left[\!\!\left[\frac{\partial^2 f}{\partial \eta_\mu \partial \tau}\right]\!\!\right] = -V\xi_\mu \overline{f} - V\frac{\tilde{\partial}\hat{f}}{\partial \eta_\mu} + \xi_\mu \frac{d\hat{f}}{d\tau} + \hat{f}\frac{d\xi_\mu}{d\tau}, \qquad (7.1.22)$$

$$\left[\!\!\left[\frac{\partial^2 f}{\partial \eta_\mu \partial \eta_\delta}\right]\!\!\right] = \xi_\mu \xi_\delta \overline{f} + \xi_\delta \frac{\tilde{\partial}\hat{f}}{\partial \eta_\mu} + \xi_\mu \frac{\tilde{\partial}\hat{f}}{\partial \eta_\delta} + \hat{f}\frac{\tilde{\partial}\xi_\mu}{\partial \eta_\delta},$$

where

$$\hat{f} \equiv \left[\!\!\left[\xi_\mu \frac{\partial f}{\partial \eta_\mu}\right]\!\!\right], \quad \overline{f} \equiv \left[\!\!\left[\xi_\mu \xi_\delta \frac{\partial^2 f}{\partial \eta_\mu \partial \eta_\delta}\right]\!\!\right], \quad \frac{d}{d\tau} \equiv \frac{\partial}{\partial \tau} + V\xi_\mu \frac{\partial}{\partial \eta_\mu},$$

$$\frac{\tilde{\partial}[\![f]\!]}{\partial \eta_\mu} \equiv \left[\!\!\left[\frac{\partial f}{\partial \eta_\mu}\right]\!\!\right] - \xi_\mu \left[\!\!\left[\xi_\delta \frac{\partial f}{\partial \eta_\delta}\right]\!\!\right] \quad \text{and} \quad \frac{d\xi_\mu}{d\tau} = -\left(\frac{\tilde{\partial}V}{\partial \eta_\mu}\right)_{W=W_*}.$$

Forming jumps across $\psi=0$, in equations (7.1.17), we obtain on using (7.1.22)$_{1,2}$ that

$$\hat{\theta} = (\Gamma_*/c_*^{1/2} \sin\phi_*)\ \xi_2\ \hat{z}, \quad \hat{\phi} = -(\Gamma_*/c_*^{1/2})\ \xi_1\ \hat{z}, \quad \hat{x}_i = 0. \qquad (7.1.23)$$

Following a standard procedure [167], if we differentiate equations (7.1.17) with respect to η_μ, form jumps across $\psi = 0$ in the resulting equations, make use of the compatibility conditions (7.1.22) and the relations (7.1.23), and then eliminate quantities with overhead bars, we obtain the following Riccati type equation in $\hat{z}$ on $\psi = 0$:

$$\frac{d\hat{z}}{d\tau} + \left\{\frac{d(ln\Lambda^{1/2})}{d\tau} + P\right\}\hat{z} + Q\ \hat{z}^2 = 0, \qquad (7.1.24)$$

where

$$\Lambda = c_*^{1/2}/k_{1_*}(\sin\phi_*)^{\xi_2^2},$$

$$P = -V_*K + \frac{V_*}{2}\left\{\cot\phi_*\,\xi_2^2\left(\frac{\partial\phi}{\partial\xi}\right)_* - \cot\phi_*\,\xi_2\left(\frac{\partial\phi}{\partial\eta_2}\right)_*\right.$$

$$-\xi_1\cos\phi_*\left(\frac{\partial\theta}{\partial\eta_2}\right)_* + \left(\frac{\partial\,lnk_1}{\partial\xi}\Big|_{z_s}\right)_* + \left(\frac{\partial\,lnk_1}{\partial z_s}\Big|_\eta\right)_*\left(\frac{\partial z_s}{\partial\xi}\right)_*$$

$$-\frac{\xi_2}{k_{1_*}\sin\phi_*}\left(\frac{\partial F_1}{\partial\theta}\right)_* + \frac{\xi_1}{k_{1_*}}\left(\frac{\partial F_1}{\partial\phi}\right)_* + \Gamma_*\cot\phi_*\,\xi_1\,\xi_2\left(\frac{\partial z_s}{\partial\eta_2}\right)_*$$

$$+2\left(\frac{\partial\,ln\Gamma}{\partial z_s}\Big|_\eta\right)_*\left(\frac{\partial z_s}{\partial\xi}\right)_* - \sin\phi_*\,\xi_2\left(\frac{\partial F_2}{\partial z_s}\Big|_\eta\right)_* + \left(\frac{\partial\,ln\Gamma}{\partial\xi}\Big|_{z_s}\right)_*$$

$$+\frac{\xi_1}{\Gamma_*}\left(\frac{\partial F_3}{\partial z_s}\right)_*\right\} + \frac{1}{2}\left\{\sin\phi_*\left(\frac{\partial k_1}{\partial z_s}\right)_*\left(\frac{\partial\theta}{\partial\eta_2}\right)_* - \left(\frac{\partial k_1}{\partial z_s}\right)_*\left(\frac{\partial\phi}{\partial\eta_1}\right)_*\right.$$

$$-\xi_2^2\left(\frac{\partial F_2}{\partial\theta}\right)_* + \xi_1\xi_2\sin\phi_*\left(\frac{\partial F_2}{\partial\phi}\right)_* + \frac{\xi_1\xi_2}{\sin}\phi_*\left(\frac{\partial F_3}{\partial\theta}\right)_*$$

$$-\xi_1^2\left(\frac{\partial F_3}{\partial\phi}\right)_* - \left(\frac{\partial F_1}{\partial z_s}\right)_*\right\},$$

$$Q = \frac{V_*}{2}\left(\frac{\partial lnk_1}{\partial z_s}\Big|_\eta\right)_* - \frac{c_*^{3/2}}{2k_{1_*}U_*},$$

where $K = -(1/2)(\partial\xi_\mu/\partial\eta_\mu)$ is the mean curvature of the characteristic surface $\psi = 0$, and $(\partial f/\partial\xi) \equiv \xi_\mu(\partial f/\partial\eta_\mu)_*$ is the derivative of f in the direction normal to the characteristic surface on its upstream side.

Equation (7.1.24) is well known in the theory of acceleration waves; a thorough treatment of the properties of the solution of such an equation, within a general framework, is given in Section 4.2. Equation (7.1.24) can be integrated to yield

$$\hat{z} = \frac{\hat{z}(0)\,(\Lambda/\Lambda_0)^{-1/2}\exp(-q(\tau))}{1 + \Lambda_0^{1/2}\hat{z}(0)\,I_1(\tau)}, \tag{7.1.25}$$

where $q(\tau) = \int_0^\tau P(t)\,dt$, $I_1(\tau) = \int_0^\tau(Q/\Lambda^{1/2})\exp(-q(\tau'))\,d\tau'$, and $\hat{z}(0)$ indicates the value of $\hat{z}$ at $\tau = 0$. Equation (7.1.25) shows that if $q(\tau)$ and $I_1(\tau)$ are continuous on $[0,\infty)$ and have finite limits as $\tau \to \infty$ and if sgn $\hat{z}(0) = -\mathrm{sgn}I_1(\tau)$, then $\hat{z}$ will remain finite provided $|\hat{z}(0)| < \zeta$, where $\zeta = (\Lambda_0^{1/2}I_1(\infty))^{-1}$. But if $|\hat{z}(0)| > \zeta$ then there will exist a finite time τ^*, given by $I_1(\tau^*) = -(\Lambda_0^{1/2}\hat{z}(0))^{-1}$, such that $\hat{z} \to \infty$ as $\tau \to \tau^*$; this signifies the appearance of a shock wave at an instant τ^*. This shock discontinuity in the wave motion on the original gasdynamic shock has been referred to as "shock-shock," which corresponds physically to the formation of Mach stems in Mach reflection [210].

7.1.3   Two-dimensional configuration

Assuming the state ahead of the shock to be uniform and at rest, and setting $\phi = \pi/2$ and $\partial/\partial x_3 = 0$ in equations (7.1.17), the shock wave propagation in x_1, x_2 plane can be described by the following equations

$$\frac{\partial \theta}{\partial \eta} + \frac{U}{k_1}\frac{\partial z_s}{\partial \sigma} = -\frac{\gamma+1}{2\gamma(1+z_s)p_0}\left(\frac{\partial p}{\partial n}\right)^-,$$

$$\frac{\partial \theta}{\partial \sigma} + \frac{(\gamma+1)a_0^2}{4\gamma U^2}\frac{\partial z_s}{\partial \eta} = 0, \qquad (7.1.26)$$

$$\frac{\partial x_1}{\partial \sigma} = \cos\theta, \quad \frac{\partial x_2}{\partial \sigma} = \sin\theta,$$

where $d\sigma = U d\tau$ is the distance traveled by shock in time $d\tau$ along the shock ray, $d\eta \equiv d\eta_2$ is the distance measured along the shock front, and the derivative $(\partial p/\partial n)^-$ refers to the instantaneous value of the space derivative of p behind the shock along its normal. The successive shock positions and the rays, described by the family of curves $\sigma=$ constant and $\eta=$ constant, respectively, form an orthogonal coordinate system (σ, η), which is exactly the same as the orthogonal coordinate system (α, β) introduced in the theory of geometrical shock dynamics. Thus the line elements along the two family of curves are

$$d\sigma = M d\alpha \quad \text{and} \quad d\eta = A d\beta, \qquad (7.1.27)$$

where $A(\alpha, \beta)$ is the ray-tube area along the shock ray and $M = U/a_0$ is the shock Mach number, which in view of (7.1.4) can be expressed as

$$M = \{1 + (\gamma+1)z_s/2\gamma\}^{1/2}. \qquad (7.1.28)$$

From the definition of ray derivative, and the fact that $(n_i A)_{,i} = 0$, the entities A, n_i and U are related as $\partial A/\partial \tau = AUn_{i,i}$. This relation for the two-dimensional problem, under consideration, reduces to $\partial A/\partial \sigma = A(\partial \theta/\partial \eta)$, which, in turn, assumes the following form when expressed in the (α, β) coordinates

$$\partial \theta/\partial \beta = M^{-1}(\partial A/\partial \alpha). \qquad (7.1.29)$$

In view of (7.1.27) and (7.1.28), the shock strength equation (7.1.26)$_1$ in the (α, β) coordinates becomes

$$\frac{\partial \theta}{\partial \beta} + \frac{A\,b(M)}{M^2-1}\frac{\partial M}{\partial \alpha} = A\omega\left(\frac{\partial p}{\partial n}\right)^-, \qquad (7.1.30)$$

where $\omega = -(\gamma+1)^2\{2\rho_0 a_0^2(2\gamma M^2 - \gamma + 1)\}^{-1}$, and

$$b(M) = \frac{(\gamma+1)\{2(2\gamma-1)(M^2-1)^2 + (9\gamma+1)M^2 - 5\gamma+3\}}{(2\gamma M^2 - \gamma + 1)(2 + (\gamma-1)M^2)}.$$

If we use (7.1.29) in (7.1.30), we obtain the following differential equation connecting the variation of A and M along the rays:

$$\frac{1}{A}\frac{\partial A}{\partial \alpha} + \frac{Mb(M)}{M^2 - 1}\frac{\partial M}{\partial \alpha} = \omega M \left(\frac{\partial p}{\partial n}\right)^{-}. \qquad (7.1.31)$$

Similarly, equations $(7.1.26)_2$ – $(7.1.26)_4$, in terms of the (α, β) coordinates and the Mach number M, become

$$(\partial \theta / \partial \alpha) = -(1/A)\,(\partial M / \partial \beta), \qquad (7.1.32)$$
$$(\partial x_1 / \partial \alpha) = M\cos\theta, \qquad (7.1.33)$$
$$(\partial x_2 / \partial \alpha) = M\sin\theta. \qquad (7.1.34)$$

As mentioned earlier, the coupling term $(\partial p/\partial n)^{-}$ in the above equations, which is determined in the following section using a truncation criterion, is assumed to be known for the moment. Thus, equations (7.1.30) – (7.1.34), which form a complete set of equations to determine A, M, θ, x_1 and x_2, describe the shock motion taking into account the effect of acoustic disturbances behind the shock.

It is interesting to note that our equations (7.1.29) and (7.1.32) to (7.1.34) coincide exactly with those of geometrical shock dynamics. Thus, the basic equations of geometrical shock dynamics, which form an open system that requires an additional relation to complete it, are exact; the approximation lies only in the introduction of a relation between A and M, which serves to provide an approximate closure on the basic equations and determines the motion of a shock with good accuracy [210]. However, the present treatment leads to an equation (7.1.31), connecting the coupling term $(\partial p/\partial n)^{-}$ with the variations of A and M along the rays. It may be noticed that for a weak shock $(z_s \ll 1)$, $b(M) \to 4$, and consequently, the corresponding weak wave approximation of (7.1.30), which is obtained by neglecting terms of the size $O(z_s^2)$ and $O(z_s(\partial p/\partial n)^{-})$, coincides exactly with that of geometrical shock dynamics. It may be remarked that an area-Mach number relation similar to (7.1.31) for the case of shock propagation in an ideal gas with nonuniform propagation ahead has been considered previously by [29] and [161].

7.1.4 Transport equations for coupling terms

As noted earlier, it is possible to obtain an analytical description of the complete growth and decay history of a shock of arbitrary strength, provided we know the coupling term $[n_i p_{,i}]$; to achieve this goal, we proceed as follows. In order to make algebraic calculations less cumbersome, we consider the state ahead of the shock to be uniform and at rest. The transport equation (7.1.11) for the shock strength, then, becomes

$$\partial z_s / \partial \tau + k_1 n_{i,i} = -k_2 (n_i p_{,i})^{-}, \qquad (7.1.35)$$

where k_1 and k_2 are same as in (7.1.11) with $\rho_0^+ = \rho_0 =$ constant, $p_0^+ = p_0 =$ constant, and $u_0^+ = 0$.

When equations (7.1.3), (7.1.13) and (7.1.35) are combined and the resulting equation is used in (7.1.9), we obtain that

$$(n_j u_{,j}^i)^- = \epsilon_1 n_i (n_j p_{,j})^- + \epsilon_2 n_i n_{j,j} + \epsilon_3 \tilde{\partial}_i z_s, \tag{7.1.36}$$

where

$$\epsilon_1 = \{(8\gamma + 3(\gamma + 1)z_s)(2\gamma + (\gamma - 1)z_s)\}/2\rho_0 U\Theta,$$
$$\epsilon_2 = -\{z_s a_0^2 (1 + z_s)(4\gamma + (\gamma + 1)z_s)\}/U\Theta,$$
$$\epsilon_3 = a_0^2 (4\gamma + (\gamma - 3)z_s)\{2\gamma U(2\gamma + (\gamma - 1)z_s)\}^{-1},$$

with Θ being the same as in (7.1.11). Similarly, equations (7.1.8) and (7.1.9), in view of (7.1.3), (7.1.35) and (7.1.36), yield

$$(n_j \rho_{,j})^- = \epsilon_1^* (n_j p_{,j})^- + \epsilon_2^* n_{j,j}, \tag{7.1.37}$$

where

$$\epsilon_1^* = \{3(\gamma^2 - 1)z_s^2 + (10\gamma^2 - 6\gamma)z_s + 16\gamma^2\} 2\gamma^2 U^2 \{a_0^4 (2\gamma + (\gamma - 1)z_s)^2 \Theta\}^{-1},$$
$$\epsilon_2^* = 4\gamma(\gamma^2 - 1)\rho_0 U^2 z_s^3 \{a_0^2 (2\gamma + (\gamma - 1)z_s)^2 \Theta\}^{-1}.$$

The second order compatibility relations, which we need later, are as follows [201]

$$[\partial_t f_{,i}] = n_i \frac{\partial[n_j f_{,j}]}{\partial \tau} + [n_j f_{,j}]\frac{\partial n_i}{\partial \tau} + \frac{\partial(\tilde{\partial}_i[f])}{\partial \tau} - U n_i [n_l n_k f_{,lk}]$$
$$+ U(\tilde{\partial}_j n_i)(\tilde{\partial}_j [f]), \tag{7.1.38}$$
$$[f_{,ij}] = (\tilde{\partial}_i \tilde{\partial}_j [f]) + (\tilde{\partial}_i n_j)[n_k f_{,k}] + n_j(\tilde{\partial}_i [n_k f_{,k}]) + n_i(\tilde{\partial}_j [n_k f_{,k}])$$
$$- n_i(\tilde{\partial}_j n_k)(\tilde{\partial}_k [f]) + n_i n_j [n_l n_k f_{,lk}]. \tag{7.1.39}$$

Now we differentiate equations (7.1.1) with respect to x_k, take jump in the resulting equations across the shock, multiply these equations by n_k, and use the compatibility conditions (7.1.5), (7.1.38), and (7.1.39) to obtain

$$\frac{\partial(n_i \rho_{,i})^-}{\partial \tau} + \{[u^i]n_i - U\} (n_l n_m \rho_{,lm})^- + (\rho_0 + [\rho]) n_i(n_l n_k u_{,lk}^i)^-$$
$$+ n_k \frac{\partial(\tilde{\partial}_k [\rho])}{\partial \tau} + (n_l u_{,l}^i)^- \{n_i(n_k \rho_{,k})^- + \tilde{\partial}_i [\rho]\} + (\rho_0 + [\rho]) \{\tilde{\partial}_i(n_l u_{,l}^i)^-$$
$$- (\tilde{\partial}_i n_l)(\tilde{\partial}_l [u^i])\} + (n_l \rho_{,l})^- \{n_i(n_k u_{,k}^i)^- + \tilde{\partial}_i [u^i]\}$$
$$+ [u^i] \left(\tilde{\partial}_i(n_l \rho_{,l})^- - (\tilde{\partial}_i n_l)(\tilde{\partial}_l [\rho])\right) = 0, \tag{7.1.40}$$

$$\frac{\partial(n_l u^i_{,l})^-}{\partial\tau} + \{[u^j]n_j - U\}(n_l n_m u^i_{,lm})^- + (\rho_0 + [\rho])^{-1} n_i (n_l n_k p_{,lk})^-$$

$$+ n_k \frac{\partial(\tilde\partial_k[u^i])}{\partial\tau} + (\rho_0 + [\rho])^{-1}\left\{\tilde\partial_i(n_l p_{,l})^- - (\tilde\partial_i n_l)(\tilde\partial_l[p])\right\}$$

$$+ [u^j]\left\{\tilde\partial_j(n_l u^i_{,l})^- - (\tilde\partial_j n_l)(\tilde\partial_l[u^i])\right\} + (n_l u^j_{,l})^-\left\{n_j(n_k u^i_{,k})^- + \tilde\partial_j[u^i]\right\}$$

$$- (\rho_0 + [\rho])^{-2}(n_l \rho_{,l})^-\left\{n_i(n_k p_{,k})^- + \tilde\partial_i[p]\right\} = 0, \tag{7.1.41}$$

$$\frac{\partial(n_i p_{,i})^-}{\partial\tau} + \{[u^i]n_i - U\}(n_l n_m p_{,lm})^- + \gamma(p_0 + [p])n_i(n_l n_k u^i_{,lk})^-$$

$$+ n_k \frac{\partial(\tilde\partial_k[p])}{\partial\tau} + [u^i]\left\{\tilde\partial_i(n_l p_{,l})^- - (\tilde\partial_i n_l)(\tilde\partial_l[p])\right\}$$

$$+ \gamma(p_0 + [p])\left\{\tilde\partial_i(n_l u^i_{,l})^- - (\tilde\partial_i n_l)(\tilde\partial_l[u^i])\right\}$$

$$+ (n_l u^i_{,l})^-\left\{n_i(n_k p_{,k})^- + \tilde\partial_i[p]\right\}$$

$$+ \gamma(n_l p_{,l})^-\left\{n_i(n_k u^i_{,k})^- + \tilde\partial_i[u^i]\right\} = 0. \tag{7.1.42}$$

If we eliminate $(n_l n_k u^i_{,lk})^-$ between (7.1.41) and (7.1.42), substitute in the resulting equation the value of $\partial(n_j u^i_{,j})^-/\partial\tau$, which is obtained from (7.1.36), and make use of (7.1.3), (7.1.35), (7.1.36) and (7.1.37), we obtain the following transport equation for the coupling term $\Pi_{(1)} = (n_i p_{,i})^-$

$$\frac{\partial\Pi_{(1)}}{\partial\tau} + \phi_1\Pi_{(2)} + \phi_2\Pi^2_{(1)} + \phi_3\Pi_{(1)} + \phi_4 = 0, \tag{7.1.43}$$

where $\Pi_{(2)}$, defined as $\Pi_{(2)} = (n_i n_j p_{,ij})^-$, depends on the second order space derivatives of pressure behind the shock along its normal, and the quantities $\phi_1, \phi_2, \phi_3, \phi_4$, which are defined on the shock, have the following form

$$\phi_1 = (\gamma+1)z_s a_0^2\lambda/2\gamma U,$$

$$\lambda = (2\gamma + (\gamma-1)z_s)a_0^2\left\{2\gamma\rho_0 a_0^2 U(1+z_s)\epsilon_1 + (2\gamma + (\gamma-1)z_s)a_0^2\right\}^{-1},$$

$$\phi_2 = \lambda\left\{(\gamma+1)\epsilon_1 + \frac{2\gamma\rho_0 U(1+z_s)}{(2\gamma + (\gamma-1)z_s)}\left(-k_2\frac{d\epsilon_1}{dz_s} + \epsilon_1^2\right.\right.$$

$$\left.\left. -\frac{(2\gamma + (\gamma-1)z_s)^2\epsilon_1^*}{(2\gamma + (\gamma+1)z_s)^2\rho_0^2}\right)\right\},$$

$$\phi_3 = \lambda\left\{\frac{z_s a_0^2}{U} - \frac{2\gamma\rho_0 U(1+z_s)}{(2\gamma + (\gamma-1)z_s)}\left(k_1\frac{d\epsilon_1}{dz_s} + k_2\frac{d\epsilon_2}{dz_s}\right.\right.$$

$$\left.\left. +\frac{(2\gamma + (\gamma-1)z_s)^2\epsilon_2^*}{(2\gamma + (\gamma+1)z_s)^2\rho_0^2} - 2\epsilon_1\epsilon_2\right) + \gamma p_0(1+z_s)\epsilon_1 + (\gamma+1)\epsilon_2\right\}n_{i,i},$$

$$\phi_4 = \lambda \left\{ \gamma p_0 (1 + z_s) \frac{d\epsilon_3}{dz_s} + \epsilon_3 p_0 + \frac{(\gamma + 1) a_0^2 p_0}{4\gamma U} \right.$$

$$+ \frac{2\gamma \rho_0 U (1 + z_s)}{(2\gamma + (\gamma - 1)z_s)} \left(\frac{(\gamma + 1)^2 a_0^4 \epsilon_2}{16\gamma^2 U^3} + \frac{(\gamma + 1) a_0^2 \epsilon_3}{4\gamma U} \right.$$

$$\left. + \frac{(4\gamma + (\gamma + 1)z_s) a_0^4}{4\gamma^2 U^3} \left(\epsilon_3 + \frac{(\gamma + 1) a_0^2}{4\gamma U} \right) \right) \right\} \left(\tilde{\partial}_i z_s \right) \left(\tilde{\partial}_i z_s \right)$$

$$+ \lambda \left\{ \gamma p_0 (1 + z_s) \epsilon_3 - \frac{(\gamma + 1) a_0^2 \rho_0 (1 + z_s) \epsilon_2}{2(2\gamma + (\gamma - 1)z_s)} \right\} \left(\tilde{\partial}_i \tilde{\partial}_i z_s \right)$$

$$+ \lambda \left\{ \gamma p_0 (1 + z_s) \epsilon_2 + \frac{2\gamma \rho_0 U (1 + z_s)}{(2\gamma + (\gamma - 1)z_s)} \left(\epsilon_2^2 - k_1 \frac{d\epsilon_2}{dz_s} \right) \right\} (n_{i,i})^2$$

$$- \lambda \left\{ \frac{z_s (1 + z_s) a_0^2 p_0}{U} + \frac{2\gamma \rho_0 U^2 (1 + z_s) \epsilon_2}{(2\gamma + (\gamma - 1)z_s)} \right\} n_{i,j} n_{j,i} .$$

We note from (7.1.43) that the behavior of the coupling term $\Pi_{(1)}$ depends on another coupling term $\Pi_{(2)}$, which is again unknown. For its determination, we repeat the above procedure; and proceeding in this way, we obtain the following infinite set of transport equations for the coupling terms $\Pi_{(k)} = (n_{i_1} n_{i_2} \cdots n_{i_k} p_{,i_1 i_2 \ldots i_k})^-$; $k = 1, 2, 3, \ldots$

$$\frac{\partial \Pi_{(k)}}{\partial \tau} + \phi_k \Pi_{(k+1)} + \Phi_k(\rho_0, p_0, \vec{n}, z_s, \Pi_{(1)}, \Pi_{(2)}, \cdots, \Pi_{(k)}) = 0, \qquad (7.1.44)$$

where Φ_k are known functions of their arguments and the quantities ϕ_k are defined on the shock front.

The infinite set of equations (7.1.11), (7.1.13), (7.1.14), (7.1.43) and (7.1.44) with $k = 2, 3, \cdots$, and $u_0^i \equiv 0, p_0^+ = p_0, \rho_0^+ = \rho_0$, supplemented by the initial conditions

$$\vec{x} = \vec{x}_0, \quad \vec{n} = \vec{n}_0, \quad z_s = z_{s0}, \quad \Pi_{(k)} = \Pi_{(k)}^0, \quad k \geq 1 \text{ at } \tau = 0, \qquad (7.1.45)$$

can be conveniently used in the investigation of the shock location, orientation, strength and the values of the coupling terms $\Pi_{(k)}$ at any time t. It may be noted that the above infinite system of equations is an open system; in order to provide a natural closure on the infinite hierarchy of equations, we set $\Pi_{(k+1)} = 0$ in (7.1.44) for the largest k retained. The truncated system, which is closed, can be regarded as a good approximation of the infinite hierarchy of the system governing shock propagation ([169], [4]).

7.1.5   The lowest order approximation

The lowest order truncation approximation of the infinite hierarchial system of transport equations leads to the shock strength equation obtained from (7.1.11) by setting $(n_i p_{,i})^- = 0$. Let us consider the case of a weak shock $(z_s \ll 1)$ propagating into a constant state at rest. It then follows from (7.1.3) that

$$
\begin{aligned}
&[u^i] = (z_s a_0/\gamma)n_i + O(z_s^2),\\
&U = a_0\{1 + (\gamma + 1)z_s/4\gamma\} + O(z_s^2),\\
&[\rho] = \rho_0 z_s/\gamma + O(z_s^2),\\
&k_1 = z_s a_0/2 + O(z_s^2),\\
&k_2 = z_s(\gamma + 1)/4\rho_0 a_0 + O(z_s^2),\\
&\Gamma = (\gamma + 1)a_0\{1 - (\gamma + 1)z_s/4\gamma\}/4\gamma + O(z_s^2).
\end{aligned}
\tag{7.1.46}
$$

In the lowest order truncation approximation, the basic equations (7.1.26), on using (7.1.46), reduce to the following system in the weak shock limit

$$
\begin{aligned}
&\frac{\partial z_s}{\partial \tau} + \frac{z_s a_0}{2}\frac{\partial \theta}{\partial \eta} = 0,\\
&\frac{\partial \theta}{\partial \tau} + \frac{(\gamma + 1)a_0}{4\gamma}\frac{\partial z_s}{\partial \eta} = 0,\\
&\frac{\partial x_1}{\partial \tau} = a_0\left\{1 + \frac{(\gamma + 1)z_s}{4\gamma}\right\}\cos\theta,\\
&\frac{\partial x_2}{\partial \tau} = a_0\left\{1 + \frac{(\gamma + 1)z_s}{4\gamma}\right\}\sin\theta.
\end{aligned}
\tag{7.1.47}
$$

These equations represent a wave motion for disturbances propagating on the weak shock; the speeds with which the disturbances propagate turn out to be $V = \pm a_0\{(\gamma + 1)z_s/8\gamma\}^{1/2}$. These waves carry the changes of the shock shape and the shock strength along the weak shock. We consider waves with $V > 0$, running into a constant state and carrying the initial discontinuity in the derivative $(\partial z_s/\partial \eta)$. Analogous to the prototype of three-dimensional wave motion, discussed in the preceding section, these waves break in a typical nonlinear fashion; indeed, it turns out that the derivative $(\partial z_s/\partial \eta)$ at any time τ on the wave front is given by $\partial z_s/\partial \eta = (\partial z_s/\partial \eta)_0\{1 + \Lambda(\partial z_s/\partial \eta)_0\tau\}^{-1}$, where $(\partial z_s/\partial \eta)_0$ is the value of $\partial z_s/\partial \eta$ at $\tau = 0$, and $\Lambda = (a_0/2)((\gamma + 1)/8\gamma)^{1/2}$. Thus, whenever $(\partial z_s/\partial \eta)_0 < 0$, the wave motion develops a shock-shock after a finite time $\tau_s = -(\Lambda(\partial z_s/\partial \eta)_0)^{-1}$. As noted earlier, in this lowest order truncation approximation, the basic equations (7.1.26), in the weak shock limit, coincide exactly with the corresponding equations of geometrical shock dynamics.

For a three-dimensional geometrical configuration, we have $n_{i,i} = -2K$, where K is the mean curvature of the shock surface, which for a planar ($m = 0$), cylindrically ($m = 1$) or spherically ($m = 2$) symmetric flow is given by $K = -m/2X$, where X is the radius of the wave front at time t. Consequently, the shock strength equation (7.1.35), in the lowest order approximation, becomes

$$
\frac{Mb(M)}{M^2 - 1}\frac{\partial M}{\partial X} + \frac{m}{X} = 0,
\tag{7.1.48}
$$

where $M = U/a_0$ is the shock Mach number, $b(M)$ is the same as in (7.1.30), and the distance traversed by the shock along its normal is $dX = U d\tau$. For weak shocks ($M \to 1$, $b \to 4$) equation (7.1.48) yields on integration $M = 1 + (M_0 - 1)(X/X_0)^{-m/2}$, $M|_{X=X_0} = M_0$, which is the exact result of geometrical acoustics for weak pulses. However, for an infinitely strong shock ($M \to \infty$, $b \to (\gamma+1)(2\gamma-1)/\gamma(\gamma-1)$), equation (7.1.48) yields on integration $M = M_0(X/X_0)^{-\beta_*}$, where $\beta_* = m\gamma(\gamma-1)/\{(\gamma+1)(2\gamma-1)\}$. Thus, in the lowest order approximation, the values of the exponent β_* for cylindrical ($m = 1$) and spherical ($m = 2$) shocks for $\gamma = 1.4$ turn out to be 0.130 and 0.259 respectively, whilst the corresponding numerical values obtained from the Guderley's similarity solution for imploding shocks are respectively 0.197 and 0.394 (see [66]). The relative error in the values of the exponent β_* suggest that one should take into account higher approximations.

7.1.6   First order approximation

To assess the error involved in the lowest order approximation and also to give some confidence in the validity of the proposed truncation criterion, we consider the first order ($k = 1$) approximation; to this approximation, the motion of a shock running into a constant state at rest is described by setting $\Pi_{(2)} = 0$ in (7.1.43). Thus, to this approximation, the set of governing equations is the following closed system

$$\frac{\partial z_s}{\partial \tau} + k_1 n_{i,i} = -k_2 \Pi_{(1)},$$

$$\frac{\partial n_i}{\partial \tau} + \frac{(\gamma+1)a_0^2}{4\gamma U} \, \tilde{\partial}_i z_s = 0, \qquad (7.1.49)$$

$$\frac{\partial x_i}{\partial \tau} = U n_i,$$

$$\frac{\partial \Pi_{(1)}}{\partial \tau} + \phi_2 \Pi_{(1)}^2 + \phi_3 \Pi_{(1)} + \phi_4 = 0,$$

which together with the initial conditions (7.1.45) for $k = 1$ can be integrated numerically to determine the shock motion.

However, if the shock amplitude is weak ($z_s \ll 1$), equations (7.1.49) reduce to

$$\frac{\partial z_s}{\partial X} + \frac{(\gamma+1)z_s}{4\rho_0 a_0^2} \, \Pi_{(1)} + \frac{z_s}{2} \, n_{i,i} = 0,$$

$$\frac{\partial n_i}{\partial X} + \frac{\gamma+1}{4\gamma} \, \tilde{\partial}_i z_s = 0, \qquad (7.1.50)$$

$$\frac{\partial n_i}{\partial X} - \left\{ 1 + \frac{(\gamma+1)z_s}{4\gamma} \right\} n_{,i} = 0,$$

$$\frac{\partial \Pi_{(1)}}{\partial X} + \frac{\gamma+1}{2\rho_0 a_0^2} \, \Pi_{(1)}^2 + \frac{1}{2} \, n_{i,i} \Pi_{(1)} = 0,$$

where the distance traversed by the weak wave along its normal is $dX = a_0\, d\tau$. It is interesting to note that equation $(7.1.50)_4$, which governs the amplitude of a weak discontinuity or an acceleration wave, is in full agreement with the results obtained in [205].

Equations $(7.1.50)_1$ and $(7.1.50)_4$, together with the initial conditions (7.1.45), yield on integration

$$z_s = z_{s0} J(X) \exp(-\hat{I}(X)), \tag{7.1.51}$$

$$\Pi_{(1)} = \Pi^0_{(1)} J(X) \left\{ 1 + \frac{\gamma+1}{2\rho_0 a_0^2} \Pi^0_{(1)} J_1(X) \right\}^{-1}, \tag{7.1.52}$$

where z_{s0} and $\Pi^0_{(1)}$ are the values of z_s and $\Pi_{(1)}$ at $X = X_0$,

$$\hat{I}(X) = \frac{\gamma+1}{4\rho_0 a_0^2} \int_{X_0}^{X} \Pi_{(1)}(t)\, dt, \qquad J(X) = \exp\left(\int_{X_0}^{X} K(t)\, dt \right),$$

and $J_1(X) = \int_{X_0}^{X} J(t)\, dt$, with K as the mean curvature of the weak wave. For a planar $(m = 0)$, cylindrically $(m = 1)$ or spherically $(m = 2)$ symmetric flow, equations (7.1.51) and (7.1.52) yield

$$z_s = z_{s0}(X/X_0)^{-m/2} \left\{ 1 + \frac{\gamma+1}{2\rho_0 a_0^2} \Pi^0_{(1)} J_1(X) \right\}^{-1/2}, \tag{7.1.53}$$

$$\Pi_{(1)} = \Pi^0_{(1)}(X/X_0)^{-m/2} \left\{ 1 + \frac{\gamma+1}{2\rho_0 a_0^2} \Pi^0_{(1)} J_1(X) \right\}^{-1}, \tag{7.1.54}$$

where

$$J_1(X) = \begin{cases} X - X_0, & \text{(plane)} \\ 2X_0\{(X/X_0)^{1/2} - 1\}, & \text{(cylindrical)} \\ X_0\, ln(X/X_0), & \text{(spherical)}. \end{cases}$$

It can be seen from (7.1.53) and (7.1.54) that for $\Pi^0_{(1)} \geq 0$, both the shock strength z_s and the precursor disturbance $\Pi_{(1)}$ eventually decay to zero. Indeed, the shock wave decays like

$$z_s = \begin{cases} X^{-1/2}, & \text{(plane)} \\ X^{-3/4}, & \text{(cylindrical)} \\ X^{-1}(\ln X)^{-1/2}, & \text{(spherical)}, \end{cases} \quad \text{as} \quad X \to \infty,$$

while the precursor disturbance $\Pi_{(1)}$ decays like

$$\Pi_{(1)} = \begin{cases} X^{-1}, & \text{(plane and cylindrical)} \\ (X \ln X)^{-1}, & \text{(spherical)}, \end{cases} \quad \text{as} \quad X \to \infty.$$

However, if $\Pi^0_{(1)} < 0$, both the shock strength and the precursor disturbance behind the shock grow without bound after a finite running length X_* given by the solution of $J_1(X_*) = -2\rho_0 a_0^2/\{(\gamma + 1)\Pi^0_{(1)}\}$. The occurrence of this unbounded behavior of the shock strength, or equivalently that of the shock speed, can be attributed to the quasilinear hyperbolic nature of this governing system of equations whose solutions generally exist only in a finite interval.

For a strong shock ($M \gg 1$), we have the following asymptotic formulae for the functions $k_1, k_2, \phi_1, \phi_2, \phi_3$ and ϕ_4 in (7.1.49)

$$k_1 \sim 4(\gamma - 1)\gamma^2 M^3 a_0 / \{(2\gamma - 1)(\gamma + 1)^2\},$$

$$k_2 \sim \gamma(\gamma - 1)M / \{\rho_0 a_0(2\gamma - 1)\},$$

$$\phi_2 \sim 3(\gamma + 1)(5\gamma^2 - 9\gamma + 2) / \{4\rho_0 a_0 M(2\gamma - 1)(5\gamma - 1)\}, \qquad (7.1.55)$$

$$\phi_3 \sim \gamma a_0 M(14\gamma^2 - 41\gamma + 11)n_{i,i} \{(\gamma + 1)(2\gamma - 1)(5\gamma - 1)\}^{-1} n_{i,i},$$

$$\phi_4 \sim \frac{(3\gamma^4 + 8\gamma^3 - 48\gamma^2 + 36\gamma - 7)\rho_0 a_0^3}{4\gamma^2(5\gamma - 1)(\gamma - 1)^2 M} \left(\tilde{\partial}_i z_s\right) \left(\tilde{\partial}_i z_s\right)$$

$$+ \frac{3(\gamma - 1)\rho_0 a_0^3 M}{(5\gamma - 1)(\gamma + 1)} \left(\tilde{\partial}_i \tilde{\partial}_i z_s\right)$$

$$- \frac{4\gamma(\gamma^2 - 4\gamma + 1)\rho_0 a_0^3 M^3}{(\gamma - 1)(5\gamma - 1)(\gamma + 1)^2} \left\{n_{i,j} n_{j,i} + \gamma(2\gamma - 1)^{-1}(n_{i,i})^2\right\}.$$

Making use of (7.1.55) in equations (7.1.49)$_1$ and (7.1.49)$_4$, and then eliminating $\Pi_{(1)}$ from the resulting equations, we obtain the following transport equation for the shock Mach number M for plane ($m = 0$), cylindrical ($m = 1$) and spherical ($m = 2$) wave fronts

$$\frac{\partial^2 M}{\partial X^2} - \frac{(10\gamma^2 - 21\gamma + 5)}{(\gamma - 1)(5\gamma - 1)M} \left(\frac{\partial M}{\partial X}\right)^2 - \frac{m\gamma(3\gamma + 1)}{(\gamma + 1)(5\gamma - 1)X} \left(\frac{\partial M}{\partial X}\right)$$

$$- \frac{2m\gamma^2(2m + \gamma + 1)M}{(\gamma + 1)^2(5\gamma - 1)X^2} = 0.$$

This equation has the solution

$$M = B_* X^{-\beta_2} / (1 - BX^\delta)^{1/q_2}, \qquad (7.1.56)$$

where B and B_* are integration constants, $\beta_2 = (q_1 - \delta)/2q_2$, $\delta = (q_1^2 - 4q_2 q_3)$, $q_1 = 1 + m\gamma(3\gamma + 1)(5\gamma^2 + 4\gamma - 1)^{-1}$, $q_2 = (5\gamma^2 - 15\gamma + 4)/(5\gamma^2 - 6\gamma + 1)$ and $q_3 = 2m\gamma^2\{1 + 2m(\gamma + 1)^{-1}\}/(5\gamma^2 + 4\gamma - 1)$. It then follows by using (7.1.55) and (7.1.56) in (7.1.49)$_4$ that

$$\Pi_{(1)} = -\frac{4\rho_0 a_0^2(2\gamma - 1)M^2}{(\gamma^2 - 1)X} \left\{\frac{\gamma(\gamma - 1)m}{(\gamma + 1)(2\gamma - 1)} + \frac{B\beta_1 X^\delta - \beta_2}{1 - BX^\delta}\right\}, \qquad (7.1.57)$$

where $\beta_1 = (q_1 + \delta)/2q_2$. The integration constants B and B_* can be obtained from (7.1.56) and (7.1.57) by using the initial conditions $M = M_0, \Pi_{(1)} = \Pi_{(1)}^0$ at $X = X_0$.

It may be noticed that for γ lying in the interval $1 < \gamma < 2$, we have $q_1 > 0, \delta > 0, \beta_2 > 0$, while $\beta_1 < 0, q_2 < 0$. Thus, for converging waves, equation (7.1.56) yields the propagation law $M \sim B_* X^{-\beta_2}$ as $X \to 0$, where the values of the exponent β_2 for cylindrical $(m = 1)$ and spherical $(m = 2)$ shocks for $\gamma = 1.4$ turn out to be respectively 0.228 and 0.437, which are close to the numerical values obtained from the Guderley's similarity solution for imploding shocks.

7.2 An Alternative Approach Using the Theory of Distributions

In the preceding section, we have used the theory of singular surfaces to derive an infinite system of transport equations along the shock rays, which enable us to investigate the dynamical coupling between shock fronts and the flow behind them. A different method, using the theory of generalized functions, has been proposed by Maslov [119] to arrive at an infinite system of compatibility conditions that hold on the shock front. Although, the infinite system of equations arising from this method appear to be quite different from the system which is found by using the singular surface theory, we show here in this section that the two seemingly different approaches yield the same system of equations.

The basic equations (7.1.1) can be written in the following conservation forms in R^4:

$$\partial_t \rho + (\rho u^i)_{,i} = 0,$$
$$\partial_t(\rho u^i) + (\rho u^i u^j)_{,j} + p_{,i} = 0, \qquad (7.2.1)$$
$$\partial_t E + Q^i_{,i} = 0,$$

where $E = (\gamma - 1)^{-1} p + \rho u^i u^i / 2$ and $Q^i = u^i \left\{ \gamma(\gamma - 1)^{-1} p + \rho u^j u^j / 2 \right\}$. Let Ω, defined by an equation of the form $S(t, \vec{x}) = 0$, be a smooth three-dimensional surface in R^4 on which a piecewise smooth solution of (7.2.1) undergoes a discontinuity of first kind. At any fixed time t, the surface Ω is a smooth 2-dimensional surface, called a shock surface or a shock front, and it is denoted by Ω_t. Unlike Maslov, we consider a general situation where $|\nabla S|$ need not be unity.

Any piecewise smooth vector solution $f(t, \vec{x}) \equiv (\rho, \vec{u}, p)^{t_r}$ of (7.2.1), which suffers a discontinuity of first kind on Ω, can be represented in the form

$$f(t, \vec{x}) = f_0(t, \vec{x}) + \theta\left(S(t, \vec{x})\right) f_1(t, \vec{x}), \qquad (7.2.2)$$

where $f_0, f_1, S \in C^\infty(R^4)$, and θ is the Heaviside function

$$\theta(\tau) = \begin{cases} 1, & \tau > 0 \\ 0, & \tau \leq 0. \end{cases}$$

Here, f_0 refers to the known state ahead of the shock, whereas $f_0 + f_1$ refers to the state behind the shock. To the piecewise smooth solution f, we assign a set of functions $f_{\alpha\beta}$ ($\alpha = 0, 1$ and $\beta = 0, 1, 2, \cdots,$) defined on Ω and having the interpretation of derivatives along the normal $\vec{n} = (n_1, n_2, n_3)$ to the shock front Ω_t at time t, i. e.,

$$f_{\alpha\beta}|_\Omega = \left(n_i \frac{\partial}{\partial x_i}\right)^\beta f_\alpha\bigg|_\Omega. \tag{7.2.3}$$

It is assumed that the functions $f_{\alpha\beta}|_\Omega$ have been extended in a smooth manner from Ω to the space R^4 so that they are constant along a normal to Ω_t. Thus, we have

$$n_i \left(\partial f_{\alpha\beta}/\partial x_i\right)|_\Omega = 0. \tag{7.2.4}$$

Let the generalized function $h(t, \vec{x})$, representable as a finite sum of functions $h^{(r)}$ be defined as

$$h(t, \vec{x}) = \sum_{r=1}^N h^{(r)}, \qquad h^{(r)} = e_m^{(r)} \frac{\partial \hat{f}^{(r)}}{\partial x_m} ; \ m = 1, 2, 3, 4 \ (r - \text{unsummed}) \tag{7.2.5}$$

where $\hat{f}^{(r)}$ is a function of ρ, u^i and p, which can be represented in the form (7.2.2) with $\hat{f}_\alpha^{(r)}$ as functions in $C^\infty(R^4)$ and $\hat{f}_{\alpha\beta}^{(r)}$ having the definitions outlined as in (7.2.3) and (7.2.4); $e^{(r)}$ are constant vectors in R^4 having components $e_m^{(r)}$; x_4 stands for time t; and the derivatives are to be interpreted in the generalized sense. We assign to h a set of functions $h_{\alpha\beta}$, $\alpha = 0, 1$ and $\beta = -1, 0, 1, 2, 3, \ldots$ such that

$$h_{1,-1}|_\Omega = \sum_{r=1}^N \hat{f}_{10}^{(r)} e_m^{(r)} \frac{\partial S}{\partial x_m}\bigg|_\Omega, \tag{7.2.6}$$

$$h_{\alpha\beta}|_\Omega = \sum_{r=1}^N \left(\hat{f}_{\alpha,\beta+1}^{(r)} e_m^{(r)} \frac{\partial S}{\partial x_m} + e_m^{(r)} \frac{\partial \hat{f}_{\alpha\beta}^{(r)}}{\partial x_m}\right)\bigg|_\Omega, \ \alpha = 0, 1; \beta = 0, 1, 2, \cdots \tag{7.2.7}$$

Now, if $h = \sum_{r=1}^N e_m^{(r)} \frac{\partial \hat{f}^{(r)}}{\partial x_m} = 0$ with the usual summation convention on m, then it has been shown by Maslov [119] that

$$h_{1,-1}|_\Omega = 0, \quad h_\alpha = \sum_{r=1}^N e_m^{(r)} \frac{\partial \hat{f}_\alpha^{(r)}}{\partial x_m} = 0, \quad \alpha = 0, 1. \tag{7.2.8}$$

Further, since $h = 0$, all derivatives of the functions h_α vanish on Ω, i.e.,

$$\left(n_i \frac{\partial}{\partial x_i}\right)^\beta h_\alpha\bigg|_\Omega = 0, \quad \alpha = 0, 1, \beta = 0, 1, 2, \ldots \tag{7.2.9}$$

Maslov has inferred from (7.2.9) that $h_{\alpha\beta}|_\Omega = 0$, which, in fact, may not be true for all β as has been pointed out in [150]. However his relation (7.2.9) is absolutely right on Ω, and it shall be exploited here in the present context. One can easily check that for $\beta = 0$, the relation (7.2.9) yields

$$h_{\alpha 0}|_\Omega = 0, \quad \alpha = 0, 1, \tag{7.2.10}$$

while for $\beta = 1$, it yields

$$h_{\alpha 1}|_\Omega + \sum_{r=1}^{N} \frac{1}{|\nabla_x S|} \left(\hat{f}_{\alpha 1}^{(r)} e_m^{(r)} n_i \frac{\partial^2 S}{\partial x_i x_m} + e_m^{(r)} n_i \frac{\partial^2 \hat{f}_{\alpha 0}^{(r)}}{\partial x_i x_m} \right)\Bigg|_\Omega = 0, \tag{7.2.11}$$

where ∇_x is the spatial gradient operator.

Let $h(t, \vec{x})$ be a generalized function which is the result of substituting (7.2.2) into the left-hand side of the system (7.2.1). Thus, equation (7.2.8)$_1$, together with (7.2.6), yields

$$\begin{aligned} \rho_{10} (\partial_t S) + (\rho u^i)_{10} S_{,i} &= 0, \\ (\rho u^i)_{10} (\partial_t S) + (\rho u^i u^j)_{10} S_{,j} + p_{10} S_{,i} &= 0, \\ E_{10} (\partial_t S) + (Q^i)_{10} S_{,i} &= 0, \end{aligned} \tag{7.2.12}$$

where

$$(\rho u^i)_{10} = (\rho_{00} + \rho_{10}) u_{10}^i + \rho_{10} u_{00}^i,$$

$$(\rho u^i u^j)_{10} = (\rho_{00} + \rho_{10}) \left\{ u_{10}^i u_{00}^j + u_{10}^j (u_{00}^i + u_{10}^i) \right\} + \rho_{10} u_{00}^i u_{00}^j,$$

$$E_{10} = (\gamma - 1)^{-1} p_{10} + \left\{ (\rho_{00} + \rho_{10})(2 u_{10}^i u_{00}^i + u_{10}^i u_{10}^i) + \rho_{10} u_{00}^i u_{00}^i \right\}/2,$$

$$Q_{10}^i = \gamma(\gamma - 1)^{-1} \left\{ p_{10} u_{00}^i + u_{10}^i (p_{00} + p_{10}) \right\} + \left\{ u_{00}^j u_{00}^j (\rho_{10}(u_{00}^i + u_{10}^i)\right.$$

$$\left. + \rho_{00} u_{10}^i) + (\rho_{00} + \rho_{10})(u_{00}^i + u_{10}^i)(u_{10}^j u_{10}^j + 2 u_{10}^j u_{00}^j) \right\}/2.$$

Eliminating ρ_{10} and u_{10}^i from equations (7.2.12)$_1$ to (7.2.12)$_3$, we obtain the equation of motion of the shock front in the following form

$$\partial_t S + u_{00}^i S_{,i} + U |\nabla_x S| = 0, \quad \text{on} \quad \Omega \tag{7.2.13}$$

where

$$U^2|_\Omega = \left\{ a_{00}^2 (1 + (\gamma + 1) z/2\gamma) \right\} |_\Omega, \tag{7.2.14}$$

with $z = (p_{10}/p_{00})$ as the shock strength and $a_{00}^2|_\Omega = \gamma(p_{00}/\rho_{00})|_\Omega$ as the square of sound speed. Consequently, we get from (7.2.12) and (7.2.13), the following Rankine-Hugoniot jump conditions across the shock front

$$\rho_{10}|_\Omega = 2\rho_{00}|_\Omega z_s \{2\gamma + (\gamma - 1) z_s\}^{-1}, \quad u_{10}^i|_\Omega = z_s \{a_{00}^2|_\Omega/\gamma U|_\Omega\} n_i, \tag{7.2.15}$$

where $z_s = (p_{10}/p_{00})|_\Omega$, and $n_i = S_{,i}/|\nabla_x S|$ is the normal vector to the shock surface, and $\rho_{10}|_{\Omega}, u_{10}^i|_{\Omega}$ and $p_{10}|_{\Omega}$ are the jumps across the shock front in ρ, u^i and p respectively. It is evident from (7.1.13) that the normal velocity, G, of the shock front is given by

$$G = u_{00}^i n_i + U. \tag{7.2.16}$$

The transport equation describing the behavior of shock strength z_s can be obtained from equations $(7.2.10)_1$ and (7.2.7), which imply the following set of equations on Ω

$$\partial_t \rho_{00} + \rho_{01}\,(\partial_t S) + (\rho u^i)_{00,i} + (\rho u^i)_{01} S_{,i} = 0,$$
$$\partial_t (\rho u^i)_{00} + (\rho u^i)_{01}\,(\partial_t S) + (\rho u^i u^j)_{00,j} + (\rho u^i u^j)_{01} S_{,j} + p_{00,i} + p_{01} S_{,i} = 0,$$
$$\partial_t E_{00} + E_{01}\,(\partial_t S) + Q_{00,i}^i + Q_{01}^i S_{,i} = 0,$$
$$\tag{7.2.17}$$

while $(7.2.10)_2$ implies the following set of equations on Ω

$$\partial_t \rho_{10} + \rho_{11}\,(\partial_t S) + (\rho u^i)_{10,i} + (\rho u^i)_{11} S_{,i} = 0,$$
$$\partial_t (\rho u^i)_{10} + (\rho u^i)_{11}\,(\partial_t S) + (\rho u^i u^j)_{10,j} + (\rho u^i u^j)_{11} S_{,j} + p_{10,i} + p_{11} S_{,i} = 0,$$
$$\partial_t E_{10} + E_{11}\,(\partial_t S) + Q_{10,i}^i + Q_{11}^i S_{,i} = 0,$$
$$\tag{7.2.18}$$

where

$$(\rho u^i)_{00} = \rho_{00} u_{00}^i; \quad (\rho u^i)_{01} = \rho_{00} u_{01}^i + \rho_{01} u_{00}^i,$$
$$(\rho u^i u^j)_{00} = \rho_{00} u_{00}^i u_{00}^j; \quad (\rho u^i u^j)_{01} = \rho_{00}(u_{00}^i u_{01}^j + u_{01}^i u_{00}^j) + \rho_{01} u_{00}^i u_{00}^j,$$
$$E_{00} = (\gamma - 1)^{-1} p_{00} + \rho_{00} u_{00}^i u_{00}^i/2,$$
$$E_{01} = (\gamma - 1)^{-1} p_{01} + u_{00}^i(\rho_{00} u_{01}^i + \rho_{01} u_{00}^i/2),$$
$$Q_{00}^i = \left\{ \gamma(\gamma - 1)^{-1} p_{00} + \rho_{00} u_{00}^j u_{00}^j/2 \right\} u_{00}^i,$$
$$Q_{01}^i = \gamma(\gamma - 1)^{-1}(p_{00} u_{01}^i + p_{01} u_{00}^i) + \rho_{00} u_{00}^i u_{00}^j u_{01}^j + (\rho_{01} u_{00}^i$$
$$+ \rho_{00} u_{01}^i) u_{00}^j u_{00}^j/2,$$
$$(\rho u^i)_{11} = \rho_{11}(u_{00}^i + u_{10}^i) + u_{11}^i(\rho_{00} + \rho_{10}) + \rho_{10} u_{01}^i + \rho_{01} u_{10}^i,$$
$$(\rho u^i u^j)_{11} = (\rho u^i)_{11}(u_{00}^j + u_{10}^j) + u_{11}^j \left\{ (\rho u^i)_{00} + (\rho u^i)_{10} \right\} + (\rho u^i)_{10} u_{01}^j$$
$$+ (\rho u^i)_{01} u_{10}^j,$$
$$E_{11} = (\gamma - 1)^{-1} p_{11} + (\rho u^i u^i)_{11}/2,$$
$$Q_{11}^i = \gamma(\gamma - 1)^{-1} \left\{ p_{11}(u_{00}^i + u_{10}^i) + u_{11}^i(p_{00} + p_{10}) + p_{10} u_{01}^i + p_{01} u_{10}^i \right\}$$
$$+ \left\{ (\rho u^i u^j)_{10} u_{01}^j + (\rho u^i u^j)_{11}(u_{00}^j + u_{10}^j) + u_{11}^j \left((\rho u^i u^j)_{00} \right. \right.$$
$$\left. \left. + (\rho u^i u^j)_{10} \right) + (\rho u^i u^j)_{01} u_{10}^j \right\} \Big/ 2;$$

here, the quantities $(\rho u^i)_{10}, (\rho u^i u^j)_{10}, E_{10}$ and Q_{10}^i are same as defined in equations (7.2.12).

Eliminating $\partial_t \rho_{10}$ between $(7.2.18)_1$ and $(7.2.18)_2$, we get an equation involving $\partial_t u_{10}^i$; this together with $(7.2.18)_1$ can be used in $(7.2.18)_3$ to yield an equation involving $\partial_t p_{10}$. When the u_{11}^i-eliminant of the resulting equations involving $\partial_t u_{10}^i$ and $\partial_t p_{10}$ is combined with equations (7.2.15) and (7.2.17), we obtain the following transport equation for the shock strength z_s

$$\left.\frac{\partial z_s}{\partial \tau}\right|_\Omega + k_1 \, n_{i,i} = \overline{I}|_\Omega - k_2(p_{11}|_\Omega)|\nabla_x S|, \qquad (7.2.19)$$

where
$$\partial /\partial \tau = \partial /\partial t + (u_{00}^i + U n_i)\partial /\partial x_i,$$
$$\overline{I} = k_3(\partial \rho_{00}/\partial \tau) + k_4(\partial p_{00}/\partial \tau) + k_7 u_{00,i}^i + \left(\overline{k}_8 u_{01}^i n_i + k_5 p_{01}\right)|\nabla_x S|,$$
$$\overline{k}_8 = -\gamma(\gamma+1)z(2+z)(2\gamma+(\gamma+1)z)/\Theta,$$

with k_1, k_2, k_3, k_4, k_6 and k_7 being the same as in (7.1.11); here p_0^+, ρ_0^+, and a_0^+ have the same meaning as $p_{00}|_\Omega$, $\rho_{00}|_\Omega$ and $a_{00}|_\Omega$, respectively.

We note the following connection between the entities that appear in the preceding section and the present section

$$[p] = p_{10}|_\Omega, \quad u_0^{i+} = u_{00}^i|_\Omega, \quad \rho_0^+ = \rho_{00}|_\Omega, \quad p_0^+ = p_{00}|_\Omega, \quad a_0^+ = a_{00}|_\Omega,$$
$$[n_i p_{,i}] = n_i \, (\partial_i p_1)|_\Omega = n_i \, (p_{10,i} + p_{11} S_{,i}) \, |_\Omega = |\nabla_x S| p_{11}|_\Omega, \qquad (7.2.20)$$
$$(p_{0,i})^+ = (p_{0,i})|_\Omega = \left(p_{00,i} + p_{01} n_i |\nabla_x S|\right)|_\Omega,$$
$$\left(u_{0,j}^i\right)^+ = \left(u_{0,j}^i\right)|_\Omega = \left(u_{00,j}^i + u_{01}^i n_j |\nabla_x S|\right)|_\Omega.$$

In view of (7.2.20), it then follows that (7.2.19) is exactly the same as (7.1.11). The bicharacteristics (or, rays) corresponding to the Hamilton-Jacobi equation (7.2.13) are given by

$$\frac{dx_i}{dt} = U n_i + u_{00}^i, \qquad \frac{dS_{,i}}{dt} = H_{,i} \qquad (7.2.21)$$

where $H \equiv \partial_t S + (U n_i + u_{00}^i)S_{,i}$. It follows from (7.2.13), (7.2.14) and the definition of $\partial /\partial \tau$ – derivative in (7.2.19), that the shock ray equations (7.2.21) are precisely the same as (7.1.13) and (7.1.14). In order to derive the transport equation for p_{11}- the jump in the pressure gradient along the normal to shock front, we need to exploit (7.2.11). For the sake of simplicity, we consider the state ahead of the shock to be uniform and at rest. Thus we have from (7.2.3), (7.2.9) and (7.2.12)

$$p_0 = p_{00}, \quad \rho_0 = \rho_{00}, \quad u_0 = u_{00},$$
$$p_{0\alpha} = \rho_{0\alpha} = u_{0\alpha} = 0, \quad \alpha = 1, 2, \cdots . \qquad (7.2.22)$$

In view of (7.2.7) and (7.2.22), the equation (7.2.11) for $\alpha = 1$, yields the following set of equations on Ω

$$|\nabla_x S|\{\partial_t \rho_{11} + \rho_{12}\,(\partial_t S) + (\rho u^i)_{11,i} + (\rho u^i)_{12}S_{,i}\} + n_i\,(\partial_t \rho_{10})_{,i}$$
$$+ \rho_{11}n_i\,(\partial_t S)_{,i} + (\rho u^i)_{11}n_j S_{,ij} = 0, \tag{7.2.23}$$
$$|\nabla_x S|\{\partial_t(\rho u^i)_{11} + (\rho u^i)_{12}\,(\partial_t S) + (\rho u^i u^j)_{11,j} + (\rho u^i u^j)_{12}S_{,j}$$
$$+ p_{11,i} + p_{12}S_{,i}\} + n_j\left(\partial_t(\rho u^i)_{10}\right)_{,j} + (\rho u^i)_{11}n_j\,(\partial_t S)_{,j}$$
$$+(\rho u^i u^j)_{11}n_k S_{,kj} = 0, \tag{7.2.24}$$
$$|\nabla_x S|\,(\partial_t E_{11} + E_{12}\,(\partial_t S) + Q^i_{11,i} + Q^i_{12}S_{,i}) + n_i\,(\partial_t E_{10})_{,i}$$
$$+ E_{11}n_i\,(\partial_t S)_{,i} + Q^i_{11}n_j S_{,ij} = 0, \tag{7.2.25}$$

where

$$
\begin{aligned}
(\rho u^i)_{12} &= \rho_{12}u^i_{10} + u^i_{12}(\rho_{00} + \rho_{10}) + 2\rho_{11}u^i_{11},\\
(\rho u^i u^j)_{11} &= (\rho_{00} + \rho_{10})(u^i_{12}u^j_{10} + u^i_{10}u^j_{12} + 2u^i_{11}u^j_{11})\\
&\quad +2\rho_{11}(u^i_{11}u^j_{10} + u^i_{10}u^j_{11}) + \rho_{12}u^i_{10}u^j_{10},\\
E_{12} &= (\gamma - 1)^{-1}p_{12} + (\rho u^i u^i)_{12}/2,\\
Q^i_{12} &= \gamma(\gamma - 1)^{-1}\{p_{12}u^i_{10} + u^i_{12}(p_{00} + p_{10}) + 2p_{11}u^i_{11}\}\\
&\quad +(\rho_{00} + \rho_{10})\{u^j_{10}u^j_{10}u^i_{12} + 2u^j_{10}u^j_{12}u^i_{10} + 4u^j_{10}u^j_{11}u^i_{11}\}/2\\
&\quad +\rho_{11}u^j_{10}\left(u^j_{10}u^i_{11} + 2u^j_{11}u^i_{10}\right) + \rho_{12}u^i_{10}u^j_{10}u^j_{10}/2,
\end{aligned}
$$

with $(\rho u^i)_{11}, (\rho u^i u^j)_{11}, E_{11}$ and Q^i_{11} being the same as defined in (75); in view of these definitions and (7.2.18), equations (7.2.23) to (7.2.25) can be rewritten as

$$|\nabla_x S|\{\partial_t \rho_{11} + \rho_{12}\,\partial_t S + (\rho_{00} + \rho_{10})(u^i_{11,i} + u^i_{12}S_{,i}) + \rho_{11}(u^i_{10,i} + u^i_{11}S_{,i})$$
$$+u^i_{10}(\rho_{11,i} + \rho_{12}S_{,i}) + u^i_{11}(\rho_{10,i} + \rho_{11}S_{,i})\} + n_i\,(\partial_t \rho_{10})_{,i}$$
$$+ \rho_{11}n_i\,(\partial_t S)_{,i} + u^i_{10}\rho_{11}n_j S_{,ij} + (\rho_{00} + \rho_{10})u^i_{11}n_j S_{,ij} = 0, \tag{7.2.26}$$
$$|\nabla_x S|\left\{\partial_t u^i_{11} + u^i_{12}\,\partial_t S + u^j_{10}(u^i_{11,j} + u^i_{12}S_{,j}) + u^j_{11}(u^i_{10,j} + u^i_{11}S_{,j})\right.$$
$$\left.+(\rho_{00} + \rho_{10})^{-1}(p_{11,i} + p_{12}S_{,i}) - \rho_{11}(\rho_{00} + \rho_{10})^{-2}(p_{10,i} + p_{11}S_{,i})\right\}$$
$$+ n_j\left(\partial_t u^i_{10}\right)_{,j} + u^i_{11}n_j\,(\partial_t S)_{,j} + u^k_{10}u^i_{11}n_j S_{,kj} \tag{7.2.27}$$
$$+ (\rho_{00} + \rho_{10})^{-1}p_{11}n_j S_{,ij} = 0,$$
$$|\nabla_x S|\{\partial_t p_{11} + p_{12}\,\partial_t S + u^i_{10}(p_{11,i} + p_{12}S_{,i}) + u^i_{11}(p_{10,i} + p_{11}S_{,i})$$
$$+ \gamma p_{11}(u^i_{10,i} + u^i_{11}S_{,i}) + \gamma(p_{00} + p_{10})(u^i_{11,i} + u^i_{12}S_{,i})\} + n_i\,(\partial_t p_{10})_{,i}$$
$$+ p_{11}n_i\,(\partial_t S)_{,i} + u^i_{10}p_{11}n_j S_{,ij} + \gamma(p_{00} + p_{10})u^i_{11}n_j S_{,ij} = 0. \tag{7.2.28}$$

According to conditions (7.2.22), equation (7.2.19) assumes the following form

$$\left.\frac{\partial z}{\partial \tau}\right|_\Omega + k_1\,n_{i,i} = -k_2(p_{11}|_\Omega)|\nabla_x S|. \tag{7.2.29}$$

In view of (7.2.15), (7.2.22) and (7.2.29), equations $(7.2.18)_{1,2}$ imply

$$u_{11}^i|_\Omega = \epsilon_1 n_i(p_{11}|_\Omega) + \left(\epsilon_2 n_i n_{j,j} + \epsilon_3\, \tilde{\partial}_i z_s\right)/|\nabla_x S|, \quad (7.2.30)$$

$$\rho_{11}|_\Omega = \epsilon_1^*\,(p_{11}|_\Omega) + \epsilon_2^* n_{j,j}/|\nabla_x S| \quad (7.2.31)$$

where $\epsilon_1, \epsilon_2, \epsilon_3, \epsilon_1^*$ and ϵ_2^* are same as in (7.1.36) and (7.1.37).

It may be noticed that the space derivative of f satisfies

$$f_{,i}|_\Omega = \tilde{\partial}_i\,(f|_\Omega)\ . \quad (7.2.32)$$

Thus, if we eliminate u_{12}^i between (7.2.27) and (7.2.28) and substitute in the resulting equation the values of u_{11}^i and ρ_{11} given by (7.2.30) and (7.2.31), we obtain the following transport equation for p_{11} involving a coupling term p_{12}

$$\left.\frac{\partial p_{11}}{\partial \tau}\right|_\Omega + \left(\phi_1(p_{12}|_\Omega) + \phi_2\left(p_{11}^2|_\Omega\right)\right)|\nabla_x S| + \chi_1 p_{11}|_\Omega + (\phi_4/|\nabla_x S| = 0, \quad (7.2.33)$$

where

$$\chi_1 = \lambda\left\{\gamma p_0(1+z_s)\epsilon_1 + (\gamma+1)\epsilon_2 + \frac{z_s a_0^2}{U} - \frac{2\gamma\rho_0 U(1+z_s)}{(2\gamma+(\gamma-1)z_s)}\left(k_1\frac{d\epsilon_1}{dz_s}\right.\right.$$

$$\left.\left. + k_2\frac{d\epsilon_2}{dz_s} + \frac{(2\gamma+(\gamma-1)z_s)^2\epsilon_2^*}{(2\gamma+(\gamma+1)z_s)^2\rho_0^2} - 2\epsilon_1\epsilon_2\right)\right\}n_{i,i} + \frac{(\gamma+1)a_{00}^2}{4\gamma U}n_j z_{s,j}$$

$$+ \frac{\phi_1}{|\nabla_x S|}n_i n_j S_{,ij},$$

with ϕ_1, ϕ_2, ϕ_4 and λ being the same as in (7.1.43). One can easily verify that

$$\epsilon_1 = \frac{1}{\gamma(1+z_s)}\left\{\frac{2\gamma+(\gamma-1)z_s}{2\rho_0 U} + k_2\right\},$$

$$\epsilon_2 = -\frac{1}{\gamma(1+z_s)}\left\{\frac{z_s(1+z_s)a_{00}^2}{U} - k_1\right\} \quad (7.2.34)$$

$$\tilde{\partial}_i|\nabla_x S| = n_k S_{,ik},$$

$$n_{i,j} = \left\{S_{,ij} - n_i n_k S_{,kj}\right\}/|\nabla_x S|.$$

Further, it may be noticed that

$$\Pi_{(1)} = p_{11}|\nabla_x S|, \quad \Pi_{(2)} = p_{12}|\nabla_x S| + p_{11}n_i n_j S_{,ij}.$$

Consequently, equation (7.1.43) can be rewritten as

$$\left.\frac{\partial p_{11}}{\partial \tau}\right|_\Omega + \phi_1 p_{12}|\nabla_x S| + \phi_2 p_{11}^2|\nabla_x S| + \tilde{\phi}_3 p_{11} + \phi_4/|\nabla_x S| = 0, \quad (7.2.35)$$

where $\tilde{\phi}_3 = \phi_3 + (\phi_1/|\nabla_x S|)n_i n_j S_{,ij} + ((\gamma+1)a_{00}^2/4\gamma U)n_i z_{s,i}$.

In view of (7.2.34), it follows that $\chi_1 = \tilde{\phi}_3$. Consequently, the transport equation (7.2.33) is identical with (7.1.43) or (7.2.35), which we have been seeking to establish.

7.3　Kinematics of a Bore over a Sloping Beach

In contrast to the gasdynamic case, considered in section 7.1, where the determination of the motion of a shock of arbitrary strength is highly complicated because of the entropy variations in the flow, the hydraulic analogy has no counterpart to entropy variations; and hence the flow behind hydraulic jumps or bores can be investigated with relatively greater ease as compared to the gasdynamic case. In this connection the work of Peregrine [142], which analyzes the problem of water wave interactions in the surf zone using the finite amplitude shallow water equations may be mentioned. The shallow water equations are a set of hyperbolic equations which approximate the full free surface gravity flow problem with viscosity and surface tension effects neglected [141]; these equations, being quasilinear and hyperbolic, admit discontinuous solutions which in water are called bores. The motion of a bore moving over a sloping beach through still water has been studied by Keller et al. [90], while an asymptotic analysis of its approach to the shoreline has been performed by Ho and Meyer [72]. An application of the weighted average flux (WAF) method for the computation of the global solution to an initial boundary value problem for the unsteady two dimensional shallow water equations has been given by Toro [199]. Using an approach based mainly on the theory of structure of the nonsingular Euler-Poisson-Darboux equation, an interesting study on the shoreward travel of a bore into a water at rest on a sloping beach has been carried out by Ho and Meyer [72]. They clarified the role of shore-singularity and used it to provide an understanding of the way the solution forgets its initial conditions as the bore approaches the shore. In their work they introduced a monotonicity assumption concerning the seaward boundary which enabled them to determine an asymptotic first order approximation for bore development near the shore, and the results for relatively weak bores with an initial bore strength $M = 1.15$ were shown to coincide with those obtained by Keller et al. [90].

In this section, we study the propagation of bore of arbitrary strength as it approaches the shoreline over a sloping beach, and investigate the dynamical coupling between the bore and the rearward flow by considering an infinite system of transport equations for the variations of the bore strength M and the jump in the space derivatives of the flow variables across the bore. The first transport equation from an infinite hierarchical system derived in this section bears a close structural resemblance with the evolutionary equation obtained by Ho and Meyer [72] for the bore strength M, which in the limit of vanishing wave strength ratio equals the result obtained by using the characteristic rule.

As the wave strength ratio in the evolutionary equation of Ho and Meyer [72] remains an unknown, their equations are unable to reveal the complete history of the evolutionary behavior of a bore of arbitrary strength. This is accomplished in the present study by examining the dynamical coupling between the bore and the rearward flow, which is investigated by considering an infinite system of transport equations for the variations of the bore strength M and the jumps in the space derivatives of the flow variables across the bore.

The implications of the first three truncation approximations are discussed for the case when the bore strength varies from a weak to the strong limit. The results of the present investigation are compared with the approximate results obtained by using the characteristic rule (see Whitham [210]).

7.3.1 Basic equations

The equations describing one dimensional flow of water in terms of the so called shallow water approximation can be written in the form (see [84], [80], and [141]).

$$\partial_t \mathbf{V} + \mathbf{A}(\partial_x \mathbf{V}) + \mathbf{B} = 0, \qquad (7.3.1)$$

where $\mathbf{V} = (c, u)^{tr}$, and $\mathbf{B} = (0, -2c_0(dc_0/dx))^{tr}$ are the column vectors, while $\mathbf{A}$ is the 2×2 matrix having the components $\mathbf{A}_{11} = u$, $\mathbf{A}_{12} = c/2$, $\mathbf{A}_{21} = 2c$ and $\mathbf{A}_{22} = u$. Here, u is the x component of velocity and $c = \sqrt{gh}$ is the speed of propagation of a surface disturbance with g the acceleration due to gravity, and h the depth of the water; $h_0(x)$ is the undisturbed water depth and $c_0 = \sqrt{gh_0}$ is the corresponding surface wave propagation speed. The eigenvalues of A are $\lambda_{1,2} = u \pm c$, and the corresponding left eigenvectors are found to be the row vectors $\mathbf{L}^{1,2} = (1, \pm 1/2)$. Thus, the system of equations (7.3.1) possesses two families of characteristics, $dx/dt = u \pm c$, representing waves propagating in the $\pm x$ direction with the characteristic speeds $u \pm c$. These waves carrying an increase in elevation break leading to the occurrence of a discontinuity known as "bore" in the flow region.

We consider the motion of a right running bore traveling in the region $x \leq x_0$ toward the shoreline at $x = x_0$. It is assumed that the water depth is uniform in $x < 0$ and that the bore is initially moving with constant speed in that region; the beach, which starts at $x = 0$ and has the shoreline at $x = x_0$, has slope $h_0'(x)$ for x lying in the interval $0 \leq x \leq x_0$.

The bore conditions, derived from the conservation form of (7.3.1), can be written in a convenient form (see [210], [90], and [141])

$$\begin{array}{ll} [u] = 2Mc_0(M^2 - 1)/\sqrt{2M^2 - 1}, & [c] = c_0\left(\sqrt{2M^2 - 1} - 1\right), \\ [h] = 2h_0\left(M^2 - 1\right), & U = Mc_0\sqrt{2M^2 - 1}, \end{array} \qquad (7.3.2)$$

where $dX/dt = U$ is the bore velocity being $X(t)$ as the location of bore, the parameter $M = U/c$ may be regarded as the bore strength, and $[f] = f - f_0$ denotes the jump in f across the bore, with f and f_0 the values of f immediately behind and ahead of the bore respectively; here the entity f represents any of the physical variables u, c and h. As the bore forms on a family of positive characteristics, we have $u_0 + c_0 < U < u + c$. In the sequel, we shall make use of the well known compatibility condition

$$\frac{d[f]}{dx} = \frac{1}{U}[\partial_t f] + [\partial_x f], \qquad (7.3.3)$$

where d/dx denotes the displacement derivative following the bore; indeed, if f_x and f_t are discontinuous at the bore $x = X(t)$, but continuous everywhere else, then using the chain rule for the total time derivative of f along the bore and forming the jumps in usual way, one obtains equation (7.3.3). It is evident that equation (7.3.3) expresses the condition for the discontinuity $[f]$ in the function $f(x,t)$ to persist during the motion of the bore $x = X(t)$; in particular if f is continuous across the bore, so that $[f] = 0$, condition (7.3.3) reduces to $[\partial_t f] = -U[\partial_x f]$, which shows that there is a discontinuity in the time derivative f_t at points of the bore whenever there is a discontinuity in the space derivative f_x, provided the velocity of the bore is different from zero. We shall see in what follows that when the dynamical conditions of compatibility, which are the jump relations resulting from the basic equations (7.3.1), are combined with the above compatibility condition (7.3.3), important information can be obtained regarding the evolutionary behavior of the bore.

Taking jump in the system (7.3.1) across the bore and using (7.3.3), we get

$$U\frac{d\,[\mathbf{V}]}{dx} + \mathbf{Q}\,[\partial_x \mathbf{V}] + [\mathbf{A}]\,(\partial_x \mathbf{V})_0 + [\mathbf{B}] = 0,$$

where $\mathbf{Q} = [\mathbf{A}] + \mathbf{A}_0 - U\mathbf{I}$ with $\mathbf{I}$ the unit 2×2 matrix. Since the matrix Q is nonsingular, we can define $\mathbf{P} = \mathbf{Q}^{-1}$ and obtain

$$\mathbf{P}U\frac{d\,[\mathbf{V}]}{dx} + \mathbf{P}\,[\mathbf{A}]\,(\partial_x \mathbf{V})_0 + \mathbf{P}\,[\mathbf{B}] + [\partial_x \mathbf{V}] = 0. \qquad (7.3.4)$$

Since the quantities $[\mathbf{V}]$, $[\mathbf{A}]$, $[\mathbf{P}]$, U and $[\mathbf{B}]$ are expressible in terms of the bore strength M and the flow quantities upstream from the bore, equation (7.3.4), on multiplying by the left eigenvector $\mathbf{L}^1$, yields the following transport equation for the bores of arbitrary strength (see Appendix – A)

$$\alpha_1(x, M)\frac{dM}{dx} + \beta_1(x, M) + \Pi_1 = 0, \qquad (7.3.5)$$

where α_1 and β_1 are known functions of x and M, and the entity

$$\Pi_1 = \mathbf{L}^1\,[\partial_x \mathbf{V}] = [\partial_x(c + u/2)]\,,$$

which provides a coupling of the bore with the flow behind the bore, may be regarded as the amplitude of induced disturbances that overtake the bore from behind. Since the coupling term Π_1 is an unknown function, equation (7.3.5) is unable to reveal the complete history of the evolutionary behavior of a bore of arbitrary strength. We, therefore, need to obtain a transport equation for the coupling term Π_1. To achieve this goal, we differentiate the vector equation (7.3.1) with respect to x, take the jumps across the bore, use the compatibility condition (7.3.3) and then multiply the resulting vector equation by the nonsingular matrix $\mathbf{P}$ to obtain

$$\mathbf{P}\{U d[\partial_x \mathbf{V}]/dx + \{[\partial_x \mathbf{V}] \cdot [\nabla_\mathbf{v} \mathbf{A}] + (\partial_x \mathbf{V})_0 \cdot [\nabla_\mathbf{v} \mathbf{A}]$$
$$+[\partial_x \mathbf{V}] \cdot (\nabla_\mathbf{v} \mathbf{A})_0\}([\partial_x \mathbf{V}] + (\partial_x \mathbf{V})_0)((\partial_x \mathbf{V}) \cdot (\nabla_\mathbf{v} \mathbf{A}))_0[\partial_x \mathbf{V}]$$
$$+[\partial_x \mathbf{V}] \cdot ([\nabla_\mathbf{v} \mathbf{B}] + (\nabla_\mathbf{v} \mathbf{B}_0) + (\partial_x \mathbf{V})_0 \cdot [\nabla_\mathbf{v} \mathbf{B}]$$
$$+[\mathbf{A}](\partial_x^2 \mathbf{V})_0\} + [\partial_x^2 \mathbf{V}] = 0, \tag{7.3.6}$$

where $\nabla_\mathbf{v}$ denotes gradient operator with respect to the elements of $\mathbf{V}$, and ∂_x^2 denotes second order partial derivative with respect to x.

Equation (7.3.6), after pre-multiplying by the left eigenvector $\mathbf{L}^1$ and using (7.3.2), (7.3.4) and (7.3.5) yields the following transport equation for the coupling term Π_1

$$\alpha_2(x, M)\frac{d\Pi_1}{dx} + \beta_2(x, M, \Pi_1) + \Pi_2 = 0, \tag{7.3.7}$$

where α_2 and β_2 are known functions of x, M and Π_1, and

$$\Pi_2 = \mathbf{L}^1 \left[\partial_x^2 \mathbf{V}\right] = \left[\partial_x^2(c + u/2)\right],$$

is another coupling term which is again unknown. We, thus, note that the behavior of the coupling term Π_1 depends on another coupling term Π_2 whose determination calls for the repetition of the above procedure; subsequently, we obtain the following transport equation for Π_2, which in turn involves the coupling term $\Pi_3 = \mathbf{L}^1 \left[\partial_x^3 \mathbf{V}\right] = \left[\partial_x^3(c + u/2)\right]$

$$\alpha_3(x, M, \Pi_1)\frac{d\Pi_2}{dx} + \beta_3(x, M, \Pi_1, \Pi_2) + \Pi_3 = 0, \tag{7.3.8}$$

where α_3 and β_3 are known functions of x, M, Π_1 and Π_2. Proceeding in this way, we obtain an infinite set of transport equations for the coupling terms Π_n, $n = 2, 3, 4, \ldots$, behind the bore

$$\alpha_n(x, M, \Pi_1, \Pi_2, \cdots, \Pi_{n-2})\frac{d\Pi_{n-1}}{dx} + \beta_n(x, M, \Pi_1, \Pi_2, \cdots, \Pi_{n-1}) + \Pi_n = 0, \tag{7.3.9}$$

where α_n and β_n are known functions of x, M, Π_1, $\Pi_2 \cdots$, Π_{n-1}, and

$$\Pi_n = \mathbf{L}^1 \left[\partial_x^n \mathbf{V}\right] = \left[\partial_x^n(c + u/2)\right].$$

This infinite system of transport equations (7.3.5), (7.3.7), (7.3.8) and (7.3.9) together with $dX/dt = U$, supplemented by the initial conditions

$$x = x_0, \quad M = M_0, \quad \Pi_n = \Pi_{n0}, \quad n \geq 1 \quad \text{at} \quad t = 0 \tag{7.3.10}$$

can be conveniently used in the investigation of the bore location $X(t)$, the bore strength M and the amplitudes Π_n of the induced disturbances at any time t. It may be noted that the above infinite system of equations is an open system; in order to provide a natural closure on the infinite hierarchy of equations we set $\Pi_n = 0$ in (7.3.9) for the largest n retained. The truncated system of $n+1$ equations involving unknowns $X, M, \Pi_1, \Pi_2 \cdots$, and Π_{n-1} can be regarded as a good approximation of the infinite hierarchy of equations (see Anile and Russo [3]).

7.3.2   Lowest order approximation

The lowest order truncation approximation of the infinite hierarchical system of transport equations leads to the bore strength equation obtained from (7.3.5) by setting $\Pi_1 = 0$; this corresponds to the situation when there is no dynamical coupling between the bore and the rearward flow. The resulting equation, in view of (7.3.2), can be written as (see Appendix – A)

$$\frac{dM}{dx} + \frac{G(M)}{c_0}\frac{dc_0}{dx} = 0, \qquad (7.3.11)$$

where

$$G(M) = \frac{(M^2-1)\left\{(2M^2-1)(2M^4+5M^3+M^2-1)+(4M^2-1)\sqrt{2M^2-1}\right\}}{M\left(8M^6+12M^5-6M^4-11M^3+3M^2+3M-1\right)}.$$

Here the range of M is $1 < M < \infty$; small values of $M - 1$ correspond to weak bores and large values of M correspond to strong bores. For weak and strong bores, equation (7.3.11), together with (7.3.2), yields

$$\begin{aligned}
&M - 1 \propto c_0^{-5/2}, \ \ [h] \propto c_0^{-1/2}, \ \text{for } M-1 \ll 1,\\
&M \propto c_0^{-1/2}, \ \ [h] \propto c_0, \ \text{for } M \gg 1.
\end{aligned} \qquad (7.3.12)$$

It is quite interesting to note that the above asymptotic results are identical with those predicted by the characteristic rule. At this juncture, it is worthwhile to consider the evolutionary behavior of bores using the characteristic rule, and compare the results with the approximation procedure employed here.

The characteristic rule, proposed by Whitham, consists in applying the differential relation which is valid along a positive characteristic to the flow quantities just behind the bore. The equations in system (7.3.1) may be combined to obtain a differential equation along positive characteristics as

$$\frac{Du}{Dx} + 2\frac{Dc}{Dx} - \frac{2c_0}{u+c}\frac{Dc_0}{Dx} = 0, \qquad (7.3.13)$$

where $D/Dx = \partial_x + (u+c)^{-1}\partial_t$ denotes the derivative along the characteristic $\frac{Dx}{Dt} = u + c$. When equation (7.3.13) is applied along the bore and bore conditions (7.3.2) are used, we obtain the following evolutionary equation for the bore (see [210] and [90])

$$\frac{DM}{Dx} + \frac{F(M)}{c_0}\frac{Dc_0}{Dx} = 0, \qquad (7.3.14)$$

where

$$F(M) = \frac{(M-1)(M^2-0.5)(M^4+3M^3+M^2-1.5M-1)}{2(M+1)(M-0.5)^2(M^3+M^2-M-0.5)}.$$

It is interesting to note that equation (11), obtained by using lowest order approximation, bears a close resemblance with the equation (7.3.14), obtained by using the characteristic rule, except for the functional forms of $G(M)$ and $F(M)$ and the difference in the differential operators D/Dx and d/dx. However, it is remarkable that in the weak or strong bore limit both the functions $F(M)$ and $G(M)$ exhibit the same asymptotic behavior, i.e., both $G, F \to M/2$ or $5(M-1)/2$ according as $M \to 1$ or ∞, respectively. For the sake of comparison of the functions F and G in the entire range $1 \leq M < \infty$, Fig.7.3.1 depicts their variations versus M^{-1}. It is evidently clear from the curves that except for a small range of M, i.e, $1 < M < 5$, the evolutionary behavior of the bores described by the lowest order approximation is almost identical with that predicted by the characteristic rule.

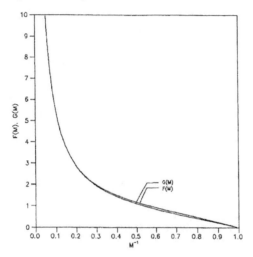

Figure 7.3.1: Functions $F(M)$ and $G(M)$, which appear in the characteristic rule (Eq.(7.3.14)) and the lowest order truncation approximation (Eq.(7.3.11)), versus M^{-1}.

In view of the relation $c_0 = \sqrt{gh_0}$, it is evident from the asymptotic results (7.3.12) that for weak bores, the bore height, $[h]$, increases as h_0 decreases, whereas for strong bores, $[h]$ decreases with h_0. It follows from (7.3.2), that the maximum of $[h]$ occurs at $x = \xi_m$ when

$$(M^2 - 1) + c_0 M \frac{dM}{dc_0} = 0, \qquad (7.3.15)$$

while, the minimum of the bore speed U occurs at $x = \xi_n$ when

$$M\sqrt{2M^2 - 1} + \frac{c_0(4M^2 - 1)}{\sqrt{2M^2 - 1}} \frac{dM}{dc_0} = 0. \qquad (7.3.16)$$

Using the lowest order approximation (i.e., equation (7.3.11)) for dM/dc_0, we find that $[h]$ attains maximum at $M = 1.3148$, while the minimum of bore speed occurs at $M = 1.3027$. These values are found to be somewhat different from those predicted by the characteristic rule, according to which the maximum of $[h]$ occurs at $M = 1.146$ and the minimum bore speed is attained at $M = 1.428$ (see [90]); strangely enough, the characteristic rule predicts the same values of M no matter what the initial bore strength and the undisturbed water depth are. The method presented here yields in a natural manner the critical values of the bore strength to a reasonably good accuracy by taking into account only the first few approximations (see below).

7.3.3   Higher order approximations

Following the truncation criterion discussed above, the evolution of bore to the first order approximation is described by the evolutionary equations (7.3.5) and (7.3.7) by setting $\Pi_2 = 0$. Then the truncated system of transport equations forms a closed system consisting of a pair of equations involving two unknowns M and Π_1; however, if we eliminate Π_1 from these equations, we obtain the following second order ordinary differential equation, which when supplemented by the initial conditions (7.3.10) with $n = 1$, can be integrated numerically to determine the evolutionary behavior of the bore to the first order

$$\frac{d^2M}{dx^2} + T_1\left(\frac{dM}{dx}\right)^2 + \frac{T_2}{c_0}\frac{dc_0}{dx}\frac{dM}{dx} + \frac{T_3}{c_0}\frac{d^2c_0}{dx^2} + \frac{T_4}{c_0^2}\left(\frac{dc_0}{dx}\right)^2 = 0, \quad (7.3.17)$$

where T_1, T_2, T_3 and T_4, which are functions of M only, are given in Appendix – B.

The evolution of a bore to the second order truncation approximation $(n = 2)$ is described by the transport equations (7.3.5), (7.3.7) and (7.3.8) by setting $\Pi_3 = 0$. This closed system, on eliminating Π_1 and Π_2, yields the following third order differential equation

$$\frac{d^3M}{dx^3} + \frac{\hat{\alpha}_1}{c_0}\frac{dc_0}{dx}\frac{d^2M}{dx^2} + \hat{\alpha}_2\frac{dM}{dx}\frac{d^2M}{dx^2} + \frac{\hat{\alpha}_3}{c_0}\frac{dc_0}{dx}\left(\frac{dM}{dx}\right)^2$$

$$+\hat{\alpha}_4\left(\frac{dM}{dx}\right)^3 + \frac{\hat{\alpha}_5}{c_0}\frac{d^2c_0}{dx^2}\frac{dM}{dx} + \frac{\hat{\alpha}_6}{c_0}\frac{d^3c_0}{dx^3} + \frac{\hat{\alpha}_7}{c_0^3}\left(\frac{dc_0}{dx}\right)^3$$

$$+\frac{\hat{\alpha}_8}{c_0^2}\frac{dc_0}{dx}\frac{d^2c_0}{dx^2} + \frac{\hat{\alpha}_9}{c_0^2}\left(\frac{dc_0}{dx}\right)^2\frac{dM}{dx} = 0, \quad (7.3.18)$$

where $\hat{\alpha}_1, \hat{\alpha}_2, \hat{\alpha}_3, \hat{\alpha}_4, \hat{\alpha}_5, \hat{\alpha}_6, \hat{\alpha}_7, \hat{\alpha}_8$ and $\hat{\alpha}_9$, which are functions of M only, are given in Appendix – B. The implications of (7.3.17) and (7.3.18) concerning the evolutionary behavior of a bore of arbitrary strength are discussed in the following section.

7.3.4 Results and discussion

A transport equation (7.3.5) for the bore strength M, which is just one of the infinite sequence of transport equations that hold on the bore front, is found to be identical with equation (53) of Ho and Meyer (see Appendix – C). The coupling term Π_1 in equation (7.3.5), is related to the wave strength ratio Γ_b of Ho and Meyer (see Appendix – C); in Ho and Meyer [72], since the quantity Γ_b remains unknown, the implications of their equation (53) are unable to reveal the evolutionary behavior of a bore of arbitrary strength. The infinite hierarchial system for the bore strength M and the coupling terms Π_k, derived in this chapter, enable us to investigate the dynamical coupling between the bore of arbitrary strength and the rearward flow. A closure on the infinite hierarchy of equations is provided by setting Π_k equal to zero for the largest k retained. Here, we present the results of the first three successive truncation approximations and compare them with the approximate results obtained by using the characteristic rule ([210] and [90]). It is quite interesting to note that the lowest order truncation approximation of the infinite hierarchical system of transport equations leads to the bore strength equation (7.3.11), which bears a close structural resemblance with the equation obtained by using the characteristic rule (see equation (7.3.14)); indeed both the equations lead to the same asymptotic behavior in the weak or strong bore limit as discussed in section - 3. It is shown in Fig. 7.3.1, that except for a small range of M, i.e $1 < M < 5$, the evolutionary behavior of the bores described by the lowest order truncation approximation is almost identical with that predicted by the characteristic rule. Equations (7.3.15) and (7.3.16), in view of the lowest order approximation — (7.3.11) (respectively, the characteristic rule — (14)), determine that the bore height attains a maximum at $M = M^* = 1.3148$ (respectively, $M = M^* = 1.146$), while the bore speed attains a minimum at $M = M_* = 1.3026$ (respectively, $M = M_* = 1.428$). It may be noticed that the values of M^* and M_* determined by using the lowest order approximation are not influenced by M_0 or the undisturbed water depth as it happens with the characteristic rule. To facilitate the comparison between the results obtained by using the truncation approximation and the characteristic rule, we follow the same notation as in [90] and render the equations (7.3.11), (7.3.14), (7.3.17) and (7.3.18) dimensionless by introducing the dimensionless quantities

$$\bar{x} = x/x_0, \quad \bar{t} = tc_0(0)/x_0, \quad \bar{u}(\bar{x},\bar{t}) = u(x,t)/c_0(0),$$
$$H(\bar{x},\bar{t}) = h(x,t)/h_0(0), \quad V = U/c_0(0), \quad N = [h]/h_0(0),$$

where $x_0, h_0(0)$ and $c_0(0)$ are known constants.

We integrate numerically the dimensionless forms of the equations (7.3.11), (7.3.14), (7.3.17), and (7.3.18) for different values of the initial bore strength M_0, and for two different dimensionless forms of the undisturbed water depth

$H_0(\overline{x})$, given by

$$H_0(\overline{x}) = \begin{cases} 1, & \text{if} \quad \overline{x} \leq 0, \\ 1 - \overline{x}, & \text{if} \quad 0 < \overline{x} \leq 1, \end{cases} \qquad (7.3.19)$$

and

$$H_0(\overline{x}) = \begin{cases} 1, & \text{if} \quad \overline{x} \leq 0, \\ 1 - \overline{x}^2, & \text{if} \quad 0 < \overline{x} \leq 1. \end{cases} \qquad (7.3.20)$$

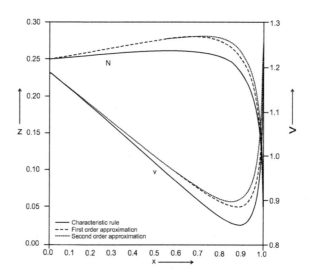

Figure 7.3.2(a): Evolutionary behavior of a bore of moderate strength described by the first and second truncation approximations and the characteristic rule. The variations of bore height N and bore velocity V $versus$ X are displayed for initial bore strength $M_0 = (1.125)^{1/2}$, which is less than the critical values M_* and M^*.

The first order approximation described by equation (7.3.17) shows that for a given M_0 and $H_0(\overline{x})$, the bore height (respectively, the bore speed) attains a maximum at $M = M_m$ (respectively, a minimum at $M = M_n$) that satisfies equations (7.3.15) and (7.3.17) (respectively, equations (7.3.16) and (7.3.17)) identically. Similar is the behavior described by the second order approximation (7.3.18); the computed results are shown in Tables 1 – 4. It may be noticed that in contrast to the characteristic rule, the higher order approximations exhibit that the bore strengths M_m and M_n for which the bore height and the bore speed attain respectively a maximum and a minimum do depend on the initial bore strength M_0 and the undisturbed water depth $H_0(\overline{x})$ as one would anticipate (see Tables 1 – 4); indeed, a feature relating to the dependence of M_n on M_0 is in agreement with the numerical results of [90], which show that the values of $U/\sqrt{gh_0(x)}$ at the points where $V = U/\sqrt{gh_0(0)}$ is minimum, are 2.462, 2.539 and 2.674 for the initial bore height $N(0) = 0.25$,

0.225 and 0.125 respectively. In order to provide a comparison between our results and the numerical results of Keller et al. [90], we have carried out the computations of η/h_0, $U/\sqrt{gh_0}$ and M_n up to second order approximation using the same values of the initial bore height $N(0)$ and undisturbed water depth $H_0(x)$, which were used in [90]; a comparison between our results and the numerical results of Keller et al. [90] is shown in Table 5. It is interesting to note that for $H_0(\overline{x})$ given by (7.3.19), which corresponds to a beach of uniform slope, the lowest order approximation yields values of M^* and M_* correct to six decimal digits as the first and second order approximations do not contribute any significant corrections to them (see Tables 1 and 2). However, if $H_0(\overline{x})$ is specified by equation (7.3.20), which corresponds to a beach of varying slope, the values of M^* and M_* determined by the lowest order approximation get modified by the first and second order approximations (see Tables 3 and 4); indeed, the values correct to four decimal digits are M^* = 1.1207 and M_* = 1.4987, which differ quantitatively from those obtained by using the characteristic rule. Further, if the initial bore strength M_0 is such that $M_0 < M^*$ (respectively, $M_0 < M_*$) then the bore height (respectively, the bore speed) exhibits a maximum (respectively, minimum) lying in the interval $0 \le \overline{x} \le 1$; however for $M_0 > M^*$(respectively, $M_0 > M_*$), the bore height (respectively, the bore speed) exhibits a monotonic decreasing (respectively, increasing) behavior.

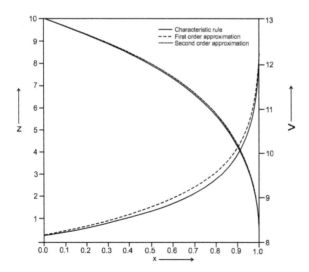

Figure 7.3.2(b): Evolutionary behavior of a bore of moderate strength described by the first and second truncation approximations and the characteristic rule. The variations of bore height N and bore velocity V versus X are displayed for initial bore strength $M_0 = (6.0)^{1/2}$, which is greater than the critical values $M*$ and M^*.

Results for the dimensionless bore height N and bore speed V, which depend on M and $\overline{x}$, through equations (7.3.2), are exhibited in Figs. 7.3.2 and

7.3.3 for typical values of M_0; indeed, for $M_0 < M^*$ (respectively, $M_0 < M_*$), the bore height N (respectively, bore speed V) exhibits a maximum (respectively, minimum) at $M = M_m$ (respectively, $M = M_n$) at $\bar{x} = \xi_m$ (respectively, $\bar{x} = \xi_n$) lying in the interval $0 \le \bar{x} \le 1$ (see Figures 7.3.2(a) and 7.3.3(a)); however, for $M_0 \ge M^*$ (respectively, $M_0 \ge M_*$) the bore height (respectively, bore speed) exhibits a monotonic decreasing (respectively, increasing) behavior over the interval $0 \le \bar{x} \le 1$ (see Figures 7.3.2(b) and 7.3.3(b)).

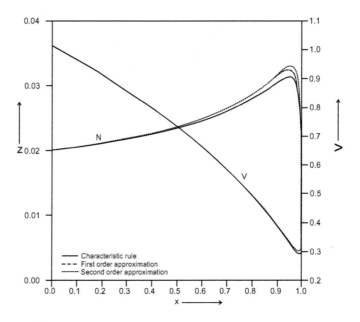

Figure 7.3.3(a): Evolutionary behavior of a bore described by the successive approximations and the characteristic rule for the initial bore strength $M_0 = 1.005$.

It is observed from Tables $1 - 4$ that as M_0 increases from 1 to M^* (respectively, M_*), the location ξ_m (respectively, ξ_n) of the maximum (respectively, minimum) bore height (respectively, bore speed) is shifted towards $\bar{x} = 0$. It may be remarked that only the first three approximations are sufficient to obtain a reasonably good accuracy in the computed values of ξ_n, ξ_m, M_m and M_n. Further, it may be remarked that the lowest order approximation describes the evolutionary behavior of a bore to a good accuracy except when the initial bore is of moderate strength, i.e., $1 < M_0 < 2$; for such a bore, only the first few approximations are sufficient to describe the evolutionary behavior to a reasonably good accuracy.

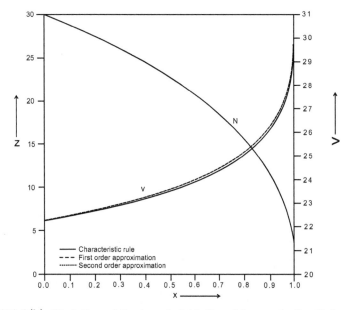

Figure 7.3.3(b): Variation of the bore height N and bore velocity V described by the successive truncation approximations and the characteristic rule for the initial bore strength $M_0 = 4.0$.

Table 1 : For a given initial bore strength M_0, and the undisturbed water depth $H_0(\bar{x})$ given by (7.3.19), the maximum bore height is attained at $\xi = \xi_m$ and $M = M_m$.

M_0	ξ_m			M_m		
	lowest order solution	first order solution	second order solution	lowest order solution	first order solution	second order solution
1.2500	0.2242	0.1549	0.1572	1.3148	1.2914	1.2922
1.3000	0.0530	0.0352	0.0353	1.3148	1.3097	1.3097
1.3027	0.0436	0.0289	0.0290	1.3148	1.3106	1.3106
1.3148	0.0000	0.0000	0.0000	1.3148	1.3148	1.3148
1.3200	0.0000	0.0000	0.0000	1.3200	1.3200	1.3200

Indeed, the improvement takes place only on a subinterval $(\bar{x}_1, \bar{x}_2)$, whose length goes on diminishing as M_0 moves closure to 1 or to 2 (see Figures 7.3.2(a) and 7.3.3(a)). In particular, for moderate strength bores with initial values $M_0 = \sqrt{1.125}$ and $M_0 = \sqrt{6}$, which correspond to the initial bore height $N(0)$ equal to 0.25 and 10, respectively, the flow variables are depicted in Figs. 7.3.2(a) and 7.3.2(b), and the values of N and V in the interval $(\bar{x}_1, \bar{x}_2)$ are improved by the higher order truncation approximations; indeed only the first three approximations are sufficient to obtain the values of N and V correct to at least three decimal digits (see Figures 7.3.2(b) and 7.3.3(b)).

Table 2 : For a given initial bore strength M_0, and the undisturbed
water depth $H_0(\overline{x})$ given by (7.3.19), the minimum bore speed
is attained at $\xi = \xi_n$ and $M = M_n$.

M_0	ξ_n			M_n		
	lowest order solution	first order solution	second order solution	lowest order solution	first order solution	second order solution
1.2500	0.1889	0.2462	0.2406	1.3027	1.3220	1.3200
1.3000	0.0099	0.0131	0.0131	1.3027	1.3035	1.3035
1.3027	0.0000	0.0001	0.0001	1.3027	1.3027	1.3027
1.3148	0.0000	0.0000	0.0000	1.3148	1.3148	1.3148
1.3200	0.0000	0.0000	0.0000	1.3200	1.3200	1.3200

Table 3 : For a given initial bore strength M_0, and the undisturbed
water depth $H_0(\overline{x})$ given by (7.3.20), the maximum bore height
is attained at $\xi = \xi_m$ and $M = M_m$.

M_0	ξ_m			M_m		
	lowest order solution	first order solution	second order solution	lowest order solution	first order solution	second order solution
1.1200	0.791	0.077	0.88	1.3148	1.1207	1.1209
1.1206	0.789	0.027	0.031	1.3148	1.1207	1.1207
1.1207	0.798	0.006	0.007	1.3148	1.1207	1.1207
1.1207	0.789	0.002	0.002	1.3148	1.1207	1.1207
1.1207	0.789	0.000	0.000	1.3148	1.1207	1.1207
1.2000	0.620	0.000	0.000	1.3148	1.2000	1.2000

Table 4 : For a given initial bore strength M_0, and the undisturbed
water depth $H_0(\overline{x})$ given by (7.3.20), the minimum bore speed
is attained at $\xi = \xi_n$ and $M = M_n$.

M_0	ξ_n			M_n		
	lowest order solution	first order solution	second order solution	lowest order solution	first order solution	second order solution
1.4500	0.0000	0.3763	0.3390	1.4500	1.4982	1.4884
1.4900	0.0000	0.1610	0.1440	1.4900	1.4986	1.4968
1.4950	0.0000	0.1060	0.0940	1.4950	1.4987	1.4979
1.4987	0.0000	0.0010	0.0010	1.4987	1.4987	1.4987
1.4987	0.0000	0.0000	0.0000	1.4987	1.4987	1.4987
1.5000	0.0000	0.0000	0.0000	1.5000	1.5000	1.5000

Table 5 : Comparison of our results, obtained using second order
approximation, with the numerical results of Keller [90]; here
η/h_0, $U/\sqrt{gh_0}$ and M_n are the values at points
where V is minimum.

M_0	Method	η/h_0	$U/\sqrt{gh_0}$	M_n
$\sqrt{1.1125}$, i.e.,	Present method	1.807	2.312	1.379
$N(0) = 0.225$	Numerical method	2.13	2.539	1.436
$\sqrt{1.125}$, i.e.,	Present method	1.8024	2.308	1.378
$N(0) = 0.25$	Numerical method	2.017	2.462	1.417
$\sqrt{1.0625}$, i.e.,	Present method	1.8595	2.3490	1.389
$N(0) = 0.125$	Numerical method	2.316	2.674	1.469

However, outside the interval $(\overline{x}_1, \overline{x}_2)$, the values of N and V obtained
by using the higher order approximations are indistinguishable from those
obtained by using the lowest order truncation approximation (see Figure
7.3.3(a)).

7.3.5 Appendices

Appendix – A

This Appendix contains the derivation of the transport equations (7.3.5)
and (7.3.11). In view of bore conditions (7.3.2), the jump quantities
$[\mathbf{V}], [\mathbf{A}], [\mathbf{P}]$ and $[\mathbf{B}]$ can be expressed in terms of the bore strength M and
the upstream flow quantities $\mathbf{V}_0(x) = (c_0(x), u_0(x))^T$; indeed,

$$\frac{d[\mathbf{V}]}{dx} = \frac{\partial[\mathbf{V}]}{\partial M}\frac{dM}{dx} + (\nabla_{\mathbf{v}_0}[\mathbf{V}])\frac{d\mathbf{V}_0}{dx}.$$

As a result, equation (7.3.4) becomes

$$\mathbf{P}U\left\{\frac{\partial[\mathbf{V}]}{\partial M}\frac{dM}{dx} + (\nabla_{\mathbf{v}_0}[\mathbf{V}])\frac{d\mathbf{V}_0}{dx}\right\} + \mathbf{P}[\mathbf{A}]\,(\partial_x\mathbf{V})_0 + \mathbf{P}[\mathbf{B}] + [\partial_x\mathbf{V}] = 0.$$

Pre-multiplication of this last result by the left eigenvector $\mathbf{L}^1$ gives the following transport equation for the bore strength

$$\alpha_1(x, M)\frac{dM}{dx} + \beta_1(x, M) + \Pi_1 = 0,$$

where

$$\alpha_1 = U\mathbf{L}^1\mathbf{P}\frac{\partial[\mathbf{V}]}{\partial M},$$

$$\beta_1 = \mathbf{L}^1\left\{U\mathbf{P}\left(\nabla_{\mathbf{v}_0}[\mathbf{V}]\right) + \mathbf{P}[\mathbf{A}]\right\}\frac{d\mathbf{V}_0}{dx} + \mathbf{L}^1\mathbf{P}[\mathbf{B}],$$

$$\Pi_1 = \mathbf{L}^1\,[\partial_x\mathbf{V}].$$

Equations (7.3.1) and (7.3.2) allow the calculation of α_1 and α_2 in terms of M and $c_0(x)$ as follows. Since

$$[\mathbf{A}] = \frac{c_0}{\sqrt{2M^2 - 1}} \begin{pmatrix} 2M(M^2 - 1) & \left(2M^2 - 1\right) - \sqrt{2M^2 - 1}\,/2 \\ 2\left(2M^2 - 1 - \sqrt{2M^2 - 1}\right) & 2M(M^2 - 1) \end{pmatrix},$$

$$[\mathbf{B}] = 0, \quad \mathbf{P} = \frac{\sqrt{2M^2 - 1}}{(4M^2 - 1)(M^2 - 1)c_0} \begin{pmatrix} M & \left(2M^2 - 1\right)/2 \\ 2\left(2M^2 - 1\right) & M \end{pmatrix},$$

$$\frac{\partial[\mathbf{V}]}{\partial M} = \begin{pmatrix} \dfrac{\partial[c]}{\partial M} \\[2mm] \dfrac{\partial[u]}{\partial M} \end{pmatrix} = \frac{2c_0}{\sqrt{2M^2 - 1}} \begin{pmatrix} M \\[2mm] \dfrac{4M^4 - 3M^2 + 1}{2M^2 - 1} \end{pmatrix},$$

$$\nabla_{\mathbf{v}_0}[\mathbf{V}] = \begin{pmatrix} \dfrac{\partial[c]}{\partial c_0} & \dfrac{\partial[c]}{\partial u_0} \\[2mm] \dfrac{\partial[u]}{\partial c_0} & \dfrac{\partial[u]}{\partial u_0} \end{pmatrix} = \begin{pmatrix} \sqrt{2M^2 - 1} - 1 & 0 \\[2mm] \dfrac{2M(M^2 - 1)}{\sqrt{2M^2 - 1}} & 0 \end{pmatrix},$$

$$\frac{d\mathbf{V}_0}{dx} = \begin{pmatrix} \dfrac{dc_0}{dx} \\[2mm] \dfrac{du_0}{dx} \end{pmatrix}, \quad \mathbf{L}^1 = \left(1, \frac{1}{2}\right), \quad \text{and} \quad U = Mc_0\sqrt{2M^2 - 1},$$

it follows that

$$\alpha_1 = U\mathbf{L}^1\mathbf{P}\frac{\partial[\mathbf{V}]}{\partial M} = \frac{(2M - 1)^3(M + 1)^2 Mc_0}{(4M^2 - 1)(M - 1)\sqrt{2M^2 - 1}},$$

$$\beta_1 = \mathbf{L}^1\left\{\mathbf{P}U\left(\nabla_{\mathbf{v}_0}[\mathbf{V}]\right) + \mathbf{P}[\mathbf{A}]\right\}\frac{d\mathbf{V}_0}{dx} + \mathbf{L}^1\mathbf{P}[\mathbf{B}],$$

$$= \left\{1 + \frac{(M^3 + 3M^2 + 2M + 1)\sqrt{2M^2 - 1}}{2M + 1}\right\}\frac{dc_0}{dx}.$$

The lowest order truncation approximation of the infinite system is obtained by setting $\Pi_1 = 0$ in equation (7.3.5), leading to the result

$$\alpha_1(x, M)\frac{dM}{dx} + \beta_1(x, M) = 0, \quad \text{or} \quad \frac{dM}{dx} + \frac{G(M)}{c_0}\frac{dc_0}{dx} = 0,$$

where

$$G(M) = \beta_1(x, M)c_0\left(\alpha_1(x, M)\frac{dc_0}{dx}\right)^{-1}$$

$$= \frac{(M^2 - 1)\left\{(2M^2 - 1)(2M^4 + 5M^3 + M^2 - 1) + (4M^2 - 1)\sqrt{2M^2 - 1}\right\}}{M\left(8M^6 + 12M^5 - 6M^4 - 11M^3 + 3M^2 + 3M - 1\right)}.$$

Appendix – B

This Appendix lists the coefficients in the transport equations (7.3.17) and (7.3.18).

$$T_1 = \frac{2T_{11} + T_{21}}{2T_{10} + T_{20}}, \quad T_2 = \frac{2T_{12} + T_{22}}{2T_{10} + T_{20}}, \quad T_3 = \frac{2T_{13} + T_{23}}{2T_{10} + T_{20}}, \quad T_4 = \frac{2T_{14} + T_{24}}{2T_{10} + T_{20}},$$

$$\xi = \frac{M(4M^4 - M^2 + 1)\sqrt{2M^2 - 1}}{(4M^2 - 1)(M^2 - 1)}, \quad \xi_* = \frac{2M^2(12M^4 - 11M^2 + 3)}{\sqrt{2M^2 - 1}(4M^2 - 1)(M^2 - 1)},$$

$$\chi = \frac{(4M^4 - 5M^2 + 1)}{M\sqrt{2M^2 - 1}}, \quad T_{10} = \frac{-M\xi\sqrt{2M^2 - 1}}{\chi}, \quad T_{20} = \frac{-M\xi_*\sqrt{2M^2 - 1}}{\chi},$$

$$T_{11} = \frac{80M^{11} + 8M^9 - 3M^7 - 21M^5 + 7M^3 + M}{\chi\left(4M^4 - 5M^2 + 1\right)^2},$$

$$T_{12} = \frac{M^2\left(2M^2 - 1\right)\left(56M^8 - 34M^6 - 42M^4 + 34M^2 - 14\right)}{\chi\left(4M^4 - 5M^2 + 1\right)^2},$$

$$T_{13} = -\frac{1}{\chi}\left\{\frac{M\left(2M^2 - 1\right)^2\left(M^4 - 1\right)}{4M^4 - 5M^2 + 1} + \frac{M}{\sqrt{2M^2 - 1}}\right\},$$

$$T_{14} = \frac{15M^3\left(2M^2 - 1\right)^2\left(M^2 + 1\right)}{\chi\left(4M^2 - 1\right)^2},$$

$$T_{21} = \frac{2M^2\left(64M^{12} + 96M^{10} - 76M^8 - 64M^6 + 83M^4 - 38M^2 + 7\right)}{\chi\left(2M^2 - 1\right)\left(4M^4 - 5M^2 + 1\right)^2},$$

$$T_{22} = \frac{4M\left(16M^{12} + 36M^{10} - 72M^8 + 8M^6 + 17M^4 - 4M^2 - 1\right)}{\chi\left(4M^4 - 5M^2 + 1\right)^2},$$

$$T_{23} = \frac{1}{\chi}\left\{2\sqrt{2M^2 - 1} - \frac{2\left(10M^6 - 5M^4 + 4M^2 - 1\right)}{4M^2 - 1}\right\},$$

$$T_{24} = \frac{2\left(8M^{10} + 104M^8 - 60M^6 - 17M^4 + 12M^2 - 2\right)}{\chi\left(4M^2 - 1\right)^2},$$

$$\hat{\alpha}_1 = \frac{2\hat{T}_{11} + \hat{T}_{21}}{2\hat{T}_{10} + \hat{T}_{20}}, \quad \hat{\alpha}_2 = \frac{2\hat{T}_{12} + \hat{T}_{22}}{2\hat{T}_{10} + \hat{T}_{20}}, \quad \hat{\alpha}_3 = \frac{2\hat{T}_{13} + \hat{T}_{23}}{2\hat{T}_{10} + \hat{T}_{20}}, \quad \hat{\alpha}_4 = \frac{2\hat{T}_{14} + \hat{T}_{24}}{2\hat{T}_{10} + \hat{T}_{20}},$$

$$\hat{\alpha}_5 = \frac{2\hat{T}_{15} + \hat{T}_{25}}{2\hat{T}_{10} + \hat{T}_{20}}, \quad \hat{\alpha}_6 = \frac{2\hat{T}_{16} + \hat{T}_{26}}{2\hat{T}_{10} + \hat{T}_{20}}, \quad \hat{\alpha}_7 = \frac{2\hat{T}_{17} + \hat{T}_{27}}{2\hat{T}_{10} + \hat{T}_{20}}, \quad \hat{\alpha}_8 = \frac{2\hat{T}_{18} + \hat{T}_{28}}{2\hat{T}_{10} + \hat{T}_{20}},$$

$$\hat{\alpha}_9 = \frac{2\hat{T}_{19} + \hat{T}_{29}}{2\hat{T}_{10} + \hat{T}_{20}}, \quad \hat{T}_{10} = -M\overline{T}_{10}\sqrt{2M^2 - 1}, \quad \hat{T}_{20} = -M\overline{T}_{20}\sqrt{2M^2 - 1},$$

$$\hat{T}_{11} = -M\sqrt{2M^2 - 1}\left(\overline{T}_{10} + \overline{T}_{12}\right) + 2(\psi - 1)\overline{T}_{20} + 2.5\psi_*\overline{T}_{10},$$

$$\hat{T}_{12} = -M\sqrt{2M^2 - 1}\left(\frac{d\overline{T}_{10}}{dM} + 2\overline{T}_{11}\right) + 2\xi\overline{T}_{20} + 2.5\xi_*\overline{T}_{10},$$

$$\hat{T}_{13} = -M\sqrt{2M^2 - 1}\left(\overline{T}_{11} + \frac{d\overline{T}_{12}}{dM}\right) + 2(\psi - 1)\overline{T}_{21} + 2.5\psi_*\overline{T}_{11}$$
$$+ 2\xi\overline{T}_{22} + 2.5\xi_*\overline{T}_{12},$$

$$\hat{T}_{14} = -M\sqrt{2M^2 - 1}\frac{d\overline{T}_{11}}{dM} + 2\xi\overline{T}_{21} + 2.5\xi_*\overline{T}_{11},$$

$$\hat{T}_{15} = -M\sqrt{2M^2 - 1}\left(\overline{T}_{12} + \frac{d\overline{T}_{13}}{dM}\right) + 2\xi\overline{T}_{23} + 2.5\xi_*\left(\overline{T}_{13} - 1\right),$$

$$\hat{T}_{16} = -M\overline{T}_{13}\sqrt{2M^2 - 1} + \frac{2M(M^2 - 1)}{\sqrt{2M^2 - 1}},$$

$$\hat{T}_{17} = M\overline{T}_{14}\sqrt{2M^2 - 1} + 2(\psi - 1)\overline{T}_{24} + 2.5\psi_*\overline{T}_{14},$$

$$\hat{T}_{18} = -2M\overline{T}_{14}\sqrt{2M^2 - 1} + 2(\psi - 1)\overline{T}_{23} + 2.5\psi_*\left(\overline{T}_{13} - 1\right),$$

$$\hat{T}_{19} = -M\sqrt{2M^2 - 1}\frac{d\overline{T}_{14}}{dM} + 2\xi\overline{T}_{24} + 2(\psi - 1)\overline{T}_{22} + 2.5\xi_*\overline{T}_{14} + 2.5\psi_*\overline{T}_{12},$$

$$\hat{T}_{21} = -M\sqrt{2M^2 - 1}\left(\overline{T}_{20} + \overline{T}_{22}\right) + 6(\psi - 1)\overline{T}_{10} + 3\psi_*\overline{T}_{20},$$

$$\hat{T}_{22} = -M\sqrt{2M^2 - 1}\left(\frac{d\overline{T}_{20}}{dM} + 2\overline{T}_{21}\right) + 6\xi\overline{T}_{10} + 3\xi_*\overline{T}_{20},$$

$$\hat{T}_{23} = -M\sqrt{2M^2 - 1}\left(\overline{T}_{21} + \frac{d\overline{T}_{22}}{dM}\right) + 6(\psi - 1)\overline{T}_{11} + 3\psi_*\overline{T}_{21}$$
$$+ 6\xi\overline{T}_{12} + 3\xi_*\overline{T}_{22},$$

$$\hat{T}_{24} = -M\sqrt{2M^2 - 1}\frac{d\overline{T}_{21}}{dM} + 6\xi\overline{T}_{11} + 3\xi_*\overline{T}_{21},$$

$$\hat{T}_{25} = -M\sqrt{2M^2 - 1}\left(\overline{T}_{22} + \frac{d\overline{T}_{23}}{dM}\right) + 6\xi\left(\overline{T}_{13} - 1\right) + 3\xi_*\overline{T}_{23},$$

$$\hat{T}_{26} = -M\overline{T}_{23}\sqrt{2M^2 - 1} + 2\sqrt{2M^2 - 1} - 2,$$

$$\hat{T}_{27} = M\overline{T}_{24}\sqrt{2M^2 - 1} + 6(\psi - 1)\overline{T}_{14} + 3\psi_*\overline{T}_{24},$$

$$\hat{T}_{28} = -2M\overline{T}_{24}\sqrt{2M^2 - 1} + 6(\psi - 1)\overline{T}_{13} + 3\psi_*\overline{T}_{23} - 6\psi,$$

$$\hat{T}_{29} = -M\sqrt{2M^2 - 1}\frac{d\overline{T}_{24}}{dM} + 6\xi\overline{T}_{14} + 6(\psi - 1)\overline{T}_{12} + 3\xi_*\overline{T}_{24} + 3\psi_*\overline{T}_{22},$$

$$\psi = \frac{(2M^2 - 1)^{3/2}(M^2 + 1)}{4M^2 - 1} + 1, \qquad \psi_* = \frac{10M^3\sqrt{2M^2 - 1}}{(4M^2 - 1)},$$

$$\overline{T}_{10} = T_{10} + \frac{(2M^2 - 1)T_{20}}{2M} \qquad \overline{T}_{20} = \frac{2(2M^2 - 1)T_{10}}{M} + T_{20},$$

$$\overline{T}_{11} = T_{11} + \frac{(2M^2 - 1)T_{21}}{2M} \qquad \overline{T}_{21} = \frac{2(2M^2 - 1)T_{11}}{M} + T_{21},$$

$$\overline{T}_{12} = T_{12} + \frac{(2M^2 - 1)T_{22}}{2M} \qquad \overline{T}_{22} = \frac{2(2M^2 - 1)T_{12}}{M} + T_{22},$$

$$\overline{T}_{13} = T_{13} + \frac{(2M^2 - 1)T_{23}}{2M} \qquad \overline{T}_{23} = \frac{2(2M^2 - 1)T_{13}}{M} + T_{23},$$

$$\overline{T}_{14} = T_{14} + \frac{(2M^2 - 1)T_{24}}{2M} \qquad \overline{T}_{24} = \frac{2(2M^2 - 1)T_{14}}{M} + T_{24},$$

Appendix – C

This Appendix shows that the transport equations (5) and (53) in Ho and Meyer [72] are identical, and also that the coupling term in this article is related to the wave strength ratio Γ in the work of Ho and Meyer (see their equation (54)).

Ho and Meyer derived an ordinary differential equation for bore strength M by introducing a transformation from (x, t) plane to $(\alpha, \beta)-$ characteristic plane, where $\alpha =$ constant on advancing characteristic lines determined by $dx/dt = u + c$, and $\beta =$ constant on receding characteristic lines determined by $dx/dt = u - c$. In view of the relationships $u = (\alpha - \beta)/2 - \gamma t$ and $c = (\alpha + \beta)/4$ used by Ho and Meyer [72], the coupling term in this article $\Pi_1 = [c_x + u_x/2]$ can be written as $\Pi_1 = -\left[\dfrac{1}{4ct_\alpha}\right]$, since $\alpha_x = -1/2ct_\alpha$. In addition, in the medium ahead of the bore $\alpha - \beta = 2\gamma t$, it follows that

$$\Pi_1 = \frac{\gamma}{2c_0} - \frac{1}{4c_b(t_\alpha)_b} = \frac{\gamma}{2c_0} - \frac{\gamma}{2(u_b + c_b)} - \frac{\gamma\Gamma_b}{4c_b},$$

where $\Gamma_b = \dfrac{1}{\gamma}\left\{\dfrac{1}{(t_a)_b} - \dfrac{2\gamma c_b}{u_b + c_b}\right\}$ and the suffixes b and 0 refer, respectively, to the quantities just behind and ahead of the bore.

Combining the expression for Π_1 with $h_0 = -\gamma x/g$, $\quad c_b = c_0\sqrt{2M^2 - 1}$ and $u_b = 2M(M^2-1)c_0/\sqrt{2M^2 - 1}$, it can be easily seen that equation (7.3.5) coincides with equation (53) in Ho and Meyer [72].

Bibliography

[1] W. F. Ames and A. Donato. On the evolution of weak discontinuities in a state characterized by invariant solutions. *Int. J. Nonlinear Mech.*, 23:167–174, 1988.

[2] W. F. Ames. *Nonlinear Partial Differential Equations in Engineering.* Vol. 2, Academic Press, New York, 1972.

[3] A. M. Anile and G. Russo. Generalized wavefront expansion I: Higher order corrections for the propagation of weak shock waves. *Wave Motion*, 8:243–258, 1986.

[4] A. M. Anile and G. Russo. Generalized wavefront expansion II: The propagation of step shocks. *Wave Motion*, 10:3–18, 1988.

[5] A. M. Anile, J. K. Hunter, P. Pantano, and G. Russo. *Ray Methods for Nonlinear Waves in Fluids and Plasmas.* Longman Scientific & Technical, New York, 1993.

[6] H. Ardavan-Rhad. The decay of a plane shock wave. *J. Fluid Mech.*, 43:737-751, 1970.

[7] N. Asano and T. Tanuiti. Reductive perturbation method for nonlinear wave propagation in inhomogeneous media, Part II. *J. Phys. Soc. Japan*, 29:209–214, 1970.

[8] A. V. Azevedo, D. Marchesin, B. J. Plohr, and K. Zumbrun. Nonuniqueness of non-classical solutions of Riemann problems. *ZAMP*, 47:977–998, 1996.

[9] P. B. Bailey and P. J. Chen. On the local and global behavior of acceleration waves. *Arch. Ratl. Mech. Anal.*, 41:121–131, 1971.

[10] G. I. Barenblatt. *Similarity, Self-Similarity, and Intermediate Asymptotics.* Consultants Bureau, New York, 1979.

[11] J. Batt and R. Ravindran. Calculation of shock using solutions of systems of ordinary differential equations. *Quart. Appl. Math.*, 63:721–746, 2005.

[12] E. Becker. Chemically reacting flows. *Ann. Rev. Fluid Mech.*, 4:155-194, 1972.

[13] S. Benzoni-Gavage and D. Serre. *Multidimensional Hyperbolic Partial Differential Equations: First-order Systems and Applications.* Clarendon Press, Oxford, 2007.

[14] P. Bertrand and M. R. Feix. Nonlinear electron plasma oscillation: The "Water bag." *Phys. Lett.*, 28A:68–69, 1968.

[15] J. P. Best. A generalization of the theory of geometrical shock dynamics. *Shock Waves*, 1:251–273, 1991.

[16] O. P. Bhutani and L. Roy-Chowdhury. Applications of some recent techniques for the exact solutions of the small disturbance potential flow equations of nonequilibrium transonic gasdynamics. *Comp. Maths. Appl.*, 40:1349–1361, 2000.

[17] G. W. Bluman and J. D. Cole. *Similarity Methods for Differential Equations*, Springer, Berlin, 1974.

[18] G. W. Bluman and S. Kumei. *Symmetries and Differential Equations.* Springer, New York, 1989.

[19] G. Boillat. *La Propagation des Ondes.* Gauthier-Villars, Paris, 1965.

[20] G. Boillat. *Recent Mathematical Methods in Nonlinear Wave Propagation*, Lecture Notes in Mathematics (ed., T. Ruggeri) Springer, 1995.

[21] G. Boillat and T. Ruggeri. On the evolution law of weak discontinuities for hyperbolic quasilinear systems. *Wave Motion*, 1:149–152, 1979.

[22] G. Boillat and T. Ruggeri. Reflection and transmission of discontinuity waves through a shock wave: General theory including also the case of characteristic shocks. *Proc. Royal Soc. Edinburgh Sect. A*, 83:17–24, 1979.

[23] R. M. Bowen and P. J. Chen. Some comments on the behavior of acceleration waves of arbitrary shape. *J. Math. Phys.*, 13:948–950, 1972.

[24] R. M. Bowen and C. C. Wang. Acceleration waves in inhomogeneous isotropic elastic bodies. *Arch. Ratl. Mech. Anal.*, 38:13–45, 1970.

[25] A. Bressan. *Hyperbolic Systems of Conservation Laws: The One-Dimensional Cauchy Problem.* Oxford University Press, 2000.

[26] L. Brun. Ondes de choc finies dans les solides elastiques. In *Mechanical Waves in Solids.* Ed.: J. Mandel and L. Brun, Springer, Vienna, 1975.

[27] H. Cabannes. *Theoretical Magnetofluiddynamics.* Academic Press, New York, 1970.

[28] J. E. Cates and B. Sturtevant. Shock wave focusing using geometrical shock dynamics. *Phys. Fluids*, 9:3058–3068, 1997.

[29] C. J. Catherasoo and B. Sturtevant. Shock dynamics in nonuniform media. *J. Fluid Mech.*, 127:539–561, 1983.

[30] P. Cehelsky and R. R. Rosales. Resonantly interacting weakly nonlinear hyperbolic waves in the presence of shocks: A single space variable in a homogeneous time dependent medium. *Stud. Appl. Math.*, 74:117–138, 1986.

[31] T. Chang and L. Hsiao. *The Riemann Problem and Interaction of Waves in Gas Dynamics.* Longman, Harlow, 1989.

[32] P. J. Chen. *Selected Topics in Wave Propagation.* Noordhoff, Leyden, 1976.

[33] T. H. Chong and L. Sirovich. Numerical integration of the gasdynamic equation. *Phys. Fluids*, 23:1296–1300, 1980.

[34] V. Choquet-Bruhat. Ondes asmptotiques et approchees pour systemes d'equations aux derivees partielles nonlineaires. *J. Math. Pure et Appl.*, 48:117–158, 1969.

[35] A. J. Chorin. Random choice solutions of hyperbolic systems. *J. Comp. Phys.*, 22:67–99, 1976.

[36] B. T. Chu. Weak nonlinear waves in nonequilibrium flows. In *Nonequilibrium Flows, Part II*. Ed.: P. P. Wegener. Marcel Dekker, New York, 1970.

[37] J. F. Clarke and M. McChesney. *Dynamics of Relaxing Gases.* Butterworth, London, 1976.

[38] B. D. Coleman and M. E. Gurtin. On the stability against shear waves of steady flows of nonlinear viscoelastic fluids. *J. Fluid Mech.*, 33:165–181, 1968.

[39] B. D. Coleman, J. M. Greenberg, and M. E. Gurtin. Waves in material with memory: On the amplitudes of acceleration waves and mild discontinuities. *Arch. Ratl. Mech. Anal.*, 22:333–354,1966.

[40] R. Collins and H. T. Chen. Propagation of shock wave of arbitrary strength in two half planes containing a free surface. *J. Comp. Phys.*, 5:415–422, 1970.

[41] R. Courant and K. O. Friedrichs. *Supersonic Flow and Shock Waves.* Interscience, New York, 1948.

[42] E. A. Cox and A. Kluwick. Waves and non-classical shocks in a scalar conservation law with non-convex flux. *ZAMP*, 52:924–949, 2001.

[43] M. S. Cramer and R. Sen. A general scheme for the derivation of evolution equations describing mixed nonlinearity. *Wave Motion*, 15:333–355, 1992.

[44] D. G. Crighton and J. F. Scott. Asymptotics solutions of model equations in nonlinear acoustics. *Phil. Trans. R. Soc. London A*, 292:101–134, 1979.

[45] C. M. Dafermos. *Hyperbolic Conservation Laws in Continuum Physics.* Springer, Berlin, 2000.

[46] R. C. Davidson. *Methods in Nonlinear Plasma Theory.* Academic Press, New York, 1972.

[47] R. DiPerna and A. Majda. The validity of geometrical optics for weak solutions of conservation laws. *Comm. Math. Phys.*, 98:313–347, 1985.

[48] A. Donato and F. Oliveri. How to build up variable transformations allowing one to map nonlinear hyperbolic equations into autonomous or linear ones. *Transp. Theory Stat. Phys.*, 25:303–322, 1996.

[49] A. Donato and F. Oliveri. When non-autonomous equations are equivalent to autonomous ones. *Appl. Anal.*, 58:313–323, 1995.

[50] A. Donato and T. Ruggeri. Similarity solutions and strong shocks in extended thermodynamics of rarefied gas. *J. Math. Anal. Appl.*, 251:395–405, 2000.

[51] J. Dunwoody. High frequency sound waves in ideal gases with internal dissipation. *J. Fluid Mech.*, 34:769–782, 1968.

[52] A. R. Elcrat. On the propagation of sonic discontinuities in the unsteady flow of a perfect gas. *Int. Engng. Sci.*, 15:29–34, 1977.

[53] R. Estrada and R. P. Kanwal. Applications of distributional derivatives to wave propagation. *J. Inst. Math. Appl.*, 26:39–63, 1980.

[54] L. C. Evans. *Partial Differential Equations, Graduate Studies in Mathematics.* Amer. Math. Soc., Providence, 1998.

[55] K. O. Friedrichs. Formation and decay of shock waves. *Comm. Pure Appl. Maths.*, 1:211–245, 1948.

[56] Y. B. Fu and N. H. Scott. One dimensional shock waves in simple materials with memory. *Proc. R. Soc. London A*, 428:547–571, 1990.

[57] D. Fusco and A. Palumbo. Similarity solutions and model constitutive laws to viscoelasticity with applications to nonlinear wave propagation. *ZAMP*, 40:78–92, 1989.

[58] D. Fusco and J. Engelbrecht. The asymptotic analysis of nonlinear waves in Rate dependent media. *Il Nuovo Cimento*, 80:49–61, 1984.

[59] D. Fusco and P. Palumbo. Dissipative or dispersive wave processes ruled by first order quasilinear systems involving source like terms. *Rendiconti di Matematica*, 9:661–679, 1989.

[60] D. Fusco. Some comments on wave motions described by nonhomogeneous quasilinear first order hyperbolic systems. *Meccanica*, 17:128–137, 1982.

[61] P. Germain. Progressive waves. *Jber. DGLR*, 11–30, 1971.

[62] J. Glimm. Solutions in the large for nonlinear hyperbolic systems of equations. *Comm. Pure Appl. Math.*, 18:95–105, 1965.

[63] E. Godlewski and P. A. Raviart. *Numerical Approximations of Hyperbolic Systems of Conservation Laws*. Springer, Berlin, 1996.

[64] A. A. Goldstein. Optimal temperament. In *Mathematical Modelling: Classroom Notes in Applied Mathematics*. Ed.: M. Klamkin. SIAM, Philadelphia, 1987.

[65] M. A. Grinfeld. Ray method of calculating the wavefront intensity in nonlinearly elastic materials. *PMM (J. Appl. Math. and Mech.)*, 42:958–977, 1978.

[66] G. Guderley. Starke kugelige und zylindrische Verdichtungsstosse in der Nahe des Kugelmittelpunktes bzw der Zylinderachse. *Luftfahrtforschung*, 19:302–312, 1942.

[67] R. Haberman. *Mathematical Models: Mechanical Vibrations, Population Dynamics and Traffic Flow*. SIAM, Philadelphia, 1998.

[68] P. Hafner. Strong convergent shock waves near the center of convergence: A power series solution. *SIAM J. Appl. Math.*, 48:1244–1261, 1988.

[69] W. D. Hayes. Self-similar strong shocks in an exponential medium. *J. Fluid Mech.*, 32:305–315, 1968.

[70] J. B. Helliwell. The propagation of small disturbances in radiative magnetogasdynamics. *Arch. Ratl. Mech. Anal.*, 47:380–388, 1972.

[71] W. D. Henshaw, N. F. Smyth, and D.W. Schwendeman. Numerical shock propagation using geometrical shock dynamics. *J. Fluid Mech.*, 171:519–545, 1986.

[72] D. V. Ho and R. E. Meyer. Climb of a bore on a beach (Part 1. Uniform beach slope). *J. Fluid Mech.*, 14:305–318, 1962.

[73] H. Holden and N. H. Risebro. *Front Tracking for Hyperbolic Conservation Laws*. Springer, New York, 2002.

[74] L. Hörmander. *The Analysis of Linear Partial Differential Operators I*. Second Edition. Springer, Berlin, 1990.

[75] J. K Hunter and J. B. Keller. Weakly nonlinear high frequency waves. *Comm. Pure Appl. Math.*, 36:547–569, 1983.

[76] J. K. Hunter. Nonlinear Geometrical Optics. In *Multidimensional Hyperbolic Problems and Computations*. Eds.: J. Glimm and A. Majda. IMA volumes in Mathematics and its Applications, Springer, New York, 1991.

[77] J. K. Hunter, A. Majda, and R. R. Rosales. Resonantly interacting weakly nonlinear hyperbolic waves II. Several space variables. *Stud. Appl. Math.*, 75:187–226, 1986.

[78] N. H. Ibragimov. *Transformation Groups Applied to Mathematical Physics*. Riedel, Dordrecht, 1985.

[79] A. Jeffrey and T. Tanuiti. *Nonlinear Wave Propagation with Applications to Physics and Magnetohydrodynamics*. Academic Press, New York, 1964.

[80] A. Jeffrey. *Quasilinear Hyperbolic Systems and Waves*. Pitman, London, 1976.

[81] A. Jeffrey and J. Mvungi. On the breaking of water waves in a channel of arbitrarily varying depth and width. *ZAMP*, 31:758–761, 1980.

[82] A. Jeffrey and T. Kawahara. *Asymptotic Methods in Nonlinear Wave Theory*. Pitman, London, 1982.

[83] A. Jeffrey and H. Zhao. Global existence and optimal temporal decay estimates for systems of parabolic conservation laws I: the one-dimensional case. *Appl. Anal.*, 70:175–193, 1998.

[84] A. Jeffrey. The breaking of waves on a sloping beach. *ZAMP*, 15:97–106, 1964.

[85] H. Jeffrey and B. S. Jeffrey. *Methods of Mathematical Physics*. Cambridge University Press, 1999.

[86] J. Jena and V. D. Sharma. Lie transformation group solutions of nonlinear equations describing viscoelastic materials. *Int. J. Engng. Sci.*, 35:1033–1044, 1997.

[87] N. H. Johannesen and W. A. Scott. Shock formation distances in real air. *J. Sound Vib.*, 61:169–177.

[88] J. L. Joly, G. Metivier, and J. Rauch. Coherent and focusing multidimensional nonlinear geometric optics. *Ann. Sc. ENS.*, 28:51–113.

[89] K. V. Karelsky and A. S. Petrosyan. Particular solutions and Riemann problem for modifying shallow water equations. *Fluid Dynam. Res.*, 38:339–358, 2006.

[90] H. B. Keller, D. A. Levine, and G. B. Whitham. Motion of a bore over a sloping beach. *J. Fluid Mech.*, 7:302–316, 1960.

[91] J. B. Keller. Geometrical acoustics I: the theory of weak shock waves. *J. Appl. Phys.*, 25:938–947, 1954.

[92] J. Kevorkian. *Partial Differential Equations; Analytical Solution Techniques*. Wadsworth & Brooks/Cole, 1990.

[93] B. Keyfitz and H. C. Kranzer. A viscosity approximation to a system of conservation laws with no classical Riemann solution. In *Nonlinear Hyperbolic Systems*. Ed.: C. Carasso et al., Lecture Notes in Math., Vol. 1402, Springer, New York, 1988.

[94] A. Kluwick and E. A. Cox. Nonlinear waves in materials with mixed nonlinearity. *Wave Motion*, 27:23–41, 1998.

[95] Z. Kopal. The propagation of shock waves in self-similar gas spheres. *Ap. J.*, 120:159–171, 1954.

[96] V. P. Korobeinikov. *Problems in the Theory of Point Explosion in Gases*. Amer. Math. Soc., Providence, 1976.

[97] S. N. Kruzkov. First order quasilinear equations in several independent variables. *Math. USSR Sbornik*, 10:217–243, 1970.

[98] L. Landau. and E. Lifshitz. *Fluid Mechanics*. Pergamon Press, New York, 1959.

[99] L. D. Landau. On shock waves at a large distances from their place of origin. *J. Phys. USSR*, 9:494–500, 1945.

[100] P. D. Lax. *Hyperbolic Partial Differential Equations*. Amer. Math. Soc., Providence, 2006.

[101] P. D. Lax. Hyperbolic systems of conservation laws II. *Comm. Pure Appl. Maths.*, 10:537–566, 1957.

[102] P. D. Lax. *Hyperbolic Systems of Conservation Laws on the Mathematical Theory of Shock Waves*. CBMS-NSF Regional Conference Series in Applied Mathematics, SIAM, Philadelphia, 1973.

[103] R. B. Lazarus. Self-similar solutions for converging shocks and collapsing cavities. *SIAM J. Numer. Anal.*, 18:316–371, 1981.

[104] G. Lebon, C. Perez-Garcia, and J. Casas-Vazquez. On a new thermodynamic description of viscoelastic materials. *Physica*, 137:531–545, 1986.

[105] P. G. LeFloch. *Hyperbolic Systems of Conservation Laws: The Theory of Classical and Nonclassical Shock Waves*. Birkhauser, Berlin 2002.

[106] R. J. LeVeque. *Numerical Methods for Conservation Laws*. Birkhauser, Basel, 1990.

[107] J. Li, T. Zhang, and S. Yang. *The Two Dimensional Riemann Problem in Gas Dynamics*. Pitman, London, 1998.

[108] M. J. Lighthill. A method for rendering approximate solutions to physical problems uniformly valid. *Phil. Mag.*, 40:1179–1201, 1949.

[109] M. J. Lighthill. The energy distribution behind decaying shocks. *Phil. Mag.*, 41:1101–1128, 1950.

[110] T. P. Liu. *Nonlinear stability of shock waves for viscous conservation laws*. Mem. Amer. Math. Soc., Providence, Vol. 308, 1985.

[111] T. P. Liu. *Hyperbolic and Viscous Conservation Laws*. CBMS-NSF Regional Conferences Series in Applied Mathematics, SIAM, Philadelphia, 2000.

[112] J. D. Logan and J. D. J. Perez. Similarity solutions for reactive shock hydrodynamics. *SIAM J. Appl. Math.*, 39:512–527, 1980.

[113] J. Lou and T. Ruggeri. Acceleration waves amd weak Shizuta-Kawashima condition. *Rendiconti Del Circolo Matematico Di Palermo*, 78:187-200, 2006.

[114] A. Majda and R. Pego. Stable viscosity matrices for systems of conservation laws. *Journal of Differential Equations*, 56:229–262, 1985.

[115] A. Majda and R. R. Rosales. Resonantly interacting weakly nonlinear hyperbolic waves I. A single space variable. *Stud. Appl. Math.*, 71:149–179, 1984.

[116] A. Majda. Nonlinear geometric optics for hyperbolic systems for conservation laws. In *Oscillation theory, Computation and Methods of Compensated Compactness*. Eds.: C. Dafermos et al., IMA volumes in Mathematics and its Applications, Springer, New York, 1986.

[117] S. A. V. Manickam, Ch. Radha, and V. D. Sharma. Far field behavior of waves in a vibrationally relaxing gas. *Appl. Numer. Math.*, 45:293–307, 2003.

[118] F. E. Marble. Dynamics of dusty gases. *Ann. Rev. Fluid Mech.*, 2:397–447, 1970.

[119] V. P. Maslov. Propagation of shock waves in an isentropic nonviscous gas. *J. Soviet Math.*, 13:119-163,1980.

[120] M. F. McCarthy. Singular surfaces and waves. In *Continuum Physics, Vol. II*. Ed.: A.C. Eringen, Academic Press, 1975.

[121] R. C. McOwen. *Partial Differential Equations: Methods and Applications*. Pearson Education, Singapore, 2003.

[122] V. V. Menon and V. D. Sharma. Characteristic wave fronts in magnetogasdynamics. *J. Math. Anal. Appl.*, 81:189–203, 1981.

[123] V. V. Menon, V. D. Sharma, and A. Jeffrey. On the general behavior of acceleration waves. *Appl. Anal.*. 16:101–120,1983.

[124] R. E. Meyer. Theory of characteristics of inviscid gasdynamics. In *Handbuch der Physik, Vol. IX*. Ed.: S. Flugge, Springer, New York, 1960.

[125] P. J. Montgomery and T. B. Moodie. Jump conditions for hyperbolic systems of conservation laws with an application to gravity currents. *Stud. Appl. Math.*, 106:367–392,2001.

[126] A. Morro. Interaction of acoustic waves with shock waves in elastic solids. *ZAMP*, 29:822–827, 1978.

[127] A. Morro. Interaction of waves with shock in magnetogasdynamics. *Acta Mech.*, 35:197–213, 1980.

[128] M. P. Mortell and E. Varley. Finite amplitude waves in bounded media; nonlinear free vibrations of an elastic panel. *Proc. Roy. Soc. London A*, 318:169–196, 1970.

[129] S. Murata. New Exact solution of the blast wave problem in gasdynamics. *Chaos, Solitons and Fractals*, 28:327–330, 2006.

[130] G. A. Nariboli. Some aspects of wave propagation. *Jour. Math. Phys. Sci.*, 3:294–310, 1968.

[131] J. W. Nunziato and E. K. Walsh. Propagation and growth of shock waves in inhomogeneous fluids. *Phys. Fluids*, 15:1397–1402, 1972.

[132] J. R. Ockendon, S. D. Howison, A. A. Lacey, and A. B. Movchan. *Applied Partial Differential Equations*. Oxford University Press, 2003.

[133] O. A. Oleinik. Uniqueness and stability of the generalized solutions of the Cauchy problem for a quasilinear equation. *Upeskhi Mat. Nauk*, 14:165–170, 1959.

[134] F. Oliveri and M. P. Speciale. Exact solutions to the ideal magnetogasdynamic equations of perfect gases through Lie group analysis and substitution principles. *J. Phys. A (Math. and General)*, 38:8803–8820, 2005.

[135] F. Oliveri and M. P. Speciale. Exact solutions to the unsteady equations of perfect gases through Lie group analysis and substitution principles. *Int. J. Nonlinear Mech.*, 37:257–274, 2002.

[136] P. J. Olver. *Applications of Lie groups to Differential Equations.* Springer, New York, 1986.

[137] P. J. Olver. *Equivalence, Invariants and Symmetry.* Cambridge University Press, Cambridge, 1995.

[138] L. V. Ovsiannikov. *Group Analysis of Differential Equations.* Academic, New York, 1982.

[139] S. I. Pai, V. D. Sharma, and S. Menon. Time evolution of discontinuities at the wave head in a nonequilibrium two phase flow of a mixture of gas and dust particles. *Acta Mech.*, 46:1–13, 1983.

[140] D. F. Parker. Propagation of a rapid pulse through a relaxing gas. *Phys. Fluids*, 15:256–262, 1972.

[141] D. H. Peregrine. Equations for water waves and the approximation behind them. In *Waves on Beaches*. Ed.: R. Meyer, Academic Press, New York, 1972.

[142] D. H. Peregrine. Water wave interaction in the surf zone. *Coastal Engineering Conference, Copenhagen*, 500–517, 1976.

[143] A. F. Pillow. The formation and growth of shock waves in the one dimensional motion of a gas. *Proc. Camb. Phil. Soc.*, 45:558–586, 1949.

[144] H. Poincaré. *Les Méthodes Nouvelles de la Méchanique Céleste, Vol. III.* Dover, New York, 1957.

[145] P. Prasad. *Nonlinear Hyperbolic Waves in Multi-dimensions.* Monographs and Surveys In Pure and Applied Mathematics, Chapman & Hall / CRC, 211:2001.

[146] Ch. Radha and V. D. Sharma. Propagation and Interaction of Waves in a relaxing gas. *Phil. Trans. Roy. Soc. London A*, 352:169–195, 1995.

[147] Ch. Radha, V. D. Sharma, and A. Jeffrey. An approximate analytical method for describing the kinematics of a bore over a sloping beach. *Appl. Anal.*, 81:867–892, 2002.

[148] Ch. Radha, V. D. Sharma, and A. Jeffrey. On interaction of shock waves with weak discontinuities. *Appl. Anal.*, 50:145–166, 1993.

[149] T. R. Sekhar and V. D. Sharma. Interaction of shallow water waves. *Stud. Appl. Maths.*, 121:1–25, 2008.

[150] R. Ravindran and P. Prasad. On an infinite system of compatibility conditions along a shock ray. *Q. Jl. Mech. Appl. Math.*, 46:131–140, 1993.

[151] C. Rogers and W. F. Ames. *Nonlinear Boundary Value Problems in Science and Engineering*. Academic Press, New York, 1989.

[152] M. H. Rogers. Analytic solutions for the blast wave problem with an atmosphere of varying density. *Ap. J.*, 125:478–493, 1953.

[153] R. Rosales. An introduction to weakly nonlinear geometrical optics. In *Multidimensional Hyperbolic Problems and Computations: IMA Volumes in Mathematics and Its Applications*. Eds.: J. Glimm and A. Majda, Springer, New York, 1991.

[154] R. R. Rosales and M. Shefter. Quasi-periodic solutions in weakly nonlinear gasdynamics, Part I. Numerical results in the inviscid case. *Stud. Appl. Maths.*, 103:279–337, 1999.

[155] M. Roseau. *Asymptotic Wave Theory*. North Holland Publishing Company, Amsterdam, 1976.

[156] G. Rudinger. *Fundamentals of Gas-Particle Flow*. Elsevier, Amsterdam, 1980.

[157] T. Ruggeri. Interaction between a discontinuity wave and a shock wave: critical time for the fastest transmitted wave; example of the polytropic fluid. *Appl. Anal.*, 11:103–112, 1980.

[158] T. Ruggeri. Stability and discontinuity waves for symmetric hyperbolic systems, in *Nonilinear Wave Motion*, (ed. A. Jeffrey). Longman, New York, 1989.

[159] A. Sakurai. On the problem of a shock wave arriving at the edge of a gas. *Comm. Pure Appl. Math.*, 13:353–370, 1960.

[160] H. Schmitt. Acceleration waves in condensing gases. *Acta Mech.*, 78:109–128, 1989.

[161] D. W. Schwendeman. A new numerical method for shock wave propagation on geometrical shock dynamics. *Proc. R. Soc. London. A*, 441:331–341, 1993.

[162] N. H. Scott. Acceleration waves in incompressible elastic solids. *Q. Jl. Mech. Appl. Math.*, 29:295–310, 1976.

[163] W. A. Scott and N. H. Johannesen. Spherical nonlinear wave propagation in a vibrationally relaxing gas. *Proc. R. Soc. London A*, 382:103–134, 1982.

[164] D. Serre. *Systems of Conservation Laws II.* Cambridge University Press, 2000.

[165] B. R. Seymour and M. P. Mortell. Nonlinear geometrical acoustics. In *Mechanics Today, Vol. 2.* Ed.: S. Nimat-Nasser, Pergamon Press, New York, 1975.

[166] B. R. Seymour and E. Varley. High frequency, periodic disturbances in dissipative systems. I: Small amplitude finite rate theory. *Proc. Roy. Soc. London A,* 314:387–415, 1970.

[167] V. D. Sharma and R. Shyam. Growth and Decay of Weak Discontinuities in radiative gasdynamics. *Acta Astronautica,* 8:31–45, 1981.

[168] V. D. Sharma and Ch. Radha. Similarity solutions for converging shocks in a relaxing gas. *Int. J. Engng Sci.,* 33:535–553, 1995.

[169] V. D. Sharma and Ch. Radha. One dimensional planar and nonplanar shock waves in a relaxing gas. *Phys. Fluid,* 6:2177–2190, 1994.

[170] V. D. Sharma and R. Radha. Exact solutions of Euler equations of ideal gasdynamics via Lie group analysis. *ZAMP,* 59:1029–1038, 2008.

[171] V. D. Sharma and G. K. Srinivasan. Wave interaction in a nonequilibrium gas flow. *Int. J. Nonlinear Mech.,* 40:1031–1040, 2005.

[172] V. D. Sharma, G. Madhumita, and A. V. Manickam. Dissipative waves in a relaxing gas exhibiting mixed nonlinearity. *Nonlinear Dynamics,* 33:283–299, 2003.

[173] V. D. Sharma and G. Madhumita. Nonlinear wave propagation through a stratified atmosphere. *J. Math. Anal. Appl.,* 311:13–22, 2005.

[174] V. D. Sharma and Ch. Radha. Three dimensional shock wave propagation in an ideal gas. *Int. J. Nonlinear Mech.,* 30:305–322, 1995.

[175] V. D. Sharma and V. V. Menon. Further comments on the behavior of acceleration waves of arbitrary shape. *J. Math. Phys.,* 22:683–684, 1981.

[176] V. D. Sharma. On the evolution of compression pulses in a steady magnetohydrodynamic flow over a concave wall. *Q. Jl. Mech. Appl. Math.,* 40:527–537, 1987.

[177] V. D. Sharma, B. D. Pandey, and R. R. Sharma. Weak discontinuities in a nonequilibrium flow of a dusty gas. *Acta Mech.,* 99:103–111, 1993.

[178] V. D. Sharma. Propagation of discontinuities in magnetohydrodynamics. *Phys. Fluids,* 24:1386–1387, 1981.

[179] V. D. Sharma, R. Ram, and P. L. Sachdev. Uniformly valid analytical solution to the problem of a decaying shock wave. *J. Fluid. Mech.*, 185:153-170, 1987.

[180] V. D. Sharma, R. R. Sharma, B. D. Pandey, and N. Gupta. Nonlinear wave propagation in a hot electron plasma. *Acta Mech.*, 88:141–152, 1991.

[181] V. D. Sharma, R. R. Sharma, B. D. Pandey, and N. Gupta. Nonlinear analysis of a traffic flow. *ZAMP*, 40:828–837, 1989.

[182] V. D. Sharma. The development of jump discontinuities in radiative magnetogasdynamics. *Int. J. Engng. Sci.*, 24:813–818, 1986.

[183] M. Shearer. Nonuniqueness of admissible solutions of Riemann initial value problem for a system of conservation laws of mixed type. *Arch. Ratl.Mech. Anal.*, 93:45–59, 1986.

[184] F. V. Shugaev and L. S. Shtemenko. *Propagation and Reflection of Shock Waves*. World Scientific, New Jersey, 1998.

[185] R. S. Singh and V. D. Sharma. Propagation of discontinuities along bicharacteristics in magnetohydrodynamics. *Phys. Fluids*, 23:648–649, 1980.

[186] R. S. Singh and V. D. Sharma. The growth of discontinuities in nonequilibrium flow of a relaxing gas. *Q. Jl. Mech. Appl. Math.*, 32:331–338, 1979.

[187] L. Sirovich and T. H. Chong. Approximation solution in gas dynamics. *Phys. Fluids*, 23:1291-1295, 1980.

[188] J. Smoller and B. Temple. Global solutions of relativistic Euler equations. *Comm. Math. Phys.*, 156:67–99, 1993.

[189] J. Smoller. *Shock Waves and Reaction Diffusion Equations*. Second edition, Springer, New York, 1994.

[190] G. K. Srinivasan and V. D. Sharma. On weakly nonlinear waves in media exhibiting mixed nonlinearity. *J. Math. Anal. Appl.*, 285:629–641, 2003.

[191] G. K. Srinivasan and V. D. Sharma. A note on the jump conditions for systems of conservation laws. *Stud. Appl. Math.*, 110:391–396, 2003.

[192] G. K. Srinivasan and V. D. Sharma. Modulation equations of weakly nonlinear geometrical optics in media exhibiting mixed nonlinearity. *Stud. Appl. Math.*, 110:103–122, 2003.

[193] G. W. Swan and G. R. Fowles. Shock wave stability. *Phys. Fluids.*, 18:28–35, 1975.

[194] T. Tanuiti and C. C. Wei. Reductive perturbation method in nonlinear wave propagation, Part I. *J. Phys. Soc. Japan*, 24:941–946, 1968.

[195] A. H. Taub. Relativistic Rankine–Hugoniot equations. *Physical Reviews*, 74:328–334, 1948.

[196] T. Y. Thomas. *Plastic Flow and Fracture in Solids*. Academic Press, New York, 1961.

[197] T. Y. Thomas. The growth and decay of sonic discontinuities in ideal gases. *J. Math. Mech.*, 6:311–322, 1957.

[198] T. C. T. Ting. Nonlinear effects on the propagation of discontinuities of all orders in viscoelastic media. In *Proc. IUTAM Symposia on Mechanics of Viscoelastic Media and Bodies*. Ed.: J. Hult, Springer, New York, 84–95, 1975.

[199] E. F. Toro. Riemann problems and the WAF method of solving the two dimensional shallow water equations. *Phil. Trans. Roy. Soc. London A*, 338:43–68, 1992.

[200] E. F. Toro. *Riemann Solvers for Numerical Methods for Fluid Dynamics*. Springer, Berlin, 1999.

[201] C. Truesdell and R. A. Toupin. The Classical Field Theories. In *Handbuch der Physik, Vol. III.*, Springer, Berlin, 1960.

[202] M. Van Dyke and A. J. Guttmann. The converging shock wave from a spherical or cylindrical piston. *J. Fluid Mech.*, 120:451–462, 1982.

[203] W. K. Van Moorhen and A. R. George. On the stability of plane shocks. *J. Fluid Mech.*, 68:97–108, 1975.

[204] E. Varley and E. Cumberbatch. Large amplitude waves in stratified media: acoustic pulses. *J. Fluid Mech.*, 43:513–537, 1970.

[205] E. Varley and E. Cumberbatch. Nonlinear theory of wave-front propagation. *J. Inst. Math. Appl.*, 1:101–112, 1965.

[206] E. Varley and E. Cumberbatch. Nonlinear, high frequency sound waves, *J. Inst. Math. Appl.*, 2:133–143, 1966.

[207] W. G. Vincenti and C. H. Kruger. *Introduction to Physical Gasdynamics*. Wiley, New York, 1965.

[208] N. Virgopia and F. Ferraioli. Interaction between a weak discontinuity wave and a blast wave: search for critical times for transmitted waves in self similar flows. *II Nuovo Cimento*, 69:119–135, 1982.

[209] Yu. Weinan, G. Rykov, and Ya. Sinai. Generalized variational principles, global weak solutions and behavior with random initial data for systems of conservation laws arising in adhesion particle dynamics. *Comm. Math. Phys.*, 177:349–380, 1996.

[210] G. B. Whitham. *Linear and Nonlinear Waves*. Wiley, New York, 1974.

[211] T. W. Wright. Acceleration waves in simple elastic materials. *Arch. Ratl. Mech. Anal.*, 50:237–277, 1973.

[212] T. W. Wright. An intrinsic description of unsteady shock waves. *Q. Jl. Mech. Appl. Math.*, 29:311–324, 1976.

[213] C. C. Wu. New Theory of MHD Shock Waves. In *Viscous Profiles and Numerical Methods for Shock Waves*. Ed.: M. Shearer, SIAM, Philadelphia, 1991.

[214] Y. B. Zeldovich and Y. P. Raizer. *Physics of Shock Waves and High Temperature Hydrodynamic Phenomena–II*. Academic Press, New York, 1967.

[215] Y. Zheng. *Systems of Conservation Laws: Two dimensional Riemann Problems*, Birkhauser, Basel, 2001.

Index

Milton Keynes UK
Ingram Content Group UK Ltd.
UKHW040444071024
449327UK00020B/973

9 780367 384159